M. B. Kirkham, BA, MS, PhD
Editor

Water Use
in Crop Production

Water Use in Crop Production has been co-published simultaneously as *Journal of Crop Production*, Volume 2, Number 2 (#4) 1999.

Pre-publication
REVIEWS,
COMMENTARIES,
EVALUATIONS . . .

More pre-publication
REVIEWS, COMMENTARIES, EVALUATIONS . . .

"Competition for water for domestic, municipal, industrial, environmental, and agricultural purposes is growing worldwide. How can we conserve water in agriculture so that it can be used in other sectors or used to grow more food for an ever increasing world population? What effect will rising carbon dioxide levels have on crop water use? Just how much water can be conserved without reducing yield and quality? And what means are available to achieve this goal? Answers to these questions are provided in the . . . book *Water Use in Crop Production*. . . . The editor, Prof. M. B. Kirkham, Kansas State University, has assembled an international field of experts to present current research from every continent except Antarctica. In 16 chapters, the authors show that water savings and/or higher crop yields are possible through water conservation, use and re-use of water with high salinity, improved irrigation strategies, appropriate crop water allocation, reduction in water losses, genetic improvement of crops, and increased knowledge of the factors that govern the loss of water from plant communities. The authors cover a diverse range of crops and climates: from rice production in the tropics and sub-tropics of Asia with huge demand for water, through wheat crops in temperate regions, to the drought tolerant olive tree in the dry Mediterranean climate. The use of irrigation water with high salinity and its effect on crop growth and soil physical and chemical characteristics is addressed in 5 out of 16 chapters, reflecting the increasing need to use saline water in world food production.

The wide range of topics covered in this book provides a great overview of current research, highlighting the shortcomings and constraints of current practices as well as showing the potential of new approaches to water management in agriculture. This book should appeal for both the novice and the experienced professional, from under-graduate student to professor, or anyone interested in water conservation and water resource management."

Horst Caspari, Dr. Agr.
Scientist
The Horticulture
and Food Research Institute
of New Zealand Ltd.

"*Water Use in Crop Production* is edited by Prof. M. B. Kirkham who is world-renowned for . . . outstanding contributions to the area of soil-plant water relationships. The book presents contributions from prominent international scientists covering important aspects of water relations in soil and in plant. Environmental and plant factors affecting plant water use are discussed and each topic has a comprehensive list of up-to-date literature which will prove invaluable for further consultation by readers. Crop responses to water shortage are discussed and the concepts of deficit irrigation and reusing of drainage water for irrigation are developed for important crops. The most modern methods and equipment for measuring plant water use are assessed. Risk assessment of the irrigation requirements, which is based on a sound understanding of soil and plant water relations, is introduced for making the most efficient use of available water for irrigation. Basic principles and practices for agricultural water conservation are explored contributing to sustainability of crop production. Of special interest are the possible changes in the pattern of plant water use in the new millennium in view of global warming and rising carbon dioxide level of the atmosphere. This book is a powerful education and research tool for those involved in the science of soil-plant-water relationships of economic plants."

**M. Hossein Behboudian,
PhD, MSc, BSc**
Associate Professor
Plant Physiology
Institute of Natural Resources
College of Sciences
Massey University
Palmerston North, New Zealand

Food Products Press
An Imprint of The Haworth Press, Inc.

Water Use in Crop Production

Water Use in Crop Production has been co-published simultaneously as *Journal of Crop Production*, Volume 2, Number 2 (#4) 1999.

The *Journal of Crop Production* Monographic "Separates"

Below is a list of "separates," which in serials librarianship means a special issue simultaneously published as a special journal issue or double-issue *and* as a "separate" hardbound monograph. (This is a format which we also call a "DocuSerial.")

"Separates" are published because specialized libraries or professionals may wish to purchase a specific thematic issue by itself in a format which can be separately cataloged and shelved, as opposed to purchasing the journal on an on-going basis. Faculty members may also more easily consider a "separate" for classroom adoption.

"Separates" are carefully classified separately with the major book jobbers so that the journal tie-in can be noted on new book order slips to avoid duplicate purchasing.

You may wish to visit Haworth's website at . . .

http://www.haworthpressinc.com

. . . to search our online catalog for complete tables of contents of these separates and related publications.

You may also call 1-800-HAWORTH (outside US/Canada: 607-722-5857), or Fax 1-800-895-0582 (outside US/Canada: 607-771-0012), or e-mail at:

getinfo@haworthpressinc.com

Water Use
in Crop Production

M. B. Kirkham, BA, MS, PhD
Editor

Water Use in Crop Production has been co-published simultaneously as *Journal of Crop Production*, Volume 2, Number 2 (#4) 1999.

Food Products Press
An Imprint of
The Haworth Press, Inc.
New York • London • Oxford

Published by

Food Products Press®, 10 Alice Street, Binghamton, NY 13904-1580

Food Products Press® is an imprint of The Haworth Press, Inc., 10 Alice Street, Binghamton, NY 13904-1580 USA.

Water Use in Crop Production has been co-published simultaneously as *Journal of Crop Production*, Volume 2, Number 2 (#4) 1999.

The development, preparation, and publication of this work has been undertaken with great care. However, the publisher, employees, editors, and agents of The Haworth Press and all imprints of The Haworth Press, Inc., including The Haworth Medical Press® and Pharmaceutical Products Press®, are not responsible for any errors contained herein or for consequences that may ensue from use of materials or information contained in this work. Opinions expressed by the author(s) are not necessarily those of The Haworth Press, Inc.

Cover design by Thomas J. Mayshock Jr.

Library of Congress Cataloging-in-Publication Data

Water use in crop production/M.B. Kirkham, editor.
    p. cm.
    Includes bibliographical references.
    ISBN 1-56022-068-6 (alk. paper)–ISBN 1-56022-069-4 (alk. paper)
    1. Irrigation farming. 2. Crops–Water requirements. I. Kirkham, M.B.
S613.W38 2000
631.7–dc21   99-058800

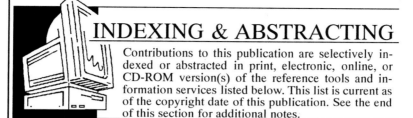

INDEXING & ABSTRACTING

Contributions to this publication are selectively indexed or abstracted in print, electronic, online, or CD-ROM version(s) of the reference tools and information services listed below. This list is current as of the copyright date of this publication. See the end of this section for additional notes.

- *AGRICOLA Database*
- *Chemical Abstracts*
- *CNPIEC Reference Guide: Chinese National Directory of Foreign Periodicals*
- *Crop Physiology Abstracts*
- *Derwent Crop Production File*
- *BIOBASE (Current Awareness in Biological Science)/Excerpta Medica*
- *Environment Abstracts*
- *Field Crop Abstracts*
- *Foods Adlibra*
- *Food Science and Technology Abstracts (FSTA)*
- *Grasslands & Forage Abstracts*
- *PASCAL, % Institute de L'Information Scientifique et Technigue*
- *Plant Breeding Abstracts*
- *Referativnyi Zhurnal (Abstracts Journal of the All-Russian Institute of Scientific and Technical Information)*
- *Seed Abstracts*
- *Soils & Fertilizers Abstracts*
- *Weed Abstracts*

(continued)

Special Bibliographic Notes related to special journal issues (separates) and indexing/abstracting:

- indexing/abstracting services in this list will also cover material in any "separate" that is co-published simultaneously with Haworth's special thematic journal issue or DocuSerial. Indexing/abstracting usually covers material at the article/chapter level.

- monographic co-editions are intended for either non-subscribers or libraries which intend to purchase a second copy for their circulating collections.

- monographic co-editions are reported to all jobbers/wholesalers/approval plans. The source journal is listed as the "series" to assist the prevention of duplicate purchasing in the same manner utilized for books-in-series.

- to facilitate user/access services all indexing/abstracting services are encouraged to utilize the co-indexing entry note indicated at the bottom of the first page of each article/chapter/contribution.

- this is intended to assist a library user of any reference tool (whether print, electronic, online, or CD-ROM) to locate the monographic version if the library has purchased this version but not a subscription to the source journal.

- individual articles/chapters in any Haworth publication are also available through the Haworth Document Delivery Service (HDDS).

Water Use in Crop Production

CONTENTS

ABOUT THE EDITOR

M. B. Kirkham, BA, MS, PhD, is Professor in the Department of Agronomy at Kansas State University. Dr. Kirkham is the author or co-author of over 180 contributions to scientific journals and is on the editorial board of *Plant and Soil* and *Soil Science*. In addition to conducting research, Professor Kirkham teaches a class about water relations, works with graduate students, and participates in national and international meetings. Dr. Kirkham is a Fellow of the American Society of Agronomy, the Soil Science Society of America, the Crop Science Society of America, the Royal Meteorological Society, and the American Association for the Advancement of Science as well as a member of thirty-four other scientific societies.

Preface

Water use in crop production is of worldwide concern, because water is the most important factor controlling plant growth. We have to make more efficient use of water to feed the earth's expanding population, which is now 5.9 billion, growing by more than 80 million a year, and projected to be 27 billion by 2150 (*National Geographic*, October, 1998). The goal of this special issue is to cover advancements in the understanding of water use in crop production around the world. The authors represent every continent except Antarctica and describe in 16 articles research in Argentina, Australia, Israel, Morocco, New Zealand, the Philippines, Spain, and the United States of America.

Paul Unger and Terry Howell (Texas, USA) give a global perspective of agricultural water conservation. They conclude that with good management, such as avoiding excessive deep percolation and reducing runoff by increased infiltration, improved use of water for greater crop production is possible under dryland and irrigated conditions. Hartwell Allen (Florida, USA) describes evapotranspiration responses of crops in the face of global climate change. He shows that rising carbon dioxide will increase water use efficiency. But if temperatures rise along with atmospheric carbon dioxide concentration, water use efficiency will decline. Peter Jamieson (New Zealand) examines the effects that water shortages have on crop production through influences on transpiration, growth rate, phenology, light interception, and biomass partitioning. He uses models to link growth and water use and finds that models can predict grain yield under drought. José Luis Costa (Argentina) studied the effect of supplementary irrigation on soil properties in Buenos Aires Province, Argentina, and shows that waters with salt concentrations even greater than 4.0 dS m^{-1} can be used for crop production.

Two contributions focus on orchard water use. Enrique Fernández (Spain) and Félix Moreno (Spain) provide a comprehensive review of water use of olive that includes information heretofore only available in Spanish or Italian. They describe why olive has an outstanding adaptation to drought. They combine their review with key results using new methods to measure water use in plants, including direct measurements of sap flow in the root and stem of olive and new techniques to optimize irrigation requirements. Tessa Mills (New Zealand), Kelly Morgan (Florida, USA), and Larry Parsons (Florida, USA) investigated responses of citrus stomata to leaf age, leaf position, canopy size, and local microclimate and show large differences due to leaf

xiii

age. They used a model of stomatal conductance to calculate total daily plant water use. The model calculated values that were close to actual values. Their results emphasize the important role of stomatal conductance in determining water use.

A series of papers documents water use in annual crops and turfgrass. Sham Goyal, Surinder Sharma, Donald Rains, and André Läuchli, all of California, USA, provide two articles concerning long-term reuse of drainage waters of varying salinites for crop irrigation in a cotton-safflower rotation. They show that irrigation waters up to 3,000 ppm salinity (about 4 dS m^{-1}) can be used without yield reductions, as long as leaching occurs through pre-plant irrigations with low salinity water. In another paper dealing with cotton, Bruria Heuer (Israel) and Arie Nadler (Israel) investigate physiological and agronomic responses of cotton to deficit irrigation and find that neither seed cotton yield nor lint quality were decreased. To Phuc Tuong (Philippines) discusses strategies for making rice more water efficient and reports that a constraint to implementing new technological innovations is lack of incentives for farmers to improve irrigation performance. Miranda Mortlock (Australia) and Graeme Hammer (Australia) studied transpiration efficiency in sorghum and learn that the best indicators of transpiration efficiency are leaf carbon concentration and leaf ash, which offer promise as an avenue for development of a selection index. Because hydraulic resistance of a C_3 and C_4 plant has not been compared, Jingxian Zhang (Texas, USA) and I (Kansas, USA) determined hydraulic resistance of sorghum (C_4) and sunflower (C_3). We found that sorghum had a constant hydraulic resistance and sunflower had a variable one and that sorghum had a higher hydraulic resistance, which may be one reason it uses less water than sunflower. Zvi Plaut (Israel) and A. Grava (Israel) look at the response of sunflower grown in a loess soil or in a clay soil to saline irrigation water. In the loess soil, sunflower extracted water to 120 cm. But sunflowers grown in the clay soil extracted limited amounts of water from deep layers due to high salinity and lack of aeration. Bingru Huang and Jack Fry, both from Kansas, USA, review evapotranspiration from turfgrass, which is a crop of increasing importance because of the interest in golf. They conclude that turfgrasses should be bred to incorporate traits that increase water use efficiency, such as selecting cultivars with extensive root systems. Ahmed Bouaziz and Hassan Chekli, both from Morocco, determined the effect of micro-basins on runoff in a dryland environment, and show that, under weeded conditions, they increase water storage and wheat yield in non-drought years.

Finally, the special issue ends with a chapter by Steve Green, Brent Clothier, Tessa Mills, and Andrew Millar, all of New Zealand. They use a simple water-balance approach to model soil-water storage changes to better estimate irrigation requirements for a range of field crops in a maritime climate.

They show how much and how often the crops should be watered. The model can be used to consider irrigation requirements for any level of prescribed risk and to quantify specific water right allocations. Their chapter reveals the power of modelling in understanding water use in crop production.

The contributions show that important advances are being made to improve water use efficiency. With the new developments, hope remains that enough water will be available to provide crops for the growing millions.

M. B. Kirkham

Agricultural Water Conservation–
A Global Perspective

Paul W. Unger
Terry A. Howell

SUMMARY. Water for agriculture generally is adequate in humid regions, but water conservation often is needed in subhumid and semi-arid regions for good crop production, even with irrigation because of limited supplies. Increasingly, urban, industrial, environmental, and recreational users compete for agricultural water supplies. Although temporally and spatially variable, annual total supplies are relatively constant. The increasing competition, therefore, makes it imperative that agriculture does its share to conserve water to achieve greater production for an ever-increasing populace. In this report, we discuss basic principles of and some practices for achieving agricultural water conservation, both under dryland (rainfed) and irrigated conditions. *[Article copies available for a fee from The Haworth Document Delivery Service: 1-800-342-9678. E-mail address: getinfo@haworthpressinc.com <Website: http://www.haworthpressinc.com>]*

KEYWORDS. Dryland agriculture, irrigated agriculture, water infiltration, evaporation, water retention, weed control, irrigation methods, irrigation water management

INTRODUCTION

All plants depend on an adequate water supply for optimum growth and development. For terrestrial plants, water stored in soil from precipitation or

Paul W. Unger is Soil Scientist, and Terry A. Howell is Agricultural Engineer, U.S. Department of Agriculture–Agricultural Research Service, Conservation and Production Research Laboratory, P.O. Drawer 10, Bushland, TX 79012 USA.

[Haworth co-indexing entry note]: "Agricultural Water Conservation–A Global Perspective." Unger, Paul W., and Terry A. Howell. Co-published simultaneously in *Journal of Crop Production* (Food Products Press, an imprint of The Haworth Press, Inc.) Vol. 2, No. 2 (#4), 1999, pp. 1-36; and: *Water Use in Crop Production* (ed: M. B. Kirkham) Food Products Press, an imprint of The Haworth Press, Inc., 1999, pp. 1-36. Single or multiple copies of this article are available for a fee from The Haworth Document Delivery Service [1-800-342-9678, 9:00 a.m. - 5:00 p.m. (EST). E-mail address: getinfo@haworthpressinc.com].

irrigation sustains plants until the next precipitation or irrigation event. In humid regions, precipitation often is frequent enough so that plants seldom experience a water deficiency, and removal of excess water sometimes is required for successful crop production. As a result, water conservation for agricultural crops often receives little attention in humid regions. Short-term droughts, however, occur and water conservation can be beneficial for crop production, even in humid regions. Water conservation in humid regions may be especially beneficial on soils with low water holding capacity because 7 to 14 days without rain often causes severe plant water deficiencies and major crop yield reductions. We will give some examples for humid conditions, but will stress agricultural water conservation in subhumid and semiarid regions for dryland agriculture and semiarid and arid regions for irrigated agriculture.

Precipitation in subhumid and semiarid regions often is limited, with periods of various duration without precipitation occurring during the growing season of most crops. During such periods, the amount of plant available water in soil greatly affects growth and yield of dryland crops. For example, winter wheat (*Triticum aestivum* L.), grain sorghum (*Sorghum bicolor* [L.] Moench), and sunflower (*Helianthus annuus* L.) yields increased 7.2, 17.0, and 7.0 kg ha^{-1}, respectively, for each additional millimeter of plant-available water in Pullman soil (Torrertic Paleustoll) at planting time (Johnson, 1964; Jones and Hauser, 1975). Obviously, water conservation is highly important for dryland crop production in subhumid and semiarid regions, and water conservation has received much attention in such regions throughout the world.

Irrigation often is used for crop production where precipitation is limited, as in semiarid regions, and to extend production into arid regions. Sometimes, irrigation is used in subhumid and humid regions to supplement water from precipitation, especially during short-term droughts.

Successful irrigated agriculture depends on a reliable water supply, with the source being streams, reservoirs, or aquifers. Development of irrigated agriculture in a region usually is based on availability of adequate water. Subsequently, however, competition for water may develop to serve needs of urban, industrial, environmental, and recreational users, resulting in less water being available for irrigation. Also, the supply may be limited and is not being replenished in some aquifers. As a result, water removed for irrigation limits the amount available for future use, and declining water levels in aquifers will result in reduced pumping rates and greater energy requirements for pumping the water. The increasing competition for and declining supplies of water clearly show that less will be available for irrigation in the future and that irrigated agriculture must participate in water conservation efforts so that the needs of all users can be met. At present, agriculture is the largest consumer of water worldwide and is deemed largely inefficient in using the water (Gleick, 1998). Postel (1992) estimated worldwide water use efficiency by agriculture at only 40%.

DRYLAND AGRICULTURE REGIONS

Dryland agriculture, also called dry farming (Cannell and Dregne, 1983), has been defined in various ways. According to the SSSA (1996), dryland farming is "crop production without irrigation (rainfed agriculture)." In the strictest sense, this definition would include farming in humid regions where precipitation may be excessive for successful crop production, at least for some crops. Although dryland agriculture is rainfed agriculture, others (Cannell and Dregne, 1983; Stewart, 1988) defined dryland agriculture (farming) as agriculture without irrigation where precipitation is low and erratic in amount and distribution, and generally less than potential evapotranspiration during a major part of the year. Use of special water-conserving practices usually is required for successful crop production under such conditions.

Dryland agriculture is practiced on all continents, except Antarctica, with about 600 million ha of land (~40% of the world's land surface) devoted to dryland agriculture (Brady, 1988). Dryland agriculture long has been a major provider of food and fiber products, and increased production of these products will be required under dryland conditions because of the ever-increasing world population. To achieve this, improved water conservation and use will be required because the total amount of water available annually is relatively constant.

IRRIGATED AGRICULTURE REGIONS

As of 1996, about 263 million ha of land were irrigated in the world, with irrigation being done in 166 countries (FAO, 1998). Most irrigated land was in India (57.0 million ha), the People's Republic of China (PRC) (49.9 million ha), and the United States of America (USA) (21.4 million ha), with the total for other countries being ~135 million ha. The irrigated area in the USA is relatively constant, but increases were shown for India, the PRC, and the total for other countries (FAO, 1998). Water available annually for irrigation and other users (competitors for water) is relatively constant. Therefore, water available for irrigation needs to be used more efficiently to achieve the increased production needed for the ever-increasing world population. As on dryland, improved water conservation on irrigated land will play a major role in assuring that adequate water will be available to produce the required agricultural products.

PRINCIPLES OF WATER CONSERVATION

Globally, individual tracts of land devoted to crops range from fractions of hectares for subsistence farmers to thousands of hectares for large private or commercial farms. Technologies involved may range from use of human or animal power to large tractors. Because of these size and technology differences, all water conservation practices are not equally applicable or adaptable for all conditions. The principles of water conservation, however, are applicable for all conditions, regardless of tract or equipment size.

Under all conditions, water conservation for agricultural crops depends first on water infiltration into soil and then its retention in that portion of the soil where it subsequently can be extracted by crop roots. Effective infiltration depends on conditions being favorable for adequate water flow into soil and on sufficiently low runoff rates that result in adequate time for water to enter soil. To retain water for later use by crops, evaporation, deep percolation, and use by weeds must be prevented or minimized. Water transport characteristics of a soil strongly influence water infiltration, evaporation, and deep percolation rates (van Bavel and Hanks, 1983).

Runoff water is of no direct value to a crop unless it is captured and used for irrigation or it enters a stream from which it can be used for irrigation at another site. To achieve maximum infiltration, runoff should be minimized or avoided. Runoff is avoided or minimized when the application rate (precipitation or irrigation) is at or below the soil's infiltration rate. Matching the water application rate to the infiltration rate is possible with many irrigation systems, but runoff often occurs with many surface irrigation methods and with mechanical move sprinkler systems. The application rate with precipitation, however, is not controllable and management practices are needed to reduce or prevent runoff, thus providing adequate time for infiltration. Soil surface and profile conditions, including the antecedent water content, influence the rate at which water infiltrates a soil.

Although runoff may be minimized or avoided, a soil often is not filled to capacity with water during one or even several precipitation events under dryland conditions in a semiarid region. Under such conditions, water harvesting may be used to supply additional water to a site or fallowing may be used to increase water storage for the next crop.

Water storage in a soil depends on such factors as its texture, organic matter content, profile depth, and horizon characteristics. Infiltrated water in excess of that needed to fill a soil to capacity is lost to deep percolation unless an impermeable layer is present. When deep percolation is hindered, runoff most likely will be greater and water-logging may occur.

Water retained in a soil is subject to evaporative loss from the surface. Loss is greatest during the first stage when the rate depends on the net effect of water transmission rate to the surface and aboveground conditions such as wind speed, temperature, relative humidity, and radiant energy. Evaporation decreases rapidly during the second stage as the soil water supply decreases and when it depends on the rate of water movement to the surface. Third stage evaporation is low and controlled by adsorptive forces at the solid-liquid interface. The potential for decreasing soil water evaporation is greatest during the first two stages (Lemon, 1956). Methods for decreasing evaporation include decreasing turbulent water vapor transfer to the atmosphere (e.g., crop residues as a mulch), decreasing soil capillary continuity or capillary water flow to the surface (shallow tillage), and decreasing water-holding capacity of surface soil layers.

Water retained in a soil is subject to loss through transpiration by weeds. Weeds present before planting decrease the amount available for crop use. Those present during the growing season directly compete with crops for water in soil; also for light and space. In most cases, crop yields are reduced when weeds are not adequately controlled. Weed control is especially important for dryland crop production and for efficient use of irrigation water.

The above factors also affect water conservation under irrigated conditions. In addition, irrigation involves water conveyance from the supply point to the application site. Unless a closed system is used, water losses due to seepage, use by non-crop vegetation, and evaporation are possible. Seepage may be especially large from non-lined ditches.

Use of high-pressure sprinkler systems can result in large evaporative losses. Also, large deep percolation losses can occur when applied amounts exceed the soil's water storage capacity in the root zone. Deep percolation losses often occur with furrow or flood irrigation.

WATER CONSERVATION PRACTICES

Agricultural water conservation involves water storage in soil, except for that stored in reservoirs for irrigation. Numerous practices have been researched and are available for increasing soil water storage. Some are widely applicable; others only to highly specific conditions. Storage in reservoirs also generally is applicable only to highly specific conditions. We will emphasize the more widely applicable practices, but also will provide information regarding some other practices.

In addition to practices based on research, numerous indigenous and relatively simple practices are used by farmers in developing countries to obtain some water conservation benefits (Critchley et al., 1994; Gallacher, 1990). Evaluation and improvement of these practices with the farmers' participation could lead to improved water conservation in countries where more elaborate conservation practices would not be adaptable or acceptable.

Relationships among those factors important for improving water storage in soil, namely, increasing infiltration (reducing runoff), reducing evaporation, eliminating or reducing water use by weeds, and eliminating or reducing deep percolation are highly complex. In many irrigation situations, deep percolation for salinity control and management is required and desired for sustainable crop production.

Infiltration and Runoff

Although runoff and infiltration are closely related (water lost as runoff cannot infiltrate) and reducing runoff is essential for increasing infiltration, all water retained on land does not necessarily infiltrate a soil. Rather, some water retained in surface depressions evaporates before infiltration occurs, especially when a surface seal or another restrictive layer is present. Also, water retained in surface soil often evaporates before it can be used by plants because it does not move deeply enough to add to the soil water supply. In some instances, infiltrated water may move laterally due to an impeding horizon and enter a stream, thereby contributing to runoff.

A soil must be receptive to applied water and sufficient time must be available for satisfactory infiltration to occur. Development of a soil surface seal (or crust) is a major deterrent to infiltration. When raindrops strike bare soil, their energy may disperse soil aggregates, thus resulting in seal development and runoff. In contrast, surface residues, as those resulting from use of conservation tillage, dissipate raindrop energy, thus preventing or reducing aggregate dispersion and seal development, and resulting in greater infiltration. Surface residues also retard the rate of water flow across the surface, thus providing more time for infiltration.

Numerous studies involving no-tillage, a type of conservation tillage, have shown the value of crop residues retained on the surface for increasing infiltration and, therefore, the potential for greater soil water storage (e.g., Cogle et al., 1996; Gilley et al., 1986; Harrold and Edwards, 1972; O'Leary and Connor, 1997; Opoku and Vyn, 1997). In general, runoff increases with increases in surface slope, especially on bare soils. With no-tillage, however, surface slope has less effect on runoff (Table 1). Although water contents were not given, reducing runoff provided the opportunity to replenish the soil water supply, which is the goal for water conservation efforts under all conditions. Besides reducing runoff, use of no-tillage also greatly reduced erosion.

Use of tillage may reduce runoff (increase infiltration) when residue amounts are low because of low production (as by dryland crops), use for other purposes, or incorporation by previous tillage. Under such conditions, tillage can disrupt a surface seal (crust), create contour ridges, and increase surface roughness and plow-layer pore space, thus retaining more water on the surface and providing more time for infiltration (Hien et al., 1997; Muchi-

TABLE 1. Tillage effects on runoff and sediment yield from watersheds planted to corn at Coshocton, Ohio, USA, during a severe storm in July 1969.[a]

Tillage	Slope (%)	Rainfall (mm)	Runoff (mm)	Sediment yield (Mg ha^{-1})
Plowed, clean tilled, sloping rows	6.6	140	112	50.7
Plowed, clean tilled, contour rows	5.8	140	58	7.2
No-tillage, contour rows	20.7	129	64	0.07

[a] Adapted from Harold and Edwards (1972).

ri and Gichuki, 1983; Rawitz et al., 1983; Stroosnijder and Hoogmoed, 1984; Willcocks, 1984). Greater infiltration and, hence, greater water storage also can be achieved by disrupting slowly permeable or compact layers in the profile (Eck and Taylor, 1969; McConkey et al., 1997; Schneider and Mathers, 1970) or loosening soils subject to freezing (Pikul et al., 1996). These practices increase the depth at which water is stored, thus enlarging the zone in which roots proliferate.

Other practices for increasing infiltration and water storage include graded furrowing (Krantz et al., 1978; Pathak et al., 1985; Richardson, 1973), terracing (Beach and Dunning, 1995; Gallacher, 1990; Jones, 1981), furrow diking (tied ridging) (Gallacher, 1990; Jones and Clark, 1987; Vogel et al., 1994), strip cropping and growing vegetative barriers (Alegre and Rao, 1996; Gallacher, 1990; Sharma et al., 1997), and using LEPA (low energy precision application) irrigation. Water application efficiencies were > 95% with LEPA irrigation (Howell et al., 1995; Lyle and Bordovsky, 1983; Schneider and Howell, 1990). With LEPA irrigation, water often is applied to alternate diked furrows to temporarily detain water on the surface. Diked furrows can be used with most sprinkler methods to temporarily impound "excess" water to provide more time for infiltration.

Snow provides much of the water for crops in the northern U.S. Great Plains, Canadian Prairie Provinces, northern Europe, and northern Asia. A special case involving crop residues or vegetative barriers is their use to trap snow, thus achieving greater and more uniform soil water storage when the snow melts (Black and Aase, 1988; Campbell et al., 1992; Cutforth and McConkey, 1997; Steppuhn and Waddington, 1996). Under some conditions, snow is "plowed" into ridges to create barriers for greater trapping of snow during subsequent storms (De Jong and Steppuhn, 1983). Soil water storage from snow is highly variable, but up to 50 mm more storage occurred with than without residues or barriers in place. Vegetative factors influencing snow trapping include stubble height, barrier spacing, and barrier orientation relative to wind direction. Greater soil water contents resulting from trapped snow permitted more intensive cropping and resulted in greater crop yields. The barriers also provided microclimate benefits for the next crop. Under some conditions, the greater soil water contents contributed to development of saline seeps (Black and Siddoway, 1976), thus reducing crop yields. Careful matching of crops to the available water supply helps avoid the saline seep problem (Brown et al., 1982).

The surface of some soils is highly unstable, and runoff is common when the soil is not protected by residues or other runoff control practices. Some materials applied directly to soils or with irrigation water have resulted in major increases in infiltration as compared to that where the materials were not applied. Runoff was reduced sixfold as compared with that from untreated soil when phosphogypsum (PG) was applied at 10 Mg ha^{-1} to a ridged sandy soil in Israel under field conditions (Agassi et al., 1989). When PG was applied to a clay loam at 3.0 Mg ha^{-1}, runoff was less than from bare soil, but still greater than where wheat straw was applied at 2.2 Mg ha^{-1} (Benyamini and Unger, 1984). Anionic polymers [polyacrylamide (PAM) or starch copolymer solutions] injected into water used for furrow irrigating a silt loam in Idaho (USA) reduced soil loss in runoff 70% when applied at 0.7 kg ha^{-1} per irrigation and 97% when applied at 10 g m^{-3} of water. The treatments also increased net infiltration and lateral infiltration, probably because of less surface sealing and sediment movement (Lentz et al., 1992; Trout et al., 1995).

Evaporation

Precipitation storage as soil water during the interval between crops in a semiarid region such as the U.S. Great Plains usually is < 50%, with amounts much below that occurring in many cases (Jones and Popham, 1997; Unger, 1978, 1994). While runoff accounts for some water loss, Bertrand (1966) indicated ~60% of average annual precipitation may be lost directly from soil by evaporation. Evaporative losses can be especially large when most precipitation occurs in relatively small storms. For example, 1522 storms occurred at Bushland, Texas (USA), in the southern Great Plains from 1960 through 1979. Precipitation occurred at ~100 mm hour^{-1} for up to 10 minutes in some storms, but total precipitation was > 50 mm for only 11 storms and > 25 mm for only 73 storms (unpublished data, Conservation and Production Research Laboratory, Bushland, TX). Small storms result in limited soil wetting and significant evaporative losses. Consequently, water storage efficiencies at Bushland generally are low because most storms occur in summer when the evaporation potential is greatest (Jones and Popham, 1997; Unger, 1978, 1994). Evaporation from fully wetted soils, however, also results in major water losses (Plauborg, 1995).

Soil water evaporation is a highly complex process because it involves water movement as liquid or vapor in response to soil water potentials, soil temperature gradients, and atmospheric conditions. In addition, deep percolation may occur while evaporation occurs from the surface. As a result, determining evaporation under field conditions is difficult because of the interacting effects of water infiltration, distribution in soil, deep percolation, and subsequent evaporation.

Effects of many surface mulch treatments on soil water evaporation have been studied (Unger, 1995), with crop residues used as the mulch in many cases. Residue characteristics affecting evaporation are orientation (flat, matted, or standing), which affects layer porosity and thickness; layer uniformity; rainfall interception; reflectivity, which affects surface radiant energy balance; and aerodynamic roughness resulting from the residues (Van Doren and Allmaras, 1978).

Although difficult to measure, results of some field studies clearly showed that retention of crop residues on the soil surface reduced evaporation. During 5 weeks without precipitation, water loss was 23 mm from bare soil, but only 20 mm with flat, 19 mm with 0.75 flat-0.25 standing, and 15 mm with 0.50 flat-0.50 standing wheat residue on the surface (Smika, 1983). Standing residue was 0.46 m tall and the amount was 4600 kg ha^{-1} in all cases. Greater wind speed was needed to initiate water loss as the amount of standing residue increased and the water loss rate decreased with increasing amounts of standing straw at a given wind speed. Residue orientation also affected average surface soil temperature (47.8, 41.7, 39.6, and 32.2°C for the respective conditions), which, in turn, affected evaporation through its effect on vapor pressure of soil water (Smika, 1983). Nielsen et al. (1997) showed that potential evaporation decreased as residue height increased, with the height effect being especially important when stem density was < 215 m^{-2}. The height effect decreased with increasing stem densities.

One day after a 13.5 mm rain, soil water contents were similar to a 15-cm depth where conventional-, minimum-, and no-tillage treatments were imposed after winter wheat harvest at Akron, Colorado (USA). Surface residue amounts were 1200, 2200, and 2700 kg ha^{-1} with the respective treatments. After 34 days without more rain, soil had dried to a < 0.1 m^3 m^{-3} water content to a depth of 12 cm with conventional tillage and 9 cm with minimum tillage. Blade tillage had been performed to those depths 8 days before the rain. With no-tillage, soil had dried to that water content only to a 5-cm depth. Some loss occurred at greater depths with each treatment, but water content was greatest with no-tillage for which the surface residue amount was greatest (Smika, 1976).

Evaporation reduction in the above studies involving crop residues resulted primarily from reduced turbulent transfer of water vapor to the atmosphere. Another means of reducing evaporation is to reduce capillary water flow to the surface. Therefore, there long has been an interest in using dust mulching (also called soil mulching) to reduce soil water evaporation. Dust mulching is essentially a clean-tillage (residue-free) system that involves producing a loose, fine granular or powdery soil layer at the surface by shallow tillage or cultivation. Dust mulching, in general, is effective for reducing evaporation of water already present in soil (Abdullah et al., 1985; Jalota, 1993; Jalota and Prihar, 1990; Papendick, 1987; Singh et al., 1997). Therefore, it is applicable mainly to regions where a distinct wet (rainy) season is followed by a distinct dry season. It usually is ineffective where precipitation mainly occurs when the potential for evaporation is greatest, as in summer in the U.S. Great Plains, because much of the water evaporates before tillage can be performed (Jacks et al., 1955). Dust mulching also is not suitable for such regions because the frequent tillage needed to maintain the mulch results in the soil being highly susceptible to erosion. Another reason for poor results with dust mulching in a summer precipitation region is that tillage brings moist soil to the surface, which increased evaporation and, in turn, resulted in less water storage where stubble mulch rather than no-tillage was used (Jones and Popham, 1997).

Water Retention

Water retention is influenced mainly by a soil's texture, structure (aggregation and porosity), depth, and organic matter content. A soil's texture is an inherent trait resulting from the conditions under which the soil developed. Sandy soils generally have lower water holding capacities than finer-textured soils (higher silt and clay contents). Deep plowing to mix profile layers or to bring finer materials to the surface increased the water holding capacity of soils initially having a surface horizon with a high sand content (Harper and Brensing, 1950; Miller and Aarstad, 1972). Besides increasing water retention in a given volume of soil, deep plowing and profile mixing also increase the depth to which plant roots can proliferate and extract water (Eck and Taylor, 1969; McConkey et al., 1997; Schneider and Mathers, 1970). These operations require special equipment; are energy-intensive, costly, and time consuming; and are not widely used, except where major benefits can be achieved. Chiseling is a less intensive operation often used to disrupt restrictive zones at relatively shallow depths, especially in irrigated soils.

Organic matter influences soil water retention through its direct affinity for water and its effect on aggregation, both of which increase with increases in organic matter content. Returning all or most crop residues to a soil helps maintain or, under some conditions, increase the soil's organic matter content. Maintaining or increasing a soil's organic matter content under dryland conditions in a semiarid region such as the southern U.S. Great Plains, however, is difficult because residue production generally is low. Rather, soil organic matter contents generally decreased with continued cropping with clean or stubble mulch (sweep) tillage and tended to be maintained, but not increased, under no-tillage conditions (Potter, 1998; Unger, 1997).

Whereas increasing a soil's organic matter content to increase water retention is difficult, there long has been an interest in adding organic substances to soils to improve water conservation (Unger and Stewart, 1983). Applying organic substances to soil resulted in less runoff (Weakly, 1960) and evaporation (Olsen et al., 1964), but the potential was limited under field conditions because the substances had limited stability in soil and little effect on crop yields.

Some recent reports indicated that adding coal-derived humic substances (Piccolo et al., 1996) and synthetic polymers (Choudhary et al., 1995) to soils significantly increased water retention. In a laboratory study, adding humic substances to soil at a 0.05 g kg^{-1} rate increased the available water content by up to 5.2% as compared with untreated soil, with no further increases when applied at rates up to 1.0 g kg^{-1}. The 0.05 and 0.10 g kg^{-1} rates of application resulted in 40 and 120% increases in soil aggregate stability, respectively, which contributed to the greater water retention. Further studies were needed to evaluate the potential of the substances under field conditions (Piccolo et al., 1996). Also under laboratory conditions, Choudhary et al. (1995) added synthetic polymers to two soils at rates of 0.2, 0.4, and 0.6% on a dry weight basis. Increases in amount of polymer applied resulted in greater water conservation by increasing the soils' water holding capacity and by decreasing evaporation as compared with that of untreated soil. The water retention benefits achieved were attributed to the hydrophilic groups in molecules of the applied polymers (Piccolo et al., 1996).

Weed Control

Where the need for water conservation for crop production is critical, it is imperative that water use by weeds be avoided or minimized. This is especially the case under dryland conditions in semiarid regions because water use by weeds reduces the amount available for the crop, thus reducing crop yields. Avoiding competition between weeds and crops for water is important not only during a crop's growing season, but also before planting when storing as much water as possible for the crop to be grown is the goal. Weeds also compete with crops for light, nutrients, and space; therefore, their control is important under all cropping conditions.

Land under dryland conditions generally must be kept free of weeds to obtain maximum soil water storage at planting time. Until herbicides became available, a major reason for tillage was to control weeds. Now, tillage and/or herbicides can be used for weed control. Under some conditions, hand weeding may be practiced. Also, use of crop rotations reduces the severity of some weed problems (Wiese, 1983).

Regardless of the method, timely control is important because uncontrolled weeds may use about 5 mm of water per day from a soil (Wicks and Smika, 1973). When using tillage, it usually can be delayed until weeds use more water than that lost by evaporation, thus avoiding frequent tillage operations and, thereby, resulting in production cost and energy savings (Lavake and Wiese, 1979). Another consideration is that each tillage operation exposes moist soil to the atmosphere, thus also contributing to evaporative soil water losses. Good and Smika (1978), for example, showed that each tillage operation resulted in losing 5 to 8 mm of water from the exposed moist soil. An advantage of using tillage for weed control is that water loss due to transpiration stops almost immediately, thus preventing continued water loss that may occur when herbicides are used. Several tillage operations may be needed to maintain weed control and to obtain optimum water conservation and crop yields (Pressland and Batianoff, 1976).

Small-scale farmers in many countries such as those in sub-Saharan Africa commonly control weeds by hand (Twomlow et al., 1997). As with tillage, repeated weeding usually is needed to achieve optimum crop production. In Zimbabwe, for example, weeding at 2, 4, and 6 weeks after corn (*Zea mays* L.) emergence resulted in greater water use efficiency and grain yield than a single weeding at 2 weeks. The unweeded control treatment resulted in the driest soil and lowest yields.

Herbicides can be applied to prevent weed seed germination or to control existing weeds. Preventing germination would be ideal for preventing soil water loss due to transpiration by weeds. However, use of such herbicides may not prevent germination of all weed seeds in a given crop because some weeds are not controlled by the herbicide. Use of "safener-treated" crop seed (seed treated to prevent action of a herbicide) has extended the use of some herbicides to prevent weed seed germination (Jones and Popham, 1997).

For established weeds, timely control is highly important for minimizing their competition with crops for water. In general, small weeds are easier to control than large weeds (Wiese et al., 1966). Weeds not killed immediately continue to use soil water. Weed control with herbicides often becomes especially difficult when plants are stressed for water. Development of herbicide-tolerant crops through genetic engineering has greatly expanded the opportunity for using highly-effective, quick-acting herbicides to control problem weeds in some crops.

Cover crops maintain a cover on the soil surface, thereby "preventing soil erosion, improving water infiltration, maintaining and increasing organic carbon levels, and possibly improving soil productivity" (Tyler, 1998). Although generally not considered to be weeds, cover crops and weeds affect the water supply for subsequent crops similarly. Use of cover crops generally had little effect on the soil water supply for the next crop in humid and subhumid regions because of generally adequate precipitation. However, where the goal is to increase soil water storage for the next crop (e.g., under dryland conditions in the semiarid portion of the U.S. southern Great Plains), growing cover crops usually is not recommended (Unger and Vigil, 1998) because their use generally reduced the soil water content at planting time and yield of the next crop.

Multiple-Factor Water Conservation Practices

For studies under field conditions, factors resulting in soil water content differences usually are not clearly differentiated. Rather, at any given time, prevailing water contents reflect the combined effects of water infiltration, runoff, evaporation, retention, and weed control, which were discussed separately in foregoing sections. The literature pertaining to soil water conservation is vast. In this section, some selected examples of the combined effects of the different factors are given and discussed.

Fallowing

Fallowing is the practice of allowing cropland to remain idle during all or part of the growing season when a crop normally would be grown. Objectives often are to control weeds, accumulate soil water, and/or accumulate plant nutrients. Fallowing often is used under dryland conditions in semiarid regions, primarily to provide more time to increase soil water storage for the next crop, thus increasing the yield potential and reducing the probability of a crop failure. Use of fallowing generally increases the soil water content at planting of the next crop, but precipitation storage as soil water (known as fallow efficiency or water storage efficiency) often is low. This is especially the case where long fallow periods are used such as those involving winter and spring wheat in the Canadian Prairies and U.S. Great Plains.

The winter wheat-fallow and spring wheat-fallow systems result in one crop in 2 years; they involve 14 to 17 and ~21 months of fallow between crops, respectively. Use of these systems improved and stabilized crop production in the Great Plains starting early in the 20th century, but water storage efficiencies generally were < 20%. Through the introduction of improved equipment, crop residue management techniques, and weed control practices (including the use of herbicides), water storage efficiencies of ~50% have been achieved under some conditions (Smika, 1986). Water storage efficiencies and crop yields resulting from use of improved practices in the U.S. central Great Plains are illustrated in Table 2.

TABLE 2. Improvements in fallow systems with respect to soil water storage and wheat grain yields at Akron, Colorado, USA.[a]

Year	Tillage during fallow	Fallow water storage		Wheat yield (Mg ha[-1])
		(mm)	(% of precip.)[b]	
1916-30	Maximum; plow, harrow (dust mulch	102	19	1.07
1931-45	Conventional; shallow disk, rod weeder	118	24	1.16
1946-60	Improved conventional; begin stubble mulch in 1957	137	27	1.73
1961-75	Stubble mulch; begin minimum with herbicides in 1969	157	33	2.16
1976-90	Projected estimate; minimum, begin no-tillage in 1983	183	40	2.69

[a] Adapted from Greb (1979).
[b] Based on percentage of soil water storage of precipitation received from wheat harvest in July to end of fallow in September (14-month period).

Storage efficiencies are highly variable among years and generally are greater in northern regions (US northern Great Plains and Canadian Prairies) than in southern regions (US southern and central Great Plains). In the northern Great Plains, average storage efficiency was 28%, but ranged from 16 to 44% from 1957 to 1970 (Black and Bauer, 1986). Water storage efficiencies resulting from use of various cropping systems and tillage methods in the southern Great Plains are given in Table 3. With improved storage efficiencies, more intensive cropping is possible and some well-adapted systems have been developed.

Crop Selection and Cropping Systems

Crops (also crop varieties or cultivars) vary in length of growing season and usually have peak growth periods at different times of the year. Therefore, for optimum production, major water requirement periods of selected crops should closely match periods of greatest potential water availability (stored soil water or precipitation). For example, winter wheat in the southern and central Great Plains is maintained during the fall and winter months mainly by water contained in soil at planting time. Although some soil water may remain for the peak demand period in spring (April till June), best yields are obtained when favorable precipitation occurs during that period, which includes the period of greatest precipitation probability in the region. Therefore, winter wheat is well adapted for that region.

TABLE 3. Cropping system and tillage method effects on average water storage efficiency during fallow before grain sorghum and winter wheat crops at Bushland, Texas, USA, 1984-1993.[a]

Cropping system and tillage method	Storage efficiency (%)[b]
Fallow before sorghum	
Continuous sorghum–stubble mulch	27.3 (4.1)
Continuous sorghum–no-tillage	32.0 (4.5)
Wheat-sorghum-fallow–stubble mulch	16.5 (2.1)
Wheat-sorghum-fallow–no-tillage	21.0 (2.3)
Fallow before wheat	
Continuous wheat–stubble mulch	13.9 (4.0)
Continuous wheat–no-tillage	19.8 (4.1)
Wheat-fallow–stubble mulch	10.6 (1.8)
Wheat-fallow–no-tillage	11.1 (2.1)
Wheat-sorghum-fallow–stubble mulch	17.0 (2.0)
Wheat-sorghum-fallow–no-tillage	16.8 (2.0)

[a] Adapted from Jones and Popham (1997).
[b] Storage efficiency = soil water storage during fallow as a percentage of fallow-season precipitation. Values in parentheses are the standard error of the mean.

Summer crops also are well adapted for the southern and central Great Plains because their growing season roughly corresponds to the time when rain is most likely to occur. Summer crops, however, differ in growing season length and, therefore, vary in adaptability. For example, sugar beet (*Beta vulgaris* L.) has a long growing season, requires a large amount of water, and generally yields poorly on dryland. Grain sorghum has a shorter growing season, requires less water than beet, and generally yields well unless water is limited during the critical grain filling period. Grain sorghum yield also is strongly influenced by the soil water content at planting (Jones and Hauser, 1975; Unger and Baumhardt, 1999). Short season crops such as millet (*Pennisetum* spp.) and some hay crops require less water and, therefore, generally produce more with a given amount of water than wheat or sorghum (Greb, 1983).

For greatest water storage after crop harvest, a crop should use most plant-available water by the time it is harvested, thus providing a soil receptive to storing water. Of course, water remaining in soil at harvest may also be available for the next crop, especially when shallow- and deep-rooted crops are grown in rotation.

The foregoing pertained mainly to crops grown annually on the same tract of land (continuous or annual cropping). The crops mentioned, along with others [corn, sunflower (*Helianthus annuus* L.), etc.], generally are well adapted for use also in crop rotations. Use of crop rotations may provide more time for soil water storage (e.g., winter wheat-grain sorghum-fallow rotation; two crops in 3 years with 10 to 11 months of fallow between crops); greater extraction of soil water (use of shallow- and deep-rooted crops, mentioned above); and better weed, insect, and disease control (use of different pesticides, tillage methods, and other management practices). All of these have water conservation ramifications, and successful dryland crop production frequently involves the use of crop rotations.

The introduction of improved equipment, crop residue management techniques, and weed control practices has resulted in greater water storage efficiencies, thus providing an opportunity for more intensive cropping. Because water storage efficiency was generally low with the wheat-fallow system, it has been replaced by more intensive cropping systems under dryland conditions in many cases. Well-adapted systems include winter wheat-fallow-grain sorghum-fallow (two crops in 3 years) in the southern and central Great Plains (Jones and Popham, 1997; Norwood, 1992; Unger, 1994) and winter wheat-corn (or grain sorghum)-millet-fallow (three crops in 4 years) in the central Great Plains (Wood et al., 1991). In the northern Great Plains, systems of spring wheat-winter wheat-fallow (two crops in 3 years); safflower (*Carthamus tinctorius*)-barley (*Hordeum vulgare* L.)-winter wheat; spring wheat-corn-peas (*Pisum sativum*); spring wheat-winter wheat-sunflower; and spring wheat in rotation with soybean (*Glycine max* L.), peas, safflower, sunflower, buckwheat (*Fagopyrum esculentum* Moench), or canola (*Brassica* spp.) are being used (Black, 1986; Black and Tanaka, 1996; Unger and Vigil, 1998). Under some conditions, use of improved management practices for continuous (annual) cropping systems has increased soil water storage, thus resulting in greater total yields than for crops grown in rotation systems (Campbell et al., 1998; Jones and Popham, 1997). More intensive cropping was reported also by Amir and Sinclair (1996), Carroll et al. (1997), and Sandal and Acharya (1997).

Mulching

Although many materials are available, crop residues usually are used as the mulch under field conditions. In essence, conservation tillage (including no-tillage) is a mulch tillage system. By definition, conservation tillage is any system that results in at least 30% residue cover on the soil surface after crop planting to control water erosion. For wind erosion control, residues equivalent to 1000 kg ha^{-1} of small grain residues should be present. Besides controlling erosion, crop residues retained on the soil surface also provide

TABLE 4. Straw mulch effects on average soil water storage during fallow, storage efficiency, dryland grain sorghum yield, and water use efficiency for grain production, Bushland, Texas, USA, 1973-1976.[a]

Mulch rate (Mg ha^{-1})	Water storage (mm)[b]	Storage effic. (%)[b]	Grain yield (kg ha^{-1})	Water use effic. (kg m^{-3})[c]
0	72 c[d]	22.6 c	1.78 c	0.56
1	99 b	31.1 b	2.41 b	0.73
2	100 b	31.4 b	2.60 b	0.74
4	116 b	35.6 b	2.98 b	0.84
8	139 a	43.7 a	3.68 a	1.01
12	147 a	46.2 a	3.99 a	1.15

[a] Adapted from Unger (1978).
[b] Water storage determined to 1.8-m depth; precipitation during fallow averaged 318 mm; fallow was 10 to 11 months.
[c] Water use efficiency based on grain produced, growing season rainfall, and soil water content changes during growing season.
[d] Column values followed by the same letter are not significantly different at the 0.05 level (Duncan's multiple range test).

water conservation benefits (Table 4). Greatest water conservation resulted from the high residue treatments, but dryland crops usually produce < 4 Mg ha^{-1} of residue. Therefore, water storage usually is lower, but still greater than where some or most crop residues are incorporated by tillage, as reported extensively in the literature. Conservation tillage, especially no-tillage, is an effective water conservation practice, even under dryland conditions.

Vertical or slot mulching is a specialized type of mulching that involves opening a slot in soil with a suitable implement (e.g., a chisel) and filling the slot with crop residues or other materials (Ramig et al., 1983; Raper et al., 1998). The mulch-filled slot provides for rapid infiltration, provided the opening to the surface is not closed by subsequent tillage. On a soil subject to freezing in the state of Washington (USA), runoff from land planted to wheat was 10 mm with slot mulching compared with 114 mm with no-tillage, resulting in the potential to increase wheat yields by 1300 to 2000 kg ha^{-1} (Ramig et al., 1983).

Water Harvesting

Ancient stone mounds and water conduits in some countries indicate water harvesting has long been used to capture or divert storm runoff for application to land where crops are grown. The water may be applied directly to cropland or retained in reservoirs for irrigating a crop at a later time. The runoff may be from natural land surfaces or from surfaces treated to enhance runoff (Abu-Awwad and Shatanawi, 1997; Frasier and Myers, 1983; Greb, 1979; Laing, 1981; Lavee et al., 1997).

Direct application of harvested water to crops generally is practiced where precipitation is limited, as in semiarid to arid regions. The goal is to capture water falling on a given area and supplement it with runoff from a contributing area. The receiving area should be capable of retaining the initial and runoff water without adversely influencing the crop being grown. Systems used for direct application of the harvested water include level pans that receive water diverted from natural waterways (Greb, 1979); conservation bench terraces for which runoff from the natural upslope area is captured on the leveled downslope area between terraces (Zingg and Hauser, 1959); level, intermittent, fish scale, and discontinuous parallel terraces for which runoff from part of the land is captured by the terraces (Unger, 1996); and various types of microbasins. Land preparation for receiving harvested water directly generally involves limited modification of the soil surface.

Where runoff is stored for later crop use, the reservoir's capacity should be adequate to hold the amount needed for irrigation, with normal frequency of runoff events and reservoir "leakage" (percolation and evaporation) influencing the capacity. Water storage in a reservoir is most frequently used in subhumid and humid regions or where distinct rainy seasons are followed by distinct dry seasons, as in parts of India (Krantz et al., 1978) and other countries.

Water conserved and crop yields resulting from water harvesting are highly variable because runoff amount and timing relative to crop requirements are highly variable (Greb, 1979; Kaushik and Gautam, 1994), especially where runoff is directly used on the land. More reliable crop yields are possible when adequate runoff is stored and used as needed for irrigation.

Crop Termination Time

When grain crops such as corn, wheat, and grain sorghum reach physiological maturity, subsequent water use does not increase their yield. Water use after physiological maturity, however, could influence a crop's harvested yield by delaying lodging until harvest is possible. Because yield potential is not increased, terminating the crop at physiological maturity would halt soil water extraction and, thereby, conserve some water for the next crop. An alternative would be to terminate plant growth immediately after harvest for crops such as grain sorghum and cotton (*Gossypium hirsutum* L.) that have an indeterminate growing season where their growth is not terminated by cold weather. Of course, second or rattoon crops are possible under some conditions [e.g., grain sorghum, sugarcane (*Saccharum* sp.), and rice (*Oryza sativa*)]. The rattoon crop most likely would require less water than the first crop because limited additional plant development would be required.

Irrigation Water Delivery Systems and Irrigation Methods

Ideally, all irrigation water would be delivered to crops without loss and at the precise time to provide the greatest benefit. Irrigation delivery may involve transporting water from a sole supply like a dedicated reservoir or a single well where one person (or company) may have complete control. Most often, however, the water is from off-site sources and its transport in conveyances varies from pipelines under various pressures (low for gravity surface flow to high for sprinklers) or canals with small head differences above the field surface itself. Sources may be streams, reservoirs, or aquifers. Irrigation water supplies often involve many institutions and legal and/or social organizations that can have a myriad of rules, regulations, and/or laws as well as varying purposes for operation.

Goals for irrigation water conservation are to achieve the greatest economic benefits (perhaps even social or political benefits) from the water applied and to provide for sustainable agriculture. The water often is a shared resource and some application and operational losses (i.e., canal spillage, return flow into streams, surface runoff, required leaching for salinity control, etc.) are regained and subsequently used by downstream irrigators. Therefore, it is difficult to characterize irrigation water conservation without defining it on a hydrologic and/or irrigation district scale (Burt et al., 1997). Even if defined precisely, it is challenging to characterize all possible components and pathways for water losses and water movements. For this report, we discuss irrigation water conservation from a field-level perspective, but we recognize the critical importance of the off-farm delivery network for achieving any water conservation goal.

Each water conservation principle discussed for dryland agriculture is equally important for irrigated agriculture. The goal of irrigation is to use the greatest fraction of the applied water to meet the crop's transpiration need. Losses to runoff (from rain or irrigation), evaporation (from plant and soil surfaces), and excess deep percolation (except that needed to maintain root zone salinity at a safe level) remain the central components of inefficient irrigation and offer the pathways for achieving enhanced irrigation water conservation. Spatial distribution of rainfall cannot be controlled, but spatial uniformity of irrigation applications remain important for successful irrigation water conservation. Irrigation spatial and temporal distribution are controlled exclusively by management and the method used. "Irrigation management consists of determining when to irrigate, the amount to apply at each irrigation and during each stage of plant growth, and the operation and maintenance of the irrigation system" (Hoffman et al., 1990; p. 9). Irrigation timing depends largely on the crop and soil water status, but the delivery schedule may be controlled by the water supplier, which can impede a producer's water conservation goals. The desired application amount also remains intertwined with the delivery schedule, crop need, and application technology. Likewise, operation and maintenance needs are directly impacted by the application method and technologies. In some cases, the crop grown may dictate using a certain application technology (i.e., to keep water off the fruit or plant).

Irrigation Technology for Conserving Water

In most countries, some form of surface irrigation technology is still used mainly because capital to acquire newer technology may be limited, skills for using the newer technology may be unavailable, or institutions desire to use manual labor (to maintain and support an agrarian populous). In the Jiftlik Valley in Jordan, switching from flood to drip-trickle irrigation (higher technology) resulted in increasing the irrigated area 10-fold while using the same amount of water. In addition, use of the drip-trickle method allowed more intensive cropping, which resulted in greater labor use, allowed repayment of loans for equipment in 3 to 4 years, increased income 13- to 15-fold, and increased off-farm benefits (commercial inputs) eight-fold (Keen, 1991). Improving irrigation application efficiency on the farm may not improve it on the larger-scale hydrologic or district level unless that change results in smaller non-recoverable losses (i.e., to non-reusable saline waters, to the vadose zone beneath the crop root zone that will not move to recoverable groundwater, etc.).

Surface irrigation often is termed "inefficient" because of large deep percolation and/or runoff losses that result from relying on soil to transport and distribute the water. Musick et al. (1983) greatly reduced deep percolation losses by using tractor traffic and wide furrow spacings (alternate furrows) on a permeable soil, and the practices did not reduce corn yields. Surface irrigation technology can be efficient on a farm or field basis when runoff water is captured and reused, and when managed to avoid or minimize percolation losses. Surface irrigation often is termed "low tech" because it mainly involves manual labor for water control, but it can involve many "high tech" (e.g., automated controls on canals and pipelines using radio, satellite, or cellular telephone communications) components.

Advanced surface irrigation technologies can range from moderately "high tech" [e.g., automated surge flow using micro-computer controlled valves to reduce field runoff while achieving a more even irrigation (Bishop et al., 1981; Kemper et al., 1988)] to precise laser-leveling of irrigated basins (Dedrick et al., 1982). Other automated devices for improving surface irrigation range from simple valves to cablegation (Kemper et al., 1981, 1985; Kruse et al., 1990). Also, as previously mentioned, treating water with PAM can enhance infiltration and reduce erosion. Achieving a high surface irrigation efficiency level requires keen management and knowledge of irrigation hydraulics, on-site soil processes (e.g., infiltration), and soil variability, regardless of system sophistication.

The main limitations and challenges for surface irrigation remain avoiding excessive deep percolation and reducing and/or eliminating runoff. Stewart et al. (1983) developed the LID (limited irrigation-dryland) furrow-irrigation system that involved limited surface applications to avoid deep percolation at the input end while not irrigating the lower end where furrow dikes were installed to impound water from rain. The system is applicable for use in continental-climate regions where some growing season rain occurs, but where the irrigation water supply is limited (e.g., many semiarid regions). Improved water use efficiency resulted from making better use of rainfall and maximizing the benefit from irrigation.

Canal linings (cement or flexible membranes) and underground pipelines [cement or polyvinyl chloride (PVC)] are highly effective for reducing irrigation water transport losses. Use of gated pipes (aluminum, PVC, or flexible materials) can reduce surface ditch seepage and spillage losses. Also, tailwater (irrigation runoff) reuse can reduce net irrigation water losses. Because most surface irrigation involves low pressures, energy required for pumping water is low, except when the source is a deep well.

Sprinkler irrigation technology can be quite varied also (Keller and Bliesner, 1990). Goals for sprinkler irrigation are to "remove" soil from its conveyance role by using pressurized pipelines and to use the kinetic force of the pressurized water to distribute the water in droplet form (like rain) directly to the crop and/or soil. System pressure and nozzle diameter affect droplet diameter and, hence, the kinetic energy that drops impart at the soil surface. Large droplets can break down surface soil aggregates, cause surface sealing, and impede water infiltration. Small drops can evaporate more quickly and drift from the target, thus reducing the amount of water reaching the crop.

Solid set and mechanical types (e.g., center pivots) can be automatically and/or remotely controlled without much difficulty. Use of sprinklers should eliminate or allow better control of deep percolation losses and practically eliminate runoff from irrigation, but uneven water distribution due to system hydraulics or wind effects on spray patterns are possible. Using lower angle and closer sprinkler or spray head spacings can reduce wind effects on water distribution.

Center pivots can be equipped with spray heads that cover a smaller wetted diameter, are closer to the ground or crop to reduce wind effects and evaporation, and operate at lower pressures (Gilley and Mielke, 1980). Use of these devices, however, can result in instantaneous water application rates that exceed the soil's infiltration rate and, therefore, surface water redistribution and/or runoff. For example, Clothier and Green (1994) reported extreme macropore flow and nonuniform soil wetting when the application rate was

102 mm hour^{-1} as compared to that when the rate was 4 mm hour^{-1}. Lyle and Bordovsky (1981, 1983) developed LEPA (low energy precision application) technology for center pivots and lateral-move machines to eliminate evaporative losses, surface redistribution, and runoff. LEPA irrigation is intended for use in conjunction with furrow dikes that impound irrigation and rain water. Users of LEPA system usually apply water to alternate furrows through furrow bubbles or drag socks (Fangmeier et al., 1990). Grain crop yields differed little when adequate water was applied using LEPA and spray irrigation at Bushland, which is at a semiarid site (Schneider and Howell, 1990, 1999). Schneider (1999) reviewed much of the literature on LEPA and spray irrigations and concluded, based on efficiency and uniformity, that neither method "could be considered inherently superior to the other." However, when irrigation capacity (flow rate per unit area) becomes low and deficit irrigation is intentionally practiced, as for cotton in the U.S. Southern High Plains, then LEPA irrigation would be preferable.

Drip and trickle irrigation, now widely called microirrigation (MI) (Camp, 1998; Kruse et al., 1990), was developed mainly in Israel and is now used worldwide. MI is used on over half of Israel's irrigated land and on over 400,000 ha in California (Gleick, 1998). With MI, objectives are to "remove" soil and air as distribution mediums, as occurs with surface and sprinkler irrigation, and to irrigate only the minimum root zone volume needed for each plant. A range of technologies encompass both surface and subsurface MI, and many types of applicators are available [drippers, line-source pipes, bubblers, small spray heads, and even small sprinklers (or rotators)].

Subsurface MI systems, called SDI (subsurface drip irrigation), are installed at soil depths ranging from a few centimeters (may be placed on the surface, then covered with cultivators) to, for example, at 30 cm (installed with special chisel shanks). Deeper placements make seed germination difficult, and water from rain or portable sprinklers may be needed for crop establishment. Lateral line spacing for SDI systems can vary, depending on the crop grown and its culture. For field crops, one line often is placed midway between two rows to reduce the cost.

The main intent of using MI is to apply the precise amount of water needed by each plant at exactly the time when it needs it (Nakayama and Bucks, 1986). MI may not result in wetting the whole soil surface as with most surface irrigation and sprinkler systems, but the area (or root volume) is irrigated more frequently. Typically, MI may involve irrigation intervals of one or a few days, but can involve multiple "pulses" in a single day.

MI involves a massive pipe network compared to that with surface or sprinkler irrigation, but the pipes usually are small because of low flow rates. Also, material (polyethylene, PE, PVC, etc.) costs are lower because a lower pressure (70 to 140 kPa) can be used. Because MI involves an extensive pipe network, it was first successfully used in orchards and vineyards with low plant densities (Abbott, 1984). Now, MI is used for many row crops, but more often for high-value vegetable crops. For higher-valued vegetables in rows, it often is used under a plastic mulch. SDI is now commonly used for row crops (e.g., cotton and corn) and for vegetables. MI systems are easily automated and controlled with devices such as simple timers and microcomputers, and they easily can be used to apply nutrients to crops.

Runoff should not occur with MI because water applications are small (ranging from 2-20 mm, but typically 5-10 mm). Deep percolation can be controlled more easily with the small applications. Also, there is less dependence on water storage in the root zone because water can be applied more frequently than where larger amounts are required to achieve uniform coverage with surface irrigation methods. Use of SDI can even reduce evaporation because the soil surface usually is not wetted. In practice, with many SDI systems, except those installed deeper than 30 cm, some surface wetting occurs due to capillary water flow and total elimination of soil water evaporation should not be expected. Also, significant evaporative losses may occur because the area is irrigated more frequently. Many times, however, the wetted area is beneath the crop canopy and evaporation might still be low.

Use of MI requires water filtration and/or chemical treatment to avoid plugging the small passageways by sand or other inorganic materials, bicarbonate (lime) and iron (ochre) deposits, and slime-forming organisms. Plugging can result in poor performance (low uniformity) or complete failures of systems in some cases. Although many water filtration and water treatment functions can be automated, careful operator attention is required.

Irrigation Management for Conserving Water

As defined previously, irrigation management encompasses more than just decisions on when and how much to irrigate. Operation and maintenance are critical elements, but they depend on the specific irrigation hardware being used. Maintenance may range from installing and maintaining a surface ditch to maintaining the intricate mechanism of the tower drive for a center pivot sprinkler. Daily maintenance may be needed in some cases; possibly only annual checking in others. Proper equipment maintenance can avoid breakdowns at critical times when a missed irrigation would be highly detrimental for a crop.

Irrigation management is broadly related to irrigation scheduling. Although simple in concept (when to irrigate and how much water to apply), it is complex for the "whole" farm decision making process that involves strategic (before the season) and tactical (on the spot day-to-day) planning. The goal is to decide how to achieve the greatest net return from the fixed and variable costs and the value of the crop produced, subject to all constraints (land, labor, water, environment, salinity, legal, etc.). Thus, irrigation water management and its conservation may not always go hand in hand.

When to use a preplant irrigation is one strategic decision that can greatly affect subsequent decisions about irrigating. A preplant irrigation may be used for weed seed germination, profile water replenishment, leaching, seed bed preparation, etc. Whereas significant profile replenishment with water is difficult with MI and most sprinkler systems, excessive infiltration rates can sometimes lead to large percolation losses for the first surface-applied irrigation after primary tillage. Generally, when rainfall before planting is near or above normal, preplant irrigations do not increase crop yields (Musick and Lamm, 1990). In some cases, early spring soil water loss rates and low rainfall may dictate that a preplant irrigation be made for a summer crop (Musick et al., 1971). When needed, it should be carefully planned and executed to minimize deep percolation (unless leaching is desired) and applied shortly before planting.

Irrigation scheduling can involve using a wide range of tools, depending on several circumstances, including the irrigation method used. For each method, an "optimum" application range may be most appropriate. With surface irrigation methods, water typically is applied more efficiently and evenly when the amounts range from 70 to 120 mm, depending on the soil type, surface slope, and field geometry (length of run, furrow spacing, border width, etc.). Traditional sprinkler methods may be more suited for applications of ~10 to 50 mm. For MI, amounts ranging from 5 to 25 mm might be better, depending on the soil. These "optimum application ranges" will be site and system specific, but a few field trials and routine evaluations (Merriam and Keller, 1978) can be used to identify the operational parameters needed for achieving the desired level of irrigation uniformity and efficiency.

The desired irrigation frequency (F, days) is a direct function of application depth and irrigation capacity (flow rate per unit area), and can by computed as

$$F = D_0/(Q \times 86,400) \qquad [1]$$

where D_0 is optimum application depth (mm), Q is irrigation capacity (L s^{-1} m^{-2}), and 86,400 (a constant) is seconds in a day. Irrigation capacity is determined by the supply rate and the area being irrigated. It is closely aligned with a crop's "peak" irrigation requirement rate (usually expressed in mm d^{-1}). This rate is largely determined by the "peak" ET rate, and is influenced by crop type, the environment, "effective" rainfall, soil type (water holding capacity and depletion permissible without reducing crop yield potential), and irrigation system efficiency.

The peak requirement rate is an irrigation system design parameter that influences many aspects of irrigation, including scheduling. It affects system fixed costs because it determines the pipe sizes needed for that flow rate and variable costs because it affects pumping costs that are a function of the flow rate. Therefore, it is desirable to keep the irrigation capacity (Q) as small as practical and at an acceptable level of risk of not being able to meet the desired crop ET rate, but as large as possible to provide the greatest flexibility in irrigation scheduling.

Irrigation timing affects water conservation in two important ways (Martin et al., 1990). One pertains to the earliest date to irrigate without having appreciable water losses (typically runoff and deep percolation); it depends largely on the irrigation system and the soil's water content and water holding capacity. The second pertains to the date for the last irrigation without inflicting a significant water deficit and "potential" yield loss on the crop. It depends on the soil, crop grown, crop growth stage, and expected ET rate. In this case, the soil profile likely will not be filled to capacity, thus providing an opportunity for water storage from rain. The scheduling decisions will be subject to weather forecasts (rain and other parameters that affect ET rates).

Irrigation timing decisions can be based on simple calendars (based on "normal" ET and precipitation) (Hill and Allen, 1996), checkbook type approaches (summations of water additions and consumptive use), tracking crop water use with computer models (based on crop ET or growth), or direct sampling of the soil or crop water status. The decision should incorporate the crop's growth stage and its sensitivity to water deficits at that particular stage. Phene et al. (1990) reviewed many techniques for sensing a crop's need for irrigation. Besides determining when to irrigate, field sampling (soil or crop water status sensing) is critical for evaluating system performance (spotting areas of poor coverage or where system errors and/or malfunctions may have occurred).

Remote sensing (aerial photography or satellite imagery) is an additional useful irrigation management tool. Good crop or soil sensing along with remote sensing can guide irrigation scheduling models as well. As such, ET modeling and sensing (remote and ground based) should be regarded as complementary rather than individual or mutually exclusive tools for irrigation scheduling.

Conserving water through irrigation management largely rests on the irrigation supply capacity (irrigation capacity and/or any legal water use constraints), crop response (yield and/or quality) to irrigation water, and irrigation economics (fixed and variable irrigation costs) (English et al., 1990). Certainly, excessive irrigations that do not contribute to meeting crop water requirements (including leaching to control salinity) should be eliminated (Clothier, 1989). Increasing an irrigation system's efficiency and enhancing its uniformity should be considered next. All irrigations involve some nonuniformity (some areas receive more water while others receive less than the mean). Small under irrigations (5-10%) may be undetectable in most cases and have not affected crop yields in most studies (in some, crop quality concerns occurred).

Water for irrigation is limited in many parts of the world (Gleick, 1998). Adequate water for fully meeting crop needs is available on few farms in the western part of the U.S. Southern High Plains (Musick et al., 1987). In India, the National Water Commission based irrigation planning on a "50% dependable" water supply (Chitale, 1987). Use of deficit irrigation can be effective for soils with plant available water contents exceeding 125 mm (Keller and Bliesner, 1990) and often results in crop yields less than the maximum attainable, but reduces irrigation water use, enhances crop water use efficiency, and improves the capture and use of rainfall. However, soil salinity levels must be monitored and appropriate leaching and reclamation measures must be implemented to protect the soil from salinization in many cases.

CONCLUSIONS

All plants depend on an adequate water for optimum growth and yield. Globally, adequate water usually is available in humid regions, but limited precipitation in subhumid and semiarid regions often limits the supply for nonirrigated crop production. When available, water from streams, reservoirs, or aquifers often is used for irrigation in subhumid and semiarid regions, and to extend crop production into arid regions. Sometimes, crops are irrigated in humid regions. Our emphasis, however, was on water conservation for dryland crops in subhumid and semiarid regions and irrigated crops in subhumid to arid regions.

The total amount of water globally available annually for all purposes is relatively constant, but highly variable temporally and spatially, especially in subhumid, semiarid, and arid regions. Also, urban, industrial, environmental, and recreational users increasingly are competing with agriculture for available supplies. The ever-increasing world population, however, requires an ever-increasing supply of food and fiber. To meet this demand, agriculture must produce more with less water, and agriculture must do its share to conserve water so that adequate water will be available for all users.

Crop production requires a large amount of water. Much water potentially available for crop use, however, is not conserved, and water initially conserved often is not used efficiently. Under all conditions, agricultural water conservation depends on water infiltration into soil and its retention for later extraction by plant roots. Water conservation on dryland and with irrigation often involves reducing water losses due to runoff, evaporation, deep percolation, and use by weeds, and increasing water retention in the soil profile. Under some conditions, however, runoff may be captured down slope for immediate use by crops, stored in reservoirs for later irrigation, or enter streams and used for other purposes, including irrigation. Also, water percolation to depths beyond the root zone is allowed to control salinity in some soils.

The principles of agricultural water conservation are discussed, and they are applicable regardless of tract and equipment size or technology level involved. Although most practices involving the principles are suitable for dryland and irrigated conditions, the size and technology level constraints sometimes limit the adaptability of some practices to achieve the conservation goals. In addition, water conservation for irrigated agriculture is influenced by the water delivery system, irrigation method, and level of technology used, and by management decisions. With good management and adoption of appropriate practices, improved agricultural water conservation and subsequent use of that water for greater crop production are possible under dryland and irrigated conditions, thus helping to meet the water needs of all users and providing for the food and fiber needs of the increasing global population.

REFERENCES

Abdullah, S.M., R. Horton, and D. Kirkham. (1985). Soil water evaporation suppression by sand mulches. *Soil Science* 139: 357-361.

Abbot, J.S. (1984). Micro irrigation–World wide usage. *ICID Bulletin* 33: 4-9.

Abu-Awwad, A.M. and M.R. Shatanawi. (1997). Water harvesting and infiltration in arid areas affected by surface crust: Examples from Jordan. *Journal of Arid Environments* 37: 443-452.

Agassi, M., I. Shainberg, D. Warrington, and M. Ben-Hur. (1989). Runoff and erosion control in potato fields. *Soil Science* 148: 149-154.

Alegre, J.C. and M.R. Rao. (1996). Soil and water conservation by contour hedging in the humid tropics of Peru. *Agriculture, Ecosystems & Environment* 57: 17-25.

Amir, J. and T.R. Sinclair. (1996). A straw mulch system to allow continuous wheat production in an arid climate. *Field Crops Research* 47: 21-31.

Beach, T. and N.P. Dunning. (1995). Ancient Maya terracing and modern conservation in the Petén rain forest of Guatemala. *Journal of Soil and Water Conservation* 50: 138-145.

Benyamini, Y. and P.W. Unger. (1984). Crust development under simulated rainfall on four soils. *Agronomy Abstracts*, pp. 243-244.

Bertrand, A.R. (1966). Water conservation through improved practices. In *Plant Environment and Efficient Water Use*, eds. W.H. Pierre, D. Kirkham, and R. Shaw. Madison, WI: American Society of Agronomy, pp. 207-235.

Bishop, A.A., W.R. Walker, N.L. Allen, and G.J. Poole. (1981). Furrow advance rates under surge flow systems. *Journal of Irrigation and Drainage Division (ASCE)* 107(IR3): 257-264.

Black, A.L. (1986). Resources and problems in the northern Great Plains area. In *Planning and Management of Water Conservation Systems in the Great Plains States, Proceedings of a Workshop*, Lincoln, NE, October 1985. Lincoln, NE: U.S. Department of Agriculture, Soil Conservation Service, Midwest National Technical Center, pp. 25-38.

Black, A.L. and J.K. Aase. (1988). The use of perennial herbaceous barriers for water conservation and the protection of soils and crops. *Agriculture, Ecosystems and Environment* 22/23: 135-148.

Black, A.L. and A. Bauer. (1986). Soil water conservation strategies for Northern Great Plains. In *Planning and Management of Water Conservation Systems in the Great Plains States, Proceedings of a Workshop*, Lincoln, NE, October 1985. Lincoln, NE: U.S. Department of Agriculture, Soil Conservation Service, Midwest National Technical Center, pp. 76-86.

Black, A.L. and F.H. Siddoway. (1976). Dryland cropping sequences within a tall wheatgrass barrier system. *Journal of Soil and Water Conservation* 31: 101-105.

Black, A.L. and D.L. Tanaka. (1996). A conservation tillage-cropping systems study in the northern Great Plains of the USA. In *Soil Organic Matter in Temperate Agroecosystems*, eds. E.A. Paul, K.A. Paustian, E.T. Elliott, and C.V. Cole. Boca Raton, FL: Lewis Publishers, pp. 335-342.

Brady, N.C. (1988). Scientific and technical challenges in dryland agriculture. In *Challenges in Dryland Agriculture–A Global Perspective*, eds. P.W. Unger, T.V. Sneed, W.R. Jordan, and R. Jensen. College Station, TX: Texas Agricultural Experiment Station, pp. 6-12.

Brown, P.L., A.D. Halvorson, F.H. Siddoway, H.F. Mayland, and M.R. Miller. (1982). Saline-seep diagnosis, control, and reclamation. *U.S. Department of Agriculture Conservation Research Report No. 30*. Washington, DC: U.S. Government Printing Office.

Burt, C.M., A.J. Clemmens, T.S. Strelkoff, K.H. Solomon, R.D. Bliesner, L.A. Hardy, T.A. Howell, and D.E. Eisenhauer (1997). Irrigation performance measures: Efficiency and uniformity. *Journal of Irrigation and Drainage Engineering* 123: 423-442.

Camp, C.R. (1998). Subsurface drip irrigation: A review. *Transactions of the ASAE* 41: 1353-1367.

Campbell, C.A., B.G. McConkey, V.O. Biederbeck, R.P. Zentner, D. Curtin, and M.R. Peru. (1998). Long-term effects of tillage and fallow-frequency on soil quality attributes in a clay soil in semiarid southwestern Saskatchewan. *Soil & Tillage Research* 46: 135-144.

Campbell, C.A., B.G. McConkey, R.P. Zentner, F. Selles, and F.B. Dyck. (1992). Benefits of wheat stubble strips for conserving snow in southwestern Saskatchewan. *Journal of Soil and Water Conservation* 47: 112-115.

Cannell, G.H. and H. E. Dregne. (1983). Regional setting. In *Dryland Agriculture, Monograph 23*, eds. H.E. Dregne and W.O. Willis. Madison, WI: American Society of Agronomy, Inc., Crop Science Society of America, Inc., and Soil Science Society of America, Inc., pp. 3-17.

Carroll, C., H. Halpin, P. Burger, K. Bell, M.M. Sallaway, and D.F. Yule. (1997). The effect of crop type, crop rotation, and tillage practice on runoff and soil loss on a Vertisol in central Queensland. *Australian Journal of Soil Research* 35: 925-938.

Chitale, M.A. (1987). Water management in drought prone areas. *Water Supply* 5:11-130. Oxford.

Choudhary, M.I., A.A. Shalaby, and A.M. Al-Omran. (1995). Water holding capacity and evaporation of calcareous soils as affected by four synthetic polymers. *Communications in Soil Science and Plant Analysis* 26(13&14): 2205-2215.

Clothier, B.E. (1989). Research imperatives for irrigation science. *Journal of Irrigation and Drainage Engineering (ASCE)* 115: 421-448.

Clothier, B.E. and S.R. Green. (1994). Rootzone processes and the efficient use of irrigation water. *Agricultural Water Management* 25: 1-12.

Cogle, A.L., M. Littleboy, K.P.C. Rao, G.D. Smith, and D.F. Yule. (1996). Soil management and production of Alfisols in the semi-arid tropics: III. Long-term effects on water conservation and production. *Australian Journal of Soil Research* 34: 103-111.

Critchley, W.R.S., C. Reij, and T.J. Willcocks. (1994). Indigenous soil and water conservation: A review of the state of knowledge and prospects for building on traditions. *Land Degradation & Rehabilitation* 5: 293-314.

Cutforth, H.W. and B.G. McConkey. (1997). Stubble height effects on microclimate, yield and water use efficiency of spring wheat grown in a semiarid climate on the Canadian prairies. *Canadian Journal of Plant Science* 77: 359-366.

de Jong, E. and H. Steppuhn. (1983). Water conservation: Canadian Prairies, In *Dryland Agriculture*, ed. H.E. Dregne and W.O. Willis. Madison, WI: American Society of Agronomy, Inc., Crop Science Society of America, Inc., and Soil Science Society of America, Inc., pp. 89-104.

Dedrick, A.R., L.J. Erie, and A.J. Clemmens. (1982). Level-basin irrigation. In *Advances in Irrigation*, Volume 1. ed. D. Hillel. New York, NY: Academic Press, Inc., pp. 105-145.

Eck, H.V. and H.M. Taylor. (1969). Profile modification of a slowly permeable soil. *Soil Science Society of America Proceedings* 33: 779-783.

English, M.J., J.T. Musick, and V.V.N. Murty. (1990). Deficit irrigation. In *Management of Farm Irrigation Systems*, eds. G.J. Hoffman, T.A. Howell, and K.H. Solomon. St. Joseph, MI: American Society of Agricultural Engineers, pp. 631-663.

Fangmeier, D.D., W.F. Voltman, and S. Eftekharzadeh. (1990). Uniformity of LEPA irrigation systems with furrow drops. *Transactions of the ASAE* 33: 1907-1912.

FAO (Food and Agriculture Organization of the United Nations, Rome, Italy). (1998). World irrigated area. *The FAOSTAT Database, FAOSTAT Agricultural Data*. http://apps.fao.org/lim500/nph-wrap.pl?Irrigation&Domain=LUI&servlet=1

Frasier, G.W. and L.E. Myers. (1983). *Handbook of Water Harvesting*, Agriculture Handbook 600. Washington, DC: U.S. Department of Agriculture, Agricultural Research Service.

Gallacher, R.N. (1990). The search for low-input soil and water conservation techniques. In *Topics in Applied Resource Management*, Volume 2, *Experiences with Available Conservation Technologies*, eds. E. Baum, P. Wolff, and M.A. Zöbisch. Witzenhausen, Federal Republic of Germany: German Institute for Tropical and Subtropical Agriculture, pp. 11-37.

Gilley, J.R. and L.N. Mielke. (1980). Conserving energy with low-pressure center pivots. *Journal of Irrigation and Drainage Division (ASCE)* 106(IR1): 49-59.

Gilley, J.E., S.C. Finkner, R.G. Spomer, and L.N. Mielke. (1986). Runoff and erosion as affected by corn residues: Part I. Total losses. *Transactions of the ASAE* 29: 157-160.

Gleick, P.H. (1998). The world's water, 1998-1999. Washington, DC: Island Press.

Good, L.G. and D.E. Smika (1978). Chemical fallow for soil and water conservation in the Great Plains. *Journal of Soil and Water Conservation* 33: 89-90.

Greb, B.W. (1983). Water conservation: Central Great Plains, In *Dryland Agriculture*, ed. H.E. Dregne and W.O. Willis. Madison, WI: American Society of Agronomy, Inc., Crop Science Society of America, Inc., and Soil Science Society of America, Inc., pp. 57-72.

Greb, B.W. (1979). Reducing drought effects on croplands in the west-central Great Plains. *U.S. Department of Agriculture Bulletin No. 420*. Washington, DC: U.S. Government Printing Office, 31 pp.

Harper, J. and O.H. Brensing. (1950). Deep plowing to improve sandy land. *Bulletin B-362*. Stillwater, OK: Oklahoma Agricultural Experiment Station.

Harrold, L.L. and W.M. Edwards. (1972). A severe test of no-till corn. *Journal of Soil and Water Conservation* 27: 30.

Hien, F.G., M. Rietkerk, and L. Stroosnijder. (1997). Soil variability and effectiveness of soil and water conservation in the Sahel. *Arid Soil Research and Rehabilitation* 11: 1-8.

Hill, R.W. and R.G. Allen. (1996). Simple irrigation scheduling calendars. *Journal of Irrigation and Drainage Engineering (ASCE)* 122: 107-111.

Hoffman, G.J., T.A. Howell, and K.H. Solomon (eds.). (1990). *Management of Farm Irrigation Systems*. St. Joseph, MI: American Society of Agricultural Engineers, 1040 p.

Howell, T.A., A. Yazar, A.D. Schneider, D.A. Dusek, and K.S. Copeland. (1995). Yield and water use efficiency of corn in response to LEPA irrigation. *Transactions of the ASAE* 38: 1737-1747.

Jacks, G.V., W.D. Brind, and R. Smith. (1955). Mulching. *Commonwealth Bureaux of Soil Science (England) Technical Communication 49*. Farnham Royal, Bucks., England: Commonwealth Agricultural Bureaux.

Jalota, S.K. (1993). Evaporation through a soil mulch in relation to mulch characteristics and evaporativity. *Australian Journal of Soil Research* 31: 131-136.

Jalota, S.K. and S.S. Prihar. (1990). Bare-soil evaporation in relation to tillage. *Advances in Soil Science* 12: 187-216.

Johnson, W.C. (1964). Some observations on the contribution of an inch of seeding time soil moisture to wheat yields in the Great Plains. *Agronomy Journal* 56: 29-35.

Jones, O.R. (1981). Land forming effects on dryland sorghum production in the southern Great Plains. *Soil Science Society of America Journal* 45: 606-611.

Jones, O.R. and R.N. Clark. (1987). Effects of furrow dikes on water conservation and dryland crop yields. *Soil Science Society of America Journal* 51: 1307-1314.

Jones, O.R. and V.L. Hauser. (1975). Runoff utilization for grain sorghum. pp. 277-283. In *Proceedings Water Harvesting Symposium*, ed. G. W. Frazier, February 1975. U.S. Department of Agriculture, ARS W-22.

Jones, O.R. and T.W. Popham. (1997). Cropping and tillage systems for dryland grain production in the Southern High Plains. *Agronomy Journal* 89: 222-232.

Kaushik, S.K. and R.C. Gautam. (1994). Response of rainfed pearl millet (*Pennisetum glaucum*) to water harvesting, moisture conservation and plant population in light soils. *Indian Journal of Agricultural Sciences* 64: 858-860.

Keen, M. (1991). Drip-trickle irrigation boosts Bedouin farmers' yields. *Ceres 130* 23(4): 10-12.

Keller, J. and R.D. Bliesner. (1990). Sprinkle and trickle irrigation. New York, NY: Van Nostrand Reinhold.

Kemper, W.D., W.H. Heinemann, D.C. Kincaid, and R.V. Worstell. (1981). Cablegation. I. Cable controlled plugs in perforated supply pipes for automated furrow irrigation. *Transactions of the ASAE* 24: 1526-1532.

Kemper, W.D., D.C. Kincaid, R.V. Worstell, W.H. Heinemann, T.J. Trout, and J.E. Chapman. (1985). Cabelgation systems for irrigation: Description, design, installation, and performance. U.S. Department of Agriculture, ARS Report 21.

Kemper, W.D., T.J. Trout, A.S. Humpherys, and M.S. Bullock. (1988). Mechanisms by which surge irrigation reduces furrow infiltration rates in a silty loam soil. *Transactions of the ASAE* 31: 821-829.

Krantz, B.A., J. Kampen, and S.M. Virmani. (1978). Soil and water conservation and utilization for increased food production in the semi-arid tropics. Hyderabad, A.P., India: International Crops Research Institute for the Semi-arid Tropics, 22 pp.

Kruse, E.G., D.A. Bucks, and R.D. von Bernuth. (1990). Comparison of irrigation systems. In *Irrigation of Agricultural Crops*. eds. B.A. Stewart and D.R. Nielsen. Madison, WI: American Society of Agronomy, Inc., Crop Science Society of America, Inc., and Soil Science Society of America, Inc., pp. 475-508.

Laing, I.A.F. (1981). Evaluation of small catchment surface treatments. Australian Water Resources Council Technical Paper 61. Canberra, Aust.: Australian Government Publishing Service.

Lavake, D.E. and A.F. Wiese. (1979). Influence of weed growth and tillage interval during fallow on water storage, soil nitrates, and yield. *Soil Science Society of America Journal* 43: 565-569.

Lavee, H., J. Poesen, and A. Yair. (1997). Evidence of high efficiency water-harvesting by ancient farmers in the Negev Desert, Israel. *Journal of Arid Environments* 35: 341-348.

Lemon, E.R. (1956). The potentialities for decreasing soil moisture evaporation loss. *Soil Science Society of America Proceedings* 20: 120-125.

Lentz, R.D., I. Shainberg, R.E. Sojka, and D.L. Carter. (1992). Preventing irrigation furrow erosion with small applications of polymers. *Soil Science Society of America Journal* 56: 1926-1932.

Lyle, W.M. and J.P. Bordovsky. (1981). Low energy precision application (LEPA) irrigation system. *Transaction of the ASAE* 30: 1071-1074.

Lyle, W.M. and J.P. Bordovsky. (1983). LEPA irrigation system evaluation. *Transactions of the ASAE* 26: 776-781.

Martin, D.L., E.C. Stegman, and E. Fereres. (1990). Irrigation scheduling principles. In *Management of Farm Irrigation Systems*, eds. G.J. Hoffman, T.A. Howell, and K.H. Solomon. St. Joseph, MI: American Society of Agricultural Engineers, pp. 155-203.

McConkey, B.G., D.J. Ulrich, and F.B. Dyck. (1997). Slope position and subsoiling effects on soil water and spring wheat yield. *Canadian Journal of Soil Science* 77: 83-90.

Merriam, J.L. and J. Keller. (1978). Farm irrigation system evaluation: A guide for management. Logan, UT: Utah State University, Department of Agriculture and Irrigation Engineering, 271 p.

Miller, D.E. and J.S. Aarstad. (1972). Effect of deep plowing on the physical characteristics of Hezel soil. *Circular 556*. Pullman, Washington: Washington Agricultural Experiment Station.

Muchiri, G. and F.N. Gichuki. (1983). Conservation tillage in semi-arid areas of Kenya. In *Soil and Water Conservation in Kenya*, eds. D.B. Thomas and W.M. Senga. Nairobi, Kenya: University of Nairobi, Institute for Development Studies and Faculty of Agriculture, pp. 395-419.

Musick, J.T. and F.R. Lamm. (1990). Preplant irrigation in the Central and Southern High Plains–A review. *Transactions of the ASAE* 33: 1834-1842.

Musick, J.T., F.B. Pringle, and P.N. Johnson. (1983). Furrow compaction for controlling excessive irrigation water intake. St. Joseph, MI: American Society of Agricultural Engineers, Paper No. 83-2575.

Musick, J.T., F.B. Pringle, and J.D. Walker. (1987). Sprinkler and furrow irrigation trends–Texas High Plains. *Applied Engineering in Agriculture* 3: 190-195.

Musick, J.T., W.H. Sletten, and D.A. Dusek. (1971). Preseason irrigation of grain sorghum in the Southern High Plains. *Transactions of the ASAE* 14: 93-97.

Nakayama, F.S. and D.A. Bucks. (1986). *Trickle Irrigation for Crop Production: Design, Operation, and Management*. Amsterdam, The Netherlands: Elsevier Science Publishers B.V., 383 p.

Nielsen, D.C., R.M. Aiken, and G.S. McMaster. (1997). Optimum wheat stubble height to reduce erosion and evaporation. *Conservation Tillage Fact Sheet #4-97*. Published by: USDA-ARS, USDA-NRCS, and Colorado Conservation Tillage Association, 2 pp.

Norwood, C.A. (1992). Tillage and cropping system effects on winter wheat and grain sorghum. *Journal of Production Agriculture* 5: 120-126.

O'Leary, G.J. and Connor, D.J. (1997). Stubble retention and tillage in a semi-arid environment: 1. Soil water accumulation during fallow. *Field Crops Research* 52: 209-219.

Olsen, S.R., F.S. Watanabe, F.E. Clark, and W.D. Kemper. (1964). Effect of hexadecanol on evaporation of water from soil. *Soil Science* 97: 13-18.

Opoku, G. and T.J. Vyn. (1997). Wheat residue management options for no-till corn. *Canadian Journal of Plant Science* 77: 207-213.

Papendick, R.I. (1987). Tillage and water conservation: experience in the Pacific Northwest. *Soil Use and Management* 3: 69-74.

Pathak, P., S.M. Miranda, and S.A. El-Swaify. (1985). Improved rainfed farming for semi-arid tropics–implications for soil and water conservation. In *Soil Erosion and Conservation*, eds. S.A. El-Swaify, W.C. Moldenhauer, and A. Lo. Ankeny, IA: Soil Conservation Society of America, pp. 338-354.

Phene, C.J., R.J. Reginato, B. Itier, and B.R. Tanner. (1990). Sensing irrigation needs. In *Management of Farm Irrigation Systems*, eds. G.J. Hoffman, T.A. Howell and K.H. Solomon. St. Joseph, MI: American Society of Agricultural Engineering, pp. 208-261.

Piccolo, A., G. Pietramellara, and J.S.C. Mbagwu. (1996). Effects of coal derived humic substances on water retention and structural stability of Mediterranean soils. *Soil Use and Management* 12: 209-213.

Pikul, J.L., D.E. Wilkins, J.K. Aase, and J.F. Zuzel. (1996). Contour ripping: A tillage strategy to improve water infiltration into frozen soil. *Journal of Soil and Water Conservation* 51: 76-83.

Plauborg, F. (1995). Evaporation from bare soil in a temperate humid climate–measurement using micro-lysimeters and time domain reflectometry. *Agricultural and Forest Meteorology* 76: 1-17.

Postel, S. (1997). *Last Oasis: Facing Water Scarcity*. New York, NY: W.W. Norton & Co.

Potter, K.N., H.A. Torbert, O.R. Jones, J.E. Matocha, J.E. Morrison, Jr., and P.W. Unger. (1998). Distribution and amount of soil organic C in long-term management systems in Texas. *Soil & Tillage Research* 47: 309-321.

Pressland, A.J. and G. N. Batianoff. (1976). Soil water conservation under cultivated fallows in clay soils of south-western Queensland. *Australian Journal of Experimental Agriculture and Animal Husbandry* 16: 564-569.

Ramig, R.E., R.R. Allmaras, and R.I. Papendick. (1983). Water conservation: Pacific Northwest, In *Dryland Agriculture*, eds. H.E. Dregne and W.O. Willis. Madison, WI: American Society of Agronomy, Inc., Crop Science Society of America, Inc., and Soil Science Society of America, Inc., pp. 105-124.

Raper, R.L., J.H. Edwards, T.R. Way, B.H. Washington, E.C. Burt, and D.T. Hill. (1998). Effect of vertical trenching of cellulose waste on crop yield and hardpan reconsolidation. *Transactions of the ASAE* 41: 11-15.

Rawitz, E., J. Morin, W.B. Hoogmoed, M. Margolin, and H. Etkin. (1983). Tillage practices for soil and water conservation in a semi-arid zone. 1: Management of fallow during the rainy season preceding cotton. *Soil & Tillage Research* 3: 211-232.

Richardson, C.W. (1973). Runoff, erosion, and tillage efficiency on graded-furrow and terraced watersheds. *Journal of Soil and Water Conservation* 28: 162-164.

Sandal, S.K. and C.L. Acharya. (1997). Effect of conservation tillage on moisture conservation, soil-physical conditions, seedling emergence and grain yield of rainfed maize (*Zea mays*) and wheat (*Triticum aestivum*). *Indian Journal of Agricultural Sciences* 67: 227-231.

Schneider, A.D. (1999). Efficiency and uniformity of LEPA and spray sprinkler methods. *Transaction of the ASAE* vol. 42 (in press).

Schneider, A.D. and T.A. Howell. (1990). Sprinkler efficiency measurement with large weighing lysimeters. In *Visions of the Future, Proceedings of the Third National Irrigation Symposium*. St. Joseph, MI: American Society of Agricultural Engineers, pp. 69-76.

Schneider, A.D. and T.A. Howell. (1999). LEPA and spray irrigation for grain crops. *Journal of Irrigation and Drainage Engineering* 42: (in press).

Schneider, A.D. and A.C. Mathers. (1970). Deep plowing for increased grain sorghum yields under limited irrigation. *Journal of Soil and Water Conservation* 25: 147-150.

Sharma, K.D, N.S. Vangani, H.P. Singh, D.N. Bohra, A.K. Kalla, and P.K. Joshi. (1997). Evaluation of contour vegetative barriers as soil and water conservation measures in Arab lands. *Annals of Arid Zone* 36: 123-127.

Singh, S., S.K. Kaushik, and R.C. Gautam. (1997). Effect of tillage and moisture-conservation practices on productivity, water use and water-use efficiency of pearl millet (*Pennisetum glaucum*) on light soils under dryland conditions. *Indian Journal of Agricultural Sciences* 67: 232-236.

Smika, D.E. (1976). Seed zone soil water conditions with reduced tillage in the semiarid central Great Plains. In *Proceeding of The 7th Conference of the International Soil Tillage Research Organization*, Sweden, 1976, 6 pp.

Smika, D.E. (1983). Soil water change as related to position of wheat straw mulch on the soil surface. *Soil Science Society of America Journal* 47: 988-991.

Smika, D.E. (1986). Resources and problems in the central Great Plains. In *Planning and Management of Water Conservation Systems in the Great Plains States, Proceedings of a Workshop*, Lincoln, NE, October 1985. Lincoln, NE: U.S. Department of Agriculture, Soil Conservation Service, Midwest National Technical Center, pp. 39-55.

SSSA (Soil Science Society of America). (1996). *Glossary of Soil Science Terms.* Madison, WI: Soil Science Society of America.

Steppuhn, H. and J. Waddington. (1996). Conserving water and increasing alfalfa production using a tall wheatgrass windbreak system. *Journal of Soil and Water Conservation* 51: 439-445.

Stewart, B.A. (1988). Dryland farming: The North American experience. In *Challenges in Dryland Agriculture–A Global Perspective*, eds. P.W. Unger, T.V. Sneed, W.R. Jordan, and R. Jensen. College Station, TX: Texas Agricultural Experiment Station, pp. 54-59

Stewart, B.A., J.T. Musick, and D.A. Dusek. (1983). Yield and water-use efficiency of grain sorghum in a limited irrigated-dryland farming system. *Agronomy Journal* 75: 629-634.

Stroosnijder, B.A. and W.B. Hoogmoed. (1984). Crust formation on sandy soils in the Sahel. II. Tillage and its effect on the water balance. *Soil & Tillage Research* 4: 321-337.

Trout, T.J., R.E. Sojka, and R.D. Lentz. (1995). Polyacrylamide effect on furrow erosion and infiltration. *Transactions of the ASAE* 38: 761-765.

Twomlow, S., R. Riches, and S. Mabasa. (1997). Weeding–Its contribution to soil water conservation in semi-arid maize production. In *Proceedings of The 1997 Brighton Crop Protection Conference-Weeds*. Brighton, UK, pp. 185-190.

Tyler, D. (1998). Special issue on cover crops (Editorial). *Journal of Soil and Water Conservation* 53: 186.

Unger, P.W. (1978). Straw mulch rate effects on soil water storage and sorghum yield. *Soil Science Society of America Journal* 42: 486-491.

Unger, P.W. (1994). Tillage effects on dryland wheat and sorghum production in the southern Great Plains. *Agronomy Journal* 86: 310-314.

Unger, P.W. (1995). Role of mulches in dryland agriculture. In *Production and Improvement of Crops for Drylands*, ed. U.S. Gupta. New Delhi, Bombay, Calcutta: Oxford & IBH Publishing Co. Pvt. Ltd., pp. 241-270.

Unger, P.W. (1996). Common soil and water conservation practices. In *Soil Erosion, Conservation, and Rehabilitation*, ed. M. Agassi. New York, Basel, Hong Kong: Marcel Dekker, Inc., pp. 239-266.

Unger, P.W. (1997). Management-induced aggregation and organic carbon concentrations in the surface layer of a Torrertic Paleustoll. *Soil & Tillage Research* 42: 185-208.

Unger, P.W. and R.L. Baumhardt. (1999). Factors related to dryland grain sorghum yield increases, 1939 through 1997. *Agronomy Journal* (in press).

Unger, P.W. and B.A. Stewart. (1983). Soil management for efficient water use: An Overview. In *Limitations to Efficient Water Use in Crop Production*, eds. H.M. Taylor, W.R. Jordan, and T.R. Sinclair. Madison, WI: American Society of Agronomy, Inc., Crop Science Society of America, Inc., and Soil Science Society of America, Inc., pp. 419-460.

Unger, P.W. and M.F. Vigil. (1998). Cover crop effects on soil water relationships. *Journal of Soil and Water Conservation* 53: 200-206.

van Bavel, C.H.M. and R.J. Hanks. (1983). Water conservation: Principles of soil water flow, evaporation, and evapotranspiration. In *Dryland Agriculture, Monograph 23*, eds. H.E. Dregne and W.O. Willis. Madison, WI: American Society of Agronomy, Inc., Crop Science Society of America, Inc., and Soil Science Society of America, Inc., pp. 25-34.

Van Doren, Jr., D.M. and R.R. Allmaras. (1978). Effect of residue management practices on the soil physical environment, microclimate, and plant growth. In *Crop Residue Management Systems, ASA Special Publication Number 31*, ed. W.R. Oschwald. Madison, WI: American Society of Agronomy, Crop Science Society of America, and Soil Science Society of America, pp. 49-83.

Vogel, H., I. Nyagumbo and K. Olsen. (1994). Effect of tied ridging and mulch ripping on water conservation in maize production on sandveld soils. *Der Tropenlandwirt* 95: 33-44.

Weakly, H.E. (1960). The effect of HPAN soil conditioner on runoff, erosion and soil aggregation. *Journal of Soil and Water Conservation* 15: 169-171.

Wicks, G.A. and D. E. Smika. (1973). Chemical fallow in a winter wheat-fallow rotation. *Journal of the Weed Science Society of America* 21: 97-102.

Wiese, A.F. (1983). Weed control. In *Dryland Agriculture, Monograph 23*, eds. H.E. Dregne and W.O. Willis. Madison, WI: American Society of Agronomy, Inc., Crop Science Society of America, Inc., and Soil Science Society of America, Inc., pp. 463-488.

Wiese, A.F., T.J. Army, and J.D. Thomas. (1966). Moisture utilization by plants after herbicide treatment. *Weeds* 14: 205-207.

Willcocks, T.J. (1984). Tillage requirements in relation to soil type in semi-arid rainfed agriculture. *Journal of Agricultural Engineering Research* 30: 327-336.

Wood, C.W., D.G. Westfall, and G.A. Peterson. (1991). Soil carbon and nitrogen changes on initiation of no-till cropping systems. *Soil Science Society of America Journal* 55: 470-476.

Zingg, A.W. and V.L. Hauser. (1959). Terrace benching to save potential runoff for semiarid land. *Agronomy Journal* 51: 289-292.

Evapotranspiration Responses of Plants and Crops to Carbon Dioxide and Temperature

L. H. Allen, Jr.

SUMMARY. Atmospheric carbon dioxide (CO_2) concentration has risen from about $270\,\mu$mol (CO_2) mol^{-1} (air) (i.e., mole fraction of dry atmospheric air basis) before 1700 to about $370\,\mu$mol mol^{-1} currently. General Circulation Models (GCM) have predicted a global temperature rise of 2.8 to 5.2°C for a doubling of CO_2. This review examines evapotranspiration and water-use efficiency responses of plants to rising CO_2 and climatic changes, especially temperature. Doubling of CO_2 will decrease leaf stomatal conductance to water vapor about 40%. However, water use by C_3 crop plants under field conditions has usually been decreased only 12% or less for two reasons. Firstly, feedbacks in the energy balance of plant foliage cause leaf temperatures to rise as stomatal conductance is decreased. Increases of leaf temperature raise the vapor pressure of water inside the leaf, which increases the leaf-to-air vapor pressure difference. This increased driving force for transpira-

L. H. Allen, Jr. is Soil Scientist, Crop Genetics and Environment Research, U.S. Department of Agriculture, Agricultural Research Service, and Courtesy Professor, Agronomy Department and Horticultural Sciences Department, Institute of Food and Agricultural Sciences, University of Florida.

Address Correspondence to: L. H. Allen, Jr., USDA-ARS, P.O. Box 110965, University of Florida, Gainesville, FL 32611-0965.

The contributions of previous work by numerous colleagues who are cited in this review are gratefully acknowledged. Also, earlier research contract funding by the U.S. Department of Energy and The U.S. Environmental Protection Agency made this information possible.

[Haworth co-indexing entry note]: "Evapotranspiration Responses of Plants and Crops to Carbon Dioxide and Temperature." Allen, L. H., Jr. Co-published simultaneously in *Journal of Crop Production* (Food Products Press, an imprint of The Haworth Press, Inc.) Vol. 2, No. 2 (#4), 1999, pp. 37-70; and: *Water Use in Crop Production* (ed: M. B. Kirkham) Food Products Press, an imprint of The Haworth Press, Inc., 1999, pp. 37-70. Single or multiple copies of this article are available for a fee from The Haworth Document Delivery Service [1-800-342-9678, 9:00 a.m. - 5:00 p.m. (EST). E-mail address: getinfo@haworthpressinc.com].

tion offsets in large part the decreased leaf conductance caused by elevated CO_2. Secondly, CO_2 enrichment tends to cause leaf area to increase more rapidly in many crops. This increased leaf surface area for transpiration also offsets part of the decreased stomatal conductance per unit leaf area on the whole canopy evapotranspiration, but the energy budget feedbacks are more important.

Experiments point to a yield enhancement of 30 to 35% for C_3 crops for the direct effects a doubling of CO_2 (without ancillary climate change). If temperature rises, this yield enhancement may be greater for vegetative growth but less for seed grain yield. Experiments on both ambient and elevated CO_2 treatments in sunlit growth chambers showed that transpiration rates increased 20% when air temperature was changed from 28 to $33°C$ and increased 30% when temperature was increased from 28 to $35°C$. Thus, under well-watered conditions, evapotranspiration will increase about 4 to 5% per $1°C$ rise in temperature.

Crop model predictions of yields of soybean and maize showed a reduction due to temperature increases by two GCM models. Under Southeastern USA conditions, doubling CO_2 in the Goddard Institute for Space Studies (GISS) climate change scenario resulted in an 12% increase in yields, but yields decreased 50% in the Geophysical Fluids Dynamics Laboratory (GFDL) climate change scenario. Optimum irrigation for both models gave yield increases of about 10%. These model results illustrate the critical requirement of water for production of crops. Under rainfed conditions, crop yields could suffer tremendously if growing season precipitation is decreased, but yields could increase moderately if growing season precipitation is increased. Under the high growing season rainfall scenario (GISS), irrigation requirements for optimum soil water were increased 22%, but under the low rainfall scenario (GFDL), irrigation requirements were increased 111%.

Without the effects of climate change, rising CO_2 will cause an increase in crop water-use efficiency (WUE). Most of the increases in WUE will be due to increases in dry matter, with little or no contribution from decreases in water use per unit land area. Growers could produce higher yields per unit land area with higher total production, or maintain the same total production with less land and less total water use. However, if temperatures rise, transpirational water use will increase, and WUE will decline. Higher temperatures, and especially less rainfall, would raise the irrigation requirements of crops. Competition for water resources from other uses could result in less water available for irrigation. *[Article copies available for a fee from The Haworth Document Delivery Service: 1-800-342-9678. E-mail address: getinfo@haworthpressinc. com <Website: http://www.haworthpressinc.com>]*

KEYWORDS. Transpiration, photosynthesis, water-use efficiency, stomata, yield, biomass growth ratio, carbon dioxide, temperature, climate change

INTRODUCTION

Atmospheric Carbon Dioxide

Carbon dioxide (CO_2) concentration of the earth's atmosphere has varied throughout geologic time (Figure 1). Analysis of entrapped air bubbles in ice cores from Antarctica and Greenland show CO_2 and methane concentrations of the atmosphere back to 160,000 years before present (Barnola et al., 1987; Lorius et al., 1990). Changes of deuterium content within the ice crystals have been used to establish temperature changes over this same time period (Jouzel et al., 1987). Concentration of CO_2 was as low as 180 to 200 μmol mol^{-1} during the time periods 13,000 to 30,000 and 140,000 to 160,000 years before present during the coldest parts of the last two ice ages (Barnola et al., 1987). Carbon dioxide concentrations rose to about 270 μmol mol^{-1} during the last interglacial period (116,000 to 140,000 years before present) and the current interglacial period (beginning about 13,000 years before present). Ice core data since about 1700 AD and direct atmospheric sampling data since 1958 show that CO_2 increased to 315 μmol mol^{-1} by 1958 and to about 355 μmol mol^{-1} by 1990 (Keeling et al., 1989), which should be about 370 μmol mol^{-1} by 2000. The rate of increase of CO_2 in the atmosphere is about 0.5% per year, which means that the change of actual CO_2 concentration is accelerating.

These changes in atmospheric CO_2 have important implications not only for climate but also for plants and the global carbon cycle. As the substrate for terrestrial green plant photosynthesis, atmospheric CO_2 represents the first molecular link in the food chain of almost all life on earth.

Transpiration, Biomass Productivity, and Climate

Green plants require water for biochemical life processes, for translocation of nutrients and ions, and for hydrogen which is incorporated in plant tissues in photosynthesis and is used in forming reductants for driving biochemical processes. Terrestrial and emergent aquatic plants also require water for transpiration. Of these requirements, transpiration is several hundredfold greater than the water use for photosynthesis. For example, the ratio of transpiration:dry matter was 614 and 539 g (transpired water) g^{-1} (dry matter) for oat and barley, respectively, (Chang, 1968; Arkley, 1963; Briggs and Shantz, 1913a; 1913b; 1914) from one set of measurements.

Stomata on leaves allow both entry of CO_2 for photosynthesis and escape of water vapor. Thus, the transpiration process is very closely coupled to photosynthetic CO_2 exchange processes. The transpiration stream may be necessary for other plant growth processes, such as (1) nutrient uptake from the soil and transport to other organs, especially the leaves, and (2) transpirational cooling which can be important in hot, arid environments.

FIGURE 1. Carbon dioxide concentration changes for the past 160,000 years as measured from Vostok station, Antarctica ice cores (top panel). Carbon dioxide changes since 1775 and predictions into the 21st century (bottom panel). The insert shows the measurements since 1958 at Mauna Loa, Hawaii. (Adapted from Barnola et al., 1994; Neftel et al., 1994; Keeling and Whorf, 1994.)

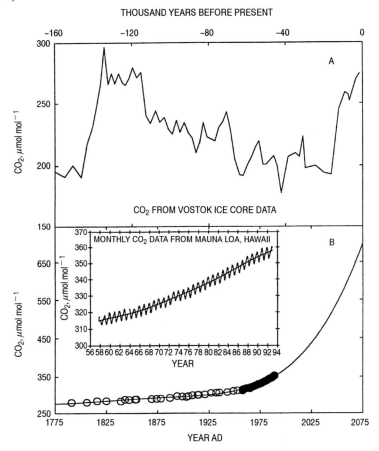

During a series of studies beginning in 1910 by USDA at Akron, Colorado, Briggs and Shantz (1913a; 1913b; 1914) showed that the biomass production of plants was linearly related to transpiration. This linear relationship was established by growing plants in metal containers filled with soil, covered to prevent evaporation from the soil and entry of rainfall. Water was added at prescribed rates to give a range of treatments from "fully" watered through several degrees of water limitation. Figure 2 shows the linear yield of oat and barley vs. water use. Grain yields in the warmer climate of Texas required more water than the same yields in Colorado.

The findings by Briggs and Shantz (1913a; 1913b; 1914) have been confirmed repeatedly (e.g., Allison et al., 1958; Arkley, 1963; Chang, 1968; Hanks et al. 1969; Stanhill, 1960). Figure 3 (left panel) shows the linear relationship between biomass yield and rainfall plus irrigation used by Sart sorghum and Starr millet in Alabama, adapted from Bennett et al. (1964).

De Wit (1958) examined the relationships among climatic factors, yield, and water use by crops, and found the following general linear relationship to be true, especially in semiarid climates with high solar radiation.

$$Y/T = m/T_{max} \qquad [1]$$

where Y = yield component (e.g., total above-ground biomass or seed)
T = cumulative actual transpiration
T_{max} = maximum possible cumulative transpiration
m = constant dependent on yield component and species, especially on differences among photosynthetic mechanisms.

T_{max} can be represented by pan evaporation, which is proportional to climatic factors, especially air vapor pressure deficit (VPD or $e_s - e_a$). Therefore:

$$T_{max} \, \alpha \, VPD = (e_s - e_a) \qquad [2]$$

FIGURE 2. Linear relationship between biomass accumulation and water use of oat and barley grown in closed soil containers at Akron, Colorado. (Adapted from Chang, 1968; Arkley, 1963; Briggs and Shantz, 1913a; 1913b; 1914.)

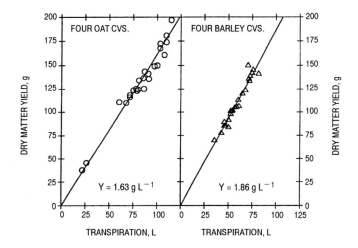

FIGURE 3. Linear relationship between biomass production and water use for two forage crops in 1956 and 1957 at Thorsby, Alabama (Left Panel). Adapted from Bennett et al. (1964). Cumulative dry matter yield vs. cumulative potential ET of pastures under a range of climatic regimes (Right Panel). Open circle: Denmark. Filled circle: The Netherlands. Open triangle: England. Filled triangle: New Jersey, USA. Open square: Toronto, Canada. Filled square: Gilat, Israel. Open inverted triangle: Trinidad, West Indies. Adapted from Stanhill (1960).

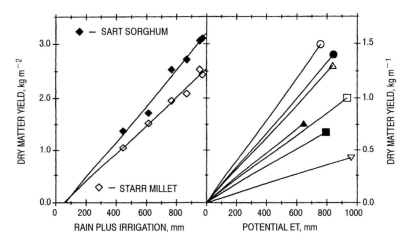

where e_s is the saturation vapor pressure at a given air temperature, and e_a is the actual vapor pressure that exists in the ambient air. This proportionality certainly holds within latitude bands where solar radiation levels would be similar. Combining relationships,

$$Y/T = k/(e_s - e_a)$$ [3a]

or,

$$Y = kT/(VPD)$$ [3b]

where k is mb (VPD) g (dry matter) g^{-1} (water). Like m, k is dependent on yield component, species, photosynthetic mechanisms, and method of aggregating an effective climatological VPD.

Thus, theory predicts that yield will be proportional to cumulative transpiration divided by VPD. Vapor pressure deficit is the driving force for the molecular process of vaporization, and it integrates climatological factors of temperature, solar radiation, and wind. Air vapor pressure deficit can be expressed several ways, ranging from an aggregation of seasonal daytime average VPD to daily values. Regardless of the method used to compute a representative VPD, the slope of yield vs. cumulative transpiration linear relationships vary with the aridity of the climate, specifically with the temperature and vapor pressure regime under which the crop is grown. Figure 3 (right panel, modified from Stanhill, 1960) shows water used vs. dry matter yield of pastures from the cool, humid latitude of Denmark to a hot, dry atmosphere of Trinidad. Comparisons among existing climates indicate that transpirational water requirements of plants per unit biomass produced will increase if climates get warmer in response to the enhanced greenhouse effect.

Atmospheric CO_2 is known to affect plant yield. Kimball (1983) reviewed 430 observations of CO_2 enrichment studies conducted prior to 1982 and reported an average yield increase of 33 ± 6% for a doubling of CO_2 concentration. This value has been generally confirmed by many other subsequent studies (e.g., Kimball, 1986; Poorter, 1993).

Transpirational water use is related to ground cover (Jensen, 1974; Doorenbos and Pruitt, 1977). Daily water use soon after planting is typically only 10 to 20% of water use after full ground cover is reached. Water use rises sharply as crop leaf area index (LAI, the ratio of leaf area per unit ground area) increases. Similarly, water use is decreased to about 30 to 40% when hay crops are cut. Transpiration rates recover rapidly as the ground cover of leaves is restored. Any CO_2-induced stimulation of early growth of leaf area and/or increase of total LAI may increase transpiration.

Increased CO_2 is known to cause smaller stomatal apertures and hence decrease the leaf conductance for water vapor (Morison, 1987). This is the primary mechanism whereby increased CO_2 may affect plant transpiration.

Another effect of rising CO_2 is the change in water-use efficiency (WUE). Water-use efficiency has a range of definitions. For whole season processes, it is defined as the ratio of total dry matter (or seed yield) produced (e.g., Y) to the amount of water used (e.g., T) by crops (e.g., WUE = Y/T). For shorter-term whole canopy processes, it is defined as the ratio of photosynthetic CO_2 uptake rate per unit land area to transpiration rate per unit land area. Figure 3 (right panel) demonstrates that warmer climates have lower WUE. Equation [3a] demonstrates that WUE decreases as VPD increases.

CARBON DIOXIDE EFFECTS ON PLANTS

Plant Photosynthetic Mechanisms

Tolbert and Zelitch (1983) reviewed responses of the three types of photosynthetic mechanisms (C_3, C_4, and CAM) of plants to CO_2. The biochemical pathway of CO_2 uptake for C_3 plant photosynthesis involves the carboxylating enzyme, ribulose 1,5-bisphosphate carboxylase/oxygenase (RuBisCO), and use and subsequent regeneration of ribulose 1,5-bisphosphate in a cyclic series of reactions. The first product of photoassimilation of CO_2 is 3-phosphoglyceric acid, a 3-carbon sugar; hence the term C_3 pathway of photosynthesis.

In C_4 plants, phosphoenol pyruvate carboxylase (PEPase) is involved in forming a 4-carbon molecule, oxalacetate, in mesophyll cells of leaves as the first step in incorporation of CO_2. This 4-carbon compound is changed into aspartic acid or malic acid, and then it is transported immediately to bundle sheath cells. Here, the CO_2 is released and utilized in the C_3 biochemical pathway. Thus, the C_4 plant mechanism first traps CO_2 in the mesophyll cells, and then transports and concentrates the CO_2 in the bundle sheath cells, where it is utilized in C_3 plant metabolism (Tolbert and Zelitch, 1983).

CAM, or crassulacean acid metabolism, is a mechanism whereby plants typically take up and store CO_2 during the night, and utilize it in photosynthetic CO_2 fixation during the day when sunlight is available. Pineapple and "air plants" such as Spanish moss and orchids have this photosynthetic mechanism. Few agricultural crops are CAM plants.

Photosynthetic rates of fully functional leaves of C_4 plants will not respond to rising CO_2 to the same extent as C_3 plants because C_4 plants have a mechanism for concentrating CO_2 in bundle sheath cells of leaves. Crop or turf plants that fit into the C_4 category include maize (corn), sorghum, millet, sugarcane, and bermudagrass. Plants that fit into the C_3 category include wheat, rice, potato, soybean, sugarbeet, alfalfa, cotton, tree and vine crops, and most vegetable crops and cool-season grasses.

Plant Growth Responses to CO_2

Increased CO_2 causes increased photosynthetic rates, biomass growth, and seed yield of all of the globally important C_3 food and feed crops (Acock and Allen, 1985; Enoch and Kimball, 1986; Warrick et al., 1986; Allen, 1990). Some plants, such as cucumber, cabbage, and tomato, have shown a tendency to first increase leaf photosynthetic rates in response to elevated CO_2, and then after several days, to decrease photosynthetic rates. In the past, this behavior was attributed to "end-product inhibition of photosynthesis" caused by the failure of translocation of photoassimilates to keep up with photosynthetic rates (Guinn and Mauney, 1980). There is evidence now that elevated CO_2 causes a decrease in the amount of messenger RNA for the synthesis of the small subunit of RuBisCO (Gesch et al., 1998).

A few experiments have been conducted with CO_2 concentration maintained across a range of 160 to 990 μmol mol^{-1}. Figure 4 shows the results of one study with soybean canopy photosynthetic rates across the range of 90

FIGURE 4. Photosynthetic CO_2 uptake rate responses of a soybean crop canopy exposed to CO_2 concentrations ranging from 110 to 990 μmol mol^{-1}. The curve is relative to the response obtained at 330 μmol mol^{-1}, indicated by the + symbol. Adapted from Allen et al. (1987).

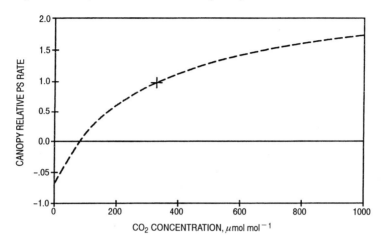

to 900 μmol mol^{-1} CO_2. A nonlinear hyperbolic model was used to fit soybean photosynthetic rate data to CO_2 concentration (Allen et al., 1987). Data sets of biomass yield and seed yield from four locations over three years, normalized to responses at 330 μmol mol^{-1} CO_2, were also fitted to the model (Allen et al., 1987).

The form of this model was:

$$R = R_{max} \times [C/(C + K_c)] + R_{int} \qquad [4]$$

where R = relative response of photosynthetic rate, biomass yield, or seed yield
   R_{max} = asymptotic upper limit for R from baseline R_{int}
   C = CO_2 concentration (μmol mol^{-1})
   K_c = Apparent Michaelis constant (μmol mol^{-1})
   R_{int} = Y-axis intercept for zero C

From the parameters of this equation, photosynthetic rate, biomass accumulation, and seed yield changes of soybean due to CO_2 changes were estimated (Allen et al., 1987). Table 1 shows the changes predicted across 3 periods of time, from the last ice age minimum of CO_2 concentration to preindustrial revolution times (about 1700), from 1700 to 1973, and from 1973 to about a century into the future. The model output indicated that there should have been large increases in plant productivity from the low CO_2

TABLE 1. Percent increases of soybean midday photosynthetic rates, biomass yield, and seed yield predicted across selected CO_2 concentration ranges associated with relevant benchmark points in time.

Period of Time	CO_2 Concentration Initial	Final	Midday Photosynthesis	Biomass Yield	Seed Yield
Years	--- μmol mol^{-1} ---		----- % increase over initial CO_2 -----		
IA-1700[1]	200	270	38	33	24
1700-1973	270	330	19	16	12
1973-20??[2]	330	660	50	41	31

[1] IA, the Ice Age about 13,000 to 30,000 years before present. The atmospheric CO_2 concentrations that prevailed during the last Ice Age, and from the end of the glacial melt until prepioneer/preindustrial revolution times, were 200 and 270 μmol mol^{-1}, respectively.

[2] The first world energy "crisis" occurred in 1973 when the CO_2 concentration was 330 μmol mol^{-1}. This CO_2 concentration is used as the basis for many CO_2-doubling studies. The CO_2 concentration is expected to double sometime within the 21st century.

The form of this model was: $R = R_{max} \times C/(C + K_c) + R_{int}$
The values determined for R_{max}, K_c, and R_{int}, respectively, were:
3.08, 279 μmol mol^{-1}, and -0.68 for relative midday photosynthesis;
3.02, 182 μmol mol^{-1}, and -0.91 for relative seasonal biomass accumulation;
2.55, 141 μmol mol^{-1}, and -0.76 for relative seed yield.

concentrations of the last ice age (200 μmol mol^{-1}) during the post-glacial period until the industrial revolution began (270 μmol mol^{-1}). Likewise, model calculations indicate a 12% increase in soybean seed yield from 1700 to 1973 when CO_2 increased from about 270 to 330 μmol mol^{-1}.

Most of the recent concerns about rising atmospheric CO_2 have been quantified by predicting changes for a doubling of CO_2 concentration. For many of the early assessment studies, doubling of CO_2 usually meant 660 versus 330 μmol mol^{-1}. Table 1 shows that soybean seed yields and biomass yields are predicted to increase 31% and 41%, respectively, from a doubling of CO_2 (from 330 to 660 μmol mol^{-1}).

The concept of a doubling of CO_2, however, has been on a sliding scale due to continuously rising concentrations, with the ambient daytime baseline having increased from about 315 μmol mol^{-1} in 1958, through steps of 330, 340, 350, 360, and 370 μmol mol^{-1} used in the literature up to the present time. Within 10 years, the ambient daytime baseline will likely be 20% greater than it was in 1958.

One method of evaluating plant responses to CO_2 while the baseline level is continually increasing is to conduct studies over wide ranges of subambient and superambient CO_2 concentrations. Table 2 shows total daytime CO_2 uptake on four days where rice was grown in CO_2 treatments ranging from 160 to 900 μmol mol^{-1} (Baker et al., 1990). These responses are similar to the predictions of the nonlinear model developed from soybean data, Equation [4].

TABLE 2. A comparison of total daily responses of rice canopies acclimated to subambient, ambient, and superambient CO_2 concentrations.[1] Adapted from Baker et al. (1990; 1997).

CO_2 treatment	Total daily photons	Total daily CO_2 uptake	Total daily H_2O loss	Water-use efficiency
μmol mol^{-1}	mol m^{-2}	mol m^{-2}	mol (H_2O)m^{-2}	m mol (CO_2) mol (H_2O)
Early planted rice, 4 March 1987 (41 DAP)				
160	36.0	0.22	418	0.53
250		0.84	599	1.40
330		1.00	548	1.82
500		1.49	585	2.55
660		1.30	555	2.34
900		1.52	491	3.09
Early planted rice, 6 April March 1987 (74 DAP)				
160	34.3	0.64	769	0.83
250		1.20	794	1.51
330		1.46	629	2.32
500		1.75	639	2.74
660		1.69	634	2.67
900		1.86	567	3.28
Late planted rice, 7 August 1987 (45 DAP)				
160	39.6	0.58	741	0.78
250		1.05	732	1.43
330		1.63	694	2.32
500		1.57	623	2.52
660		1.67	634	2.63
900		1.79	589	3.04
Late planted rice, 23 August 1987 (61 DAP)				
160	39.4	0.74	608	1.22
250		1.13	643	1.76
330		1.34	611	2.19
500		1.80	554	3.25
660		1.79	536	3.34
900		1.83	470	3.90
Rice, mean of 5 and 28 September, and 9 October 1994 (52, 75, and 86 DAP)				
350	30.1	1.08	312	346
700		1.33	282	4.72

[1] In 1987, air/dewpoint/paddy water temperature was controlled to 31/18/28°C, whereas in 1994 daytime air/dewpoint temperature was controlled to 28/16°C, with paddy water temperatures not controlled, but governed by energy balance.

Experimental studies have consistently showed a relatively lower seed yield than biomass yield for soybean when grown under doubled CO_2. If the harvest index, the ratio of seed yield to above-ground biomass yield (seed + pod walls + stems) were 0.50 for soybean grown under 330 μmol mol^{-1} CO_2 conditions, then the harvest index is predicted to be 0.46 for a doubling of

CO_2. This small decrease in soybean harvest index under elevated CO_2 conditions has been commonly observed (Allen, 1990; Jones et al., 1984). The relative midday maximum photosynthetic rates under CO_2 enrichment were consistently higher than relative biomass yields probably because the photosynthetic response to elevated CO_2 is relatively greater under high light conditions than it is under total daily solar irradiance conditions.

Transpiration and Water-Use Efficiency Responses to CO_2

The effect of CO_2 concentration on water use under field conditions has been discussed for many years. Elevated CO_2 has been labeled an ideal antitranspirant (Waggoner, 1990) because it decreases stomatal conductance without adverse effects on plants. Morison (1987) reviewed 80 observations in the literature and found that doubled CO_2 concentration decreased stomatal conductance of most plants by about $40 \pm 5\%$. Kimball and Idso (1983) calculated a 34% reduction in transpiration with doubled CO_2 from several short-term plant growth chamber experiments, which is consistent with the review by Morison (1987). However, Morison and Gifford (1984) also showed that doubled CO_2 can cause a more rapid development of leaf area of several plants, and hence cause an equal or greater plant transpiration rate in early stages of growth due to a more rapid development and exposure of transpiring surfaces. Therefore, increased rates of development of transpiring leaf surface can offset in part the decreased stomatal conductance for water vapor. Subsequently, Samarakoon and Gifford (1995) showed that, under elevated CO_2, leaf area of cotton increased with only a small decrease in stomatal conductance, so that water use rate per plant increased. On the other hand, leaf conductance of maize decreased with little leaf area response, so that water use rate per plant decreased. Wheat responded to elevated CO_2 by both increases in leaf area and decreases in stomatal conductance.

Baker et al. (1982), Allen et al. (1985), Allen (1990; 1991), and Kimball et al. (1999) also discussed the effect of reduction in stomatal conductance on foliage temperature. Any reduction in stomatal conductance due to increasing CO_2 will restrict transpiration rates per unit leaf area. A reduction in transpiration rate would result in less evaporative cooling of the leaves, and leaf temperature would rise. As leaf temperatures rise, the vapor pressure inside the leaves will increase and thus increase the leaf-to-air vapor pressure difference, which is the driving force for leaf transpiration. The increase in intercellular leaf vapor pressure will increase transpiration rates per unit leaf area and thus maintain those rates at only slightly lower values than would exist at ambient environmental CO_2 concentrations. In effect, all of the energy balance factors involved in canopy foliage energy exchange must be considered, and not just stomatal conductance alone. Kimball et al. (1999) discussed additional atmospheric feedback factors that have been raised in the literature, and concluded that transpiration reductions caused by partial stomatal closure under globally elevated CO_2 would still be significant, if not as large as measured under elevated CO_2 at the field level.

Transpiration rate changes of soybean canopies ranged from -2% (Jones et al., 1985a) to $+11\%$ (Jones et al., 1985b) for CO_2 treatments of 800 vs. 330 μmol mol^{-1} with corresponding LAI values of 6.0 vs. 3.3. These studies were conducted in controlled-environment chambers located outdoors in natural sunlight. In another experiment when differences in LAI of soybean between the 330 μmol mol^{-1} and the 660 μmol mol^{-1} treatments were small (3.36 and 3.46, respectively), the seasonal cumulative water use was decreased 12% by the doubled CO_2 treatments (Jones et al., 1985c). The percentage decrease was similar for both water stressed and nonstressed treatments of this second study. The results of these two experiments indicate that energy balance effects are greater than leaf area effects in determining canopy water use, at least when LAI is 3.3 or higher.

Field weighing lysimeter and neutron-probe water balance studies of cotton at Phoenix, Arizona have shown evapotranspiration (ET) reductions due to elevated CO_2 ranging from 0% up to 9% (Kimball et al., 1983). Later experiments on cotton over the period 1989 to 1991 showed no detectible effects of CO_2 on evapotranspiration, whether measured by sap flow gages, soil water balance, or energy balance methods (Kimball et al., 1999). However, energy balance analyses indicated a 7% decrease in daily evapotranspiration rates of wheat over four seasons of data (1992-1993, 1993-1994, 1995-1996, and 1996-1997) when grown under high nitrogen fertility, and a larger 20% decrease over the latter two seasons when grown under low nitrogen fertility.

Baker et al. (1990) measured daily evapotranspiration of rice growing at 160, 250, 330, 500, 660, and 900 μmol mol^{-1} CO_2 at a constant temperature of 31°C and a dewpoint temperature of 18°C, with paddy water temperature controlled to about 28°C (Table 2). A linear regression of daily water loss versus CO_2 concentration using the combined four days of data in Table 2 gave a slope of -0.2094 [mole (H_2O) m^{-2})]/[μmol mol^{-1} CO_2] and an intercept of 716.2 μmol mol^{-1} CO_2. Using this regression, an 11% decrease of evapotranspiration at 660 versus 330 μmol mol^{-1} CO_2 would be predicted. The evapotranspiration data of the 160 μmol mol^{-1} CO_2 treatment were not used in the regression since these ET values were smaller than those of the 250 μmol mol^{-1} CO_2 treatment, presumably because the LAI was smaller. Also, because the rice was growing in standing water, an undetermined amount of water use came from direct evaporation from the flood-water surface. Finally, crop WUE increased steadily with increasing CO_2 concentration over the range of 160 to 900 μmol mol^{-1} CO_2 (Table 2).

Rice produced later in the growing season had similar responses (Baker et al., 1997). Table 2 also shows mean values of daily evapotranspiration of rice grown in 1994 at an air temperature of 28°C and dewpoint temperature of 16°C, with paddy water temperature not being controlled but allowed to come into steady-state with the chamber energy balance. Water use was decreased almost 10% by rice grown at 700 versus 350 μmol mol^{-1} CO_2 (Table 2). Daytime water use was less in 1994 than in 1987 because plants were growing later in the season with shorter days and lower solar radiation, as well as at slightly lower temperature and without controlled paddy water temperature. When complete canopy leaf cover is achieved, midday paddy water temperature in the shade can decrease about 10°C compared with an early open paddy (Baker et al., 1994), and thus decrease direct evaporation from the water surface.

Water-use efficiencies of soybean canopies grown in outdoor, sunlit, controlled-environment chambers at 800 and 330 μmol mol^{-1} CO_2 which had LAI values of 6.0 and 3.3, respectively, were compared by Allen et al. (1985). The CO_2 exposure levels were cross-switched for one day for each of these treatments. The ratio of WUE (i.e., WUE at 800 μmol mol^{-1} CO_2 exposure/ WUE at 330 μmol mol^{-1} CO_2 exposure) for four canopies averaged 2.33. The relative contribution of photosynthesis and transpiration to the ratio of WUE was 73% and 27%, respectively. These WUE comparisons are valid for plant canopies with equal LAI only, because calculations were made on the same canopies exposed to each CO_2 levels. However, when canopies grown for the long-term at 800 μmol mol^{-1} CO_2 were compared with canopies grown for the long term at 330 μmol mol^{-1}, the ratio of WUE (800/330) was 1.80, and the relative contribution to this ratio was 104% for photosynthesis and -4% for transpiration. The negative contribution of transpiration arises from the fact that canopy transpiration rates for the 800 μmol mol^{-1} CO_2 treatments were slightly greater than the rates from the 330 μmol mol^{-1} CO_2 treatments due to the much larger LAI of the two higher CO_2 canopies. Thus, higher LAI values under elevated CO_2 could increase transpiration rates to the point where all of the improved WUE arises from increased photosynthetic rates and none from decreased water use (Allen et al., 1985).

In conclusion, although stomatal conductance may be decreased about 40% for doubled CO_2, water use by C_3 crop plants under field conditions will probably be decreased only between about 0 to 12%. If leaf area increases due to doubled CO_2 are small (or can be controlled), then the transpiration reductions would be meaningful, albeit small. If leaf area increases due to doubled CO_2 are large, then no reductions in transpiration would be expected, and even small increases might be possible.

Large increases in WUE in elevated CO_2 do not necessarily imply any meaningful reduction in crop water requirements per unit area of land. However, farmers should be able to achieve higher crop yields per unit land area with similar amounts of water. If temperatures rise, however, the overall WUE could actually decrease due to increased water requirements in warmer climates (as illustrated by Figure 3, right panel), and to possible seed yield reductions caused by higher temperatures.

TEMPERATURE EFFECTS ON PLANTS

Leaf photosynthetic rates are known to be sensitive to temperature, CO_2 concentration, and photosystem type (C_3, C_4, or CAM). Pearcy and Björkman (1983) showed an example of photosynthetic rates of a C_4 and a C_3 desert species (1 and 2) exposed to 300 μmol mol^{-1} CO_2. The C_4 species had a higher, narrow temperature optimum (about 45°C) of leaf photosynthetic rate. The C_3 species had a much broader temperature optimum with a maximum value of about 35°C. However, when exposed to 1000 μmol mol^{-1} CO_2, the C_3 temperature response curve of photosynthesis became very similar to the response curve of the C_4 plant, which scarcely changed with the increase of CO_2 concentration. Allen (1991) adapted these curves to be more representative of crop plants in temperate zones (Figure 5). This figure shows possible responses of leaf photosynthetic CO_2 uptake rates to temperature of C_3 plants (bottom curve) and C_4 plants (top curve) when grown at 330 μmol mol^{-1} CO_2 and exposed to high light levels such as midday summer conditions. This curve shows that C_4 plants have a higher maximum photosynthetic CO_2 uptake rate than C_3 plants, and also have a higher temperature maximum than C_3 plants. The relative differences are smaller at lower temperatures. The actual photosynthetic CO_2 uptake rates could be considerably different from those shown, and the temperature distribution of photosynthetic rates could be higher or lower, depending upon species, climate, or pretreatment temperature conditions (Berry and Björkman, 1980; Penning de Vries et al., 1989). In particular, the curves could be stretched to higher temperatures for species adapted to hot, desert environments (e.g., Pearcy and Björkman, 1983), and compressed to lower temperatures for species adapted to cool environments. Furthermore, photosynthetic rates of many C_3 agricultural crops are higher than shown in this illustration (e.g., Norman and Arkebauer, 1991).

Based on a biochemical model of leaf photosynthesis, Pickering et al. (1995) and Boote et al. (1997) developed curves of photosynthesis versus temperature for C_3 leaves that were somewhat similar in shape to those of Figure 5 except that they were broader with higher overall rates of photosynthetic CO_2 uptake.

Nevertheless, from Figure 5, we can conclude that C_4 plants could benefit more (or at least suffer less) than C_3 plants to an increase in global temperatures. However, the differences for a whole canopy of leaves are somewhat

FIGURE 5. Examples of maximum photosynthetic rate (PS) responses to temperature of individual leaves under high light conditions when exposed to CO_2 concentrations of 330 μmol mol^{-1} for C_3 plants (lower curve) and C_4 plants (upper curve) which have an internal mechanism for concentrating CO_2 for subsequent photoassimilation reactions. The upper curve is also similar to the maximum PS response to temperature of C_3 plant leaves grown at and exposed to 1000 μmol mol^{-1} or greater. Various species differ widely, both in maximum leaf photosynthetic rates and in the distribution of leaf photosynthetic rates with temperature. This figure was modified and adapted from the examples of Pearcy and Björkman (1983) and Norman and Arkebauer (1991). See also Berry and Björkman (1980), Penning de Vries et al. (1989), Pickering et al. (1995), and Boote et al. (1997) for further examples of the variability of response among species and among experimental conditions.

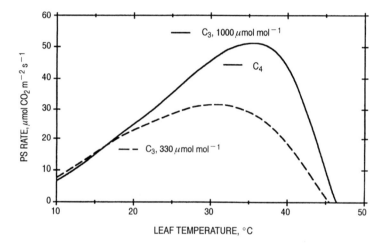

decreased from the differences of individual leaves exposed perpendicularly to high light. First, a canopy of leaves generally has leaves oriented in all directions, so that much of the total leaf area is exposed to much less irradiance than in single leaf exposure systems. Under these conditions many of the individual leaves are limited by light and the photosynthetic CO_2 uptake rates of the whole canopy becomes more similar. Secondly, solar irradiance levels are lower than midday values throughout much of the day. Nevertheless, the direction of the leaf-level differences, if not the magnitude, is maintained between C_4 and C_3 canopies.

Figure 5 also shows that C_3 plant photosynthetic rates at elevated CO_2 may increase and resemble that of C_4 plants (Pearcy and Björkman, 1983), but details will vary widely among species. Photosynthetic rates of C_3 plant leaves increase at elevated levels of CO_2 because molecules of CO_2 compete more effectively with O_2 for binding sites on rubisco, the carboxylating enzyme (Bowes and Ogren, 1972).

When plants are well-watered, leaf temperatures tend to rise slower than air temperatures throughout the daily cycle, so that the magnitude of the foliage-to-air temperature differences become greater as air temperature rises (Idso et al., 1987; Allen, 1990). For soybean, Jones et al. (1985a) found no change in crop canopy photosynthetic rates across the air temperature set-point range of 28 to 35°C. However, the canopy transpiration rates increased 30% at 35°C compared with rates at 28°C, which would lead to evaporative cooling of the leaves, and a larger magnitude of the foliage-to-air temperature differences. This 4 to 5% increase in transpiration rate per 1°C rise in temperature is close to the 6% per 1°C rise in saturation vapor pressure deficit with temperature over this temperature range.

Temperature affects growth of plants several ways. The rate of development and expression of new nodes on plants increases with increasing temperature. Leaves expressed at new nodes will grow larger, in general, if there is no concurrent water stress. Thus, plant size increases at a more rapid rate, and solar radiation capture occurs earlier in crop development. Once full ground cover is achieved, at a LAI of about 2 to 3, light capture becomes limiting and the overall temperature effects on growth are muted, but not eliminated. The duration of each ontogenic phase of plant growth decreases with increasing temperature, which is the most important effect of temperature within the upper and lower limits of survival.

INTERACTION OF CO_2
AND TEMPERATURE EFFECTS ON PLANTS

Photosynthetic and Productivity Interactions

Figure 5 showed leaf photosynthetic CO_2 uptake rate vs. temperature responses typical of C_4 plants and C_3 plants at CO_2 concentrations of 330 μmol mol^{-1}. The upper curve can also represent C_3 plants at high levels of CO_2 (1000 μmol mol^{-1} or greater). These curves suggest that a combination of rising CO_2 and rising temperature should lead to greater photosynthetic rates and hence greater biomass growth rates.

S.G. Allen et al. (1988; 1990a; 1990b) conducted experiments in Phoenix, Arizona on *Azolla*, water lily, and sorghum over the course of the season when temperatures were changing. They found that net photosynthetic rates were higher for *Azolla* and water lily during warmer times of year. Linear regressions of net photosynthetic rates of water lily vs. air temperature showed a greater increase with temperature for plants grown at 640 μmol

mol^{-1} CO_2 than those grown at 340 μmol mol^{-1} CO_2. However, there was also an interaction with solar radiation. The plants grown at 640 μmol mol^{-1} CO_2 also showed a much greater response to solar radiation than those grown at 340 μmol mol^{-1}. Although the interactions among CO_2 treatment level, air temperature, and solar radiation were not resolved, the data show that all were interrelated in the CO_2 response. S.G. Allen et al. (1990a; 1990b) also computed linear regressions of photosynthetic rate vs. previous minimum air temperatures and previous maximum air temperatures for periods of 1, 3, 6, and 9 days. They found that net photosynthetic CO_2 uptake rates were more sensitive to previous minimum air temperatures than to previous maximum air temperatures. S.G. Allen et al. (1990a; 1990b) also found that leaf photosynthetic rates were more sensitive to previous minimum temperatures than to air temperature at the time the measurements were being made.

Photosynthetic rate measurement of sorghum leaves at 640 μmol mol^{-1} CO_2 showed only small increases (about 10%) compared with leaves at 340 μmol mol^{-1} (S.G. Allen et al., 1990b). Also, the effect of temperature was quite small. These responses are to be expected since sorghum is a C_4 plant. Leaves were about 1.0 to 1.5°C warmer under the 640 μmol mol^{-1} CO_2 treatment.

Idso et al. (1987) compared growth rates (biomass accumulation rates) of carrot, radish, water hyacinth, *Azolla,* and cotton grown at 650 versus 350 μmol mol^{-1} across a seasonal range of temperatures. They found that the biomass ratios of the two CO_2 treatments increased with increasing temperature across the seasonal mean air temperature range of 12 to 36°C. The biomass growth ratio (biomass accumulation in elevated CO_2, g_{ele})/(biomass accumulation in baseline CO_2, g_{amb}) increased with temperature with a slope of 0.087 (g_{ele}/g_{amb}) °C^{-1} with a zero intercept at about 18.5°C mean daily air temperature. Under arid zone conditions at Phoenix, Arizona, daily minimum temperatures are 8 to 9°C lower than daily mean temperatures during the months of November to March when mean air temperatures are well below 18.5°C. Low nocturnal temperatures, as well as low total solar irradiance, short photoperiod (daylength), previous carbohydrate storage, and stage of growth of the plants may affect the biomass growth ratio during the winter months.

Baker et al. (1989) found a different biomass growth ratio for both final harvest dry matter and seed yield for soybean grown at 330 and 660 μmol mol^{-1} CO_2. Early vegetative growth indicated that the biomass growth ratio increased with temperature, but final harvest data showed otherwise. The experiment was conducted with day/night air temperatures of 26/19, 31/24, and 36/29°C , which gave average air temperatures of about 22.8, 27.8, and 32.8°C under the 13/11 hour thermoperiod. The biomass growth ratios for final harvest dry matter were 1.50, 1.36, and 1.24 (g_{ele}/g_{amb}) for the respective temperatures. The biomass growth ratios for final harvest seed yield were 1.46, 1.24, and 1.15 (g_{ele}/g_{amb}) for the respective temperatures. The changes in the biomass growth ratio for total dry matter and for seed yield were -0.026 (g_e/g_b) $^\circ C^{-1}$ ($r^2 = 0.98$) and -0.031 (g_e/g_b) $^\circ C^{-1}$ ($r^2 = 0.88$), respectively. This cultivar of soybean, "Bragg", has a determinate growth habit which causes vegetative growth to cease nearly when flowering begins. Elevated temperatures tend to hasten maturity and shorten the life cycle of this soybean crop. These factors were different from the Arizona study. Furthermore, the study of Baker et al. (1989) had identical light conditions for all treatments, and photoperiod interaction with temperature was minimized by two weeks of supplemental lighting at the beginning of the season.

Baker and Allen (1993) summarized total daytime CO_2 uptake as a function of CO_2 and temperature for rice, soybean and citrus treelings grown in outdoor, sunlit controlled environment growth chambers (Table 3). Rice maintained CO_2 uptake rates up to 37°C, but began to fail at 40°C. Soybean CO_2 uptake varied little with temperature over the range of 28 to 35°C. Citrus photosynthetic rates appeared to be sensitive to temperatures as low as 34°C, but elevated CO_2 alleviated the high temperature effects somewhat.

In summary, the biomass growth ratio of plants grown at elevated CO_2 may increase with increasing temperature for vegetative growth, as suggested by Figure 5. However, this response may be reversed for seed grain crops which have a determinate growth habit, such as "Bragg" soybean.

Evapotranspiration

Table 3 also included daytime total evapotranspiration and WUE of rice, soybean, and citrus (Baker and Allen, 1993). Total daytime water use increased with increasing temperature and decreased with increasing CO_2 concentration for all three species. On the other hand, WUE decreased with increasing temperature and increased with increasing CO_2 concentration. The CO_2 effect on WUE was due to both an increase in photosynthetic CO_2 uptake and to a decrease in transpiration. At the highest temperatures for rice (40°C) and citrus (34-35°C), the WUE was decreased further by a decrease in CO_2 uptake (Table 3).

Some of the best modeling studies on the simulated effects of climate change and CO_2 increase on plant canopy evapotranspiration were conducted by Rosenberg et al. (1990), using the Penman-Monteith model. A similar approach was used by Allen and Gichuki (1989) to estimate effects of CO_2-induced climate changes on evapotranspiration and irrigation water requirements in the Great Plains from Texas to Nebraska. Rosenberg et al. (1990) examined the effects of temperature, net radiation, air vapor pressure, stomatal resistance, and LAI on three types of plant canopies: wheat at Mead, NE; grassland at Konza Prairie, KS; and forest at Oak Ridge, TN. Increasing

TABLE 3. Total daytime responses of rice, soybean, and citrus treeling canopies to CO_2 and temperature. Water loss from rice included both transpiration and evaporation from the paddy water. Adapted from Baker and Allen (1993).

Crop	CO_2 Conc.	Air temperature	Daytime photons	Daytime CO_2 uptake	Daytime H_2O loss	Water-use efficiency
	μmol mol^{-1}	(°C)	mol m^{-2}	mol m^{-2}	mol m^{-2}	mmol (CO_2) mol (H_2O)
Rice[1]	330	28	23.0	0.75	495	1.51
		34	23.0	0.76	605	1.26
		40	23.1	0.34	693	0.49
	660	28	23.0	1.11	459	2.42
		34	23.0	1.13	624	1.81
		40	23.1	0.72	631	1.14
Rice[2]	330	28	26.5	0.87	613	1.42
	660	25	26.4	0.83	359	2.30
		28	26.4	1.13	476	2.37
		31	26.5	0.83	491	1.69
		34	26.5	0.89	736	1.20
		37	26.5	0.97	909	1.06
Soybean[3]	330	28	29.4	0.96	329	2.95
		31	30.6	0.96	372	2.63
		33	29.5	0.94	398	2.37
		35	29.2	0.98	414	2.33
	800	28	29.7	1.89	317	6.00
		31	30.9	1.97	387	5.11
		33	28.9	1.90	386	4.95
		35	30.1	1.87	396	4.71
Citrus[4]	330	25	51.0	0.63	274	2.30
		34	46.0	0.38	305	1.25
	840	27	49.5	1.09	202	5.40
		35	49.5	0.86	244	3.52

[1] Source: Baker et al. (1989), 14 Dec. 1988 (65 days after planting).
[2] Source: Baker et al. (1990b), 10 Sep. 1989 (58 days after planting).
[3] Source: Jones et al. (1985b), 9-19 Oct. 1982 (57-67 days after planting).
[4] Source: Brakke and Allen (1995).

temperature by 3°C gave a 6 to 8% increase in transpiration per 1°C. This compares reasonably well with 4 to 5% increase in transpiration per 1°C measured experimentally in soybean across the 28 to 35°C range by Jones et al. (1985a). Rosenberg et al. (1990) also calculated that evapotranspiration would decrease 12 to 17% for a 40% increase in stomatal resistance (a 29% decrease in stomatal conductance). This corresponds closely to a 12% decrease in seasonal transpiration obtained experimentally for soybean grown in controlled-environment chambers by Jones et al. (1985c) for doubled CO_2 conditions when leaf area index was very similar for both the ambient and doubled CO_2 treatments. In another study, Jones et al. (1985b) showed that exposure of soybean canopies to 800 μmol mol^{-1} CO_2 decreased daily total transpiration by 16% in comparison with exposure to 330 μmol mol^{-1}.

Increases in LAI of 15% caused increases in predicted evapotranspiration of about 5 to 7% according to the model of Rosenberg et al. (1990). These values were comparable to those extracted from Jones et al. (1985b) whose data showed a 33% increase in measured daily transpiration for a change in LAI from 3.3 to 6.0 (an 82% increase in LAI). However, the effect of LAI should not be linear. Using a Soil-Plant-Atmosphere Model, Shawcroft et al. (1974) showed that the effect of leaf area on canopy transpiration would be highly nonlinear across the LAI range of 0 to 8, with almost all of the effect occurring across the LAI range from 0 to 4. However, these simulations were conducted with a moist soil surface (water potential = -60 MPa) and relatively high soil surface-to-air boundary-layer conductance. Thus, predicted evapotranspiration rates at a LAI of 2 were maintained at 85% or more of the rates at a LAI of 8. Nevertheless, the modeling results of Shawcroft et al. (1974) for three solar elevation angles and three leaf elevation angle classes showed that predicted plant transpiration increased by an average of 27 $\pm$ 8% for a LAI increase from 3.3 to 6.0.

Net radiation could increase under climate change conditions both from greater downwelling thermal radiation and from increased solar radiation (decreased cloudiness), or it could decrease from increased cloudiness. Rosenberg et al. (1990) showed that evapotranspiration should change about 0.6 to 0.7% for each 1% change in net radiation. Likewise they showed that evapotranspiration should change about -0.4 to -0.8% for each 1% change in vapor pressure of the air. A combination of the following factors: temperature = +3°C, net radiation = $\pm$10%, vapor pressure = $\pm$ 10%, stomatal resistance = +40%, and leaf area index = +15%, gave changes in evapotranspiration ranging from +27% (for a case of increased net radiation and decreased vapor pressure) to -4% (for a case of decreased net radiation and increased vapor pressure). Each factor of climate change and plant response to CO_2 affects the predicted evapotranspiration.

De Bruin and Jacobs (1993) modeled the percent change in transpiration for percentage changes in canopy surface resistances ranging from -25% through +50% (which included both increased leaf area effects and increased leaf surface resistance effects). In accounting for interactions between the surface fluxes and the daytime planetary boundary layer (taken to be 1 km in depth), they found that the effects of changes in canopy surface resistances on changes of transpiration were diminished substantially. Nevertheless, they predicted that the range of surface resistance changes cited above would cause a change in transpiration ranging from +7% to -11% for aerodynamically smooth vegetation, and from +15% to -21% for aerodynamically rough surfaces. For a 40% increase in surface resistance the value used by Rosenberg et al. (1990), the model of de Bruin and Jacobs (1993) would give about -9 to -17% changes in transpiration.

PREDICTING WATER REQUIREMENTS
AND YIELDS UNDER CLIMATE CHANGE

General Circulation Models

Allen (1991) reviewed the climate change predictions generated by five general circulation models (GCM) for doubled CO_2. These models were from Oregon State University, OSU, (Schlesinger, 1984), the National Center for Atmospheric Research Community Climate Models, CCM, (Washington and Meehl 1984; 1986), the NOAA Geophysical Fluids Dynamics Laboratory, GFDL, (Manabe and Wetherald, 1987), the NASA Goddard Institute for Space Studies, GISS, (Hansen et al., 1988), and the United Kingdom Meteorological Office, UKMO, (Wilson and Mitchell, 1987; Mitchell, 1989). For a doubling of CO_2, the OSU, CCM, GFDL, GISS, and UKMO models predicted global mean surface temperature increases of 2.8, 4.0, 4.0, 4.2 and 5.2°C, and global mean precipitation increases of 7.8%, 7.1%, 8.7%, 11.0%, and 15.0%, respectively (Wilson and Mitchell, 1987).

For the USA, with a doubling of CO_2, the predicted June-July-August (JJA) median temperature increases were 3.5, 3.0, 3.8, and 5.6°C, and the predicted JJA changes in precipitation were +4, +10, +8, and -25% from the OSU, CCM, GISS, and GFDL models, respectively (Grotch, 1988). The UKMO model predictions for decreased precipitation for the USA are somewhat similar to the GFDL model scenario (Wilson and Mitchell, 1987). Higher JJA temperatures for the USA were associated with models with the lowest summer precipitation and soil wetness. The possibility of decreases in summer precipitation, coupled with temperature increases, could pose a serious problem for future agricultural productivity.

Modelling Crop Responses to CO_2 and Temperature

Growth, yield, and transpiration responses to doubled CO_2 and climate change scenarios were predicted by Peart et al. (1989) and Curry et al. (1990a; 1990b) for the Southeastern USA. Climate changes scenarios of the GISS model (Hansen et al., 1988) and the GFDL model (Manabe and Wetherald, 1987) were used to change temperatures, precipitation, and solar radiation, month by month, for 30 years of baseline weather data (1951-1980) of 19 Southeastern USA sites. The baseline, GISS, and GFDL temperature and precipitation scenarios for Memphis, Tennessee, and Columbia, South Carolina are given in Figure 6 (data provided by Roy Jenne, National Center for Atmospheric Research, Peart et al., 1989). Monthly averages of July maximum daily temperatures for baseline, GISS, and GFDL scenarios were 33.27, 35.23, and 38.19°C, respectively, for Columbia, SC, and 33.07, 35.44, and 36.03°C, respectively, for Memphis, TN. Simulations were run under

FIGURE 6. Average monthly temperature, precipitation, and potential evapo-transpiration for Memphis, Tennessee and Columbia, South Carolina for the 30-year base climate period, 1951-1980 (Top Panels), for the GISS climate change scenario for doubled atmospheric CO_2 (Middle Panels), and for the GFDL climate change scenario for doubled atmospheric CO_2 (Bottom Panels). Temperature and precipitation data adapted from Peart et al. (1989) and Roy Jenne (personal communication). Potential evapotranspiration was calculated from the method of Doorenbos and Pruitt (1977).

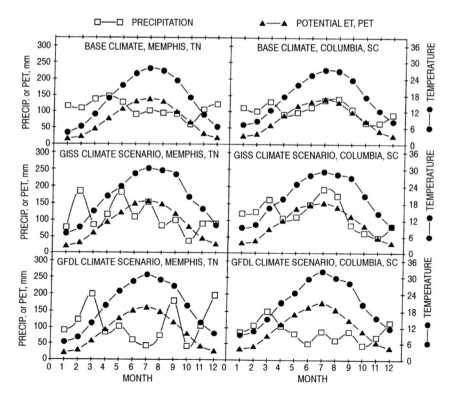

four climate change conditions: rainfed or optimum irrigation conditions, and with or without direct CO_2 fertilization effects on plant photosynthesis. Crop photosynthetic enhancement factors of 1.35 for soybean and 1.10 for maize were used (Peart et al., 1989; Curry et al., 1990a).

Table 4 shows predicted soybean seed yields averaged over 19 sites and 30 years. Under rainfed conditions without direct CO_2 fertilization effects, average soybean yield was decreased to 77% under the GISS scenario and down to 29% under the GFDL scenario, in comparison with predictions using the baseline climate data. The yields under the GFDL scenario were severely impacted because of the rainfall reductions predicted by this GCM (Figure 6).

Under optimum irrigation conditions, average yields under both the GISS and GFDL scenarios were decreased to about 80% with respect to optimum irrigation under baseline climate conditions. However, in spite of higher temperatures, the irrigated yields under the GISS and GFDL scenarios were about 32% greater than the baseline climate scenario without irrigation.

When direct CO_2 fertilization plus climate change effects were simulated (Table 4), average yields of the GISS climate scenario were increased 12% under rainfed conditions, whereas yields under the GFDL climate scenario were decreased to 50%. Under optimum irrigation with CO_2 fertilization

TABLE 4. Simulated soybean yields, cumulative evapotranspiration (ET), water-use efficiency (WUE), and irrigation for optimum soil water for BASE, GISS, and GFDL scenario climates and for two levels of CO_2. Monthly average temperatures and monthly total rainfall for the BASE, GISS, and GFDL scenarios of Memphis, TN and Charleston, SC are given in Figure 6. The values reported here are for the Southeastern USA as a whole for 30 years (1951-1980) and 19 locations from runs of the SOYGRO model. Values in parenthesis are relative to the BASE climate at $330\ \mu$mol mol^{-1} CO_2 for either the irrigated or rainfall soil water condition. Adapted from Curry et al. (1990b).

Climate Scenario	CO_2 μmol mol^{-1}	Yield kg ha^{-1}	(Rel)	Cumulative ET mm	(Rel)	WUE kg ha^{-1} mm^{-1}	Rainfall mm	(Rel)	Irrigation mm	(Rel)
IRRIGATED										
BASE	330	3810	(100%)	595	(100%)	6.40	496	(100%)	228	(100%)
GISS	330	3080	(81%)	697	(117%)	4.42	592	(119%)	290	(127%)
GFDL	330	3060	(80%)	737	(124%)	4.15	396	(80%)	486	(213%)
BASE	660	5100	(134%)	575	(97%)	8.87	497	(100%)	209	(92%)
GISS	660	4130	(108%)	683	(115%)	6.05	592	(119%)	278	(122%)
GFDL	660	4220	(111%)	729	(123%)	5.79	396	(80%)	485	(213%)
RAINFED										
BASE	330	2320	(100%)	512	(100%)	4.53	497	(100%)	0	
GISS	330	1780	(77%)	573	(112%)	3.11	593	(119%)	0	
GFDL	330	670	(29%)	468	(91%)	1.43	396	(80%)	0	
BASE	660	3520	(152%)	506	(99%)	6.96	497	(100%)	0	
GISS	660	2610	(112%)	571	(111%)	4.57	594	(120%)	0	
GFDL	660	1160	(50%)	477	(93%)	2.43	397	(80%)	0	

effects, yields under both GISS and GFDL scenarios were increased roughly 10% with respect to the irrigated baseline climate scenario.

Maize yields would be expected to decline only about 8% in the GISS climate scenario but would decline by about 75% in the GFDL scenario based on simulations at 4 Southeastern USA locations (Curry et al., 1990a). Although irrigation increased predicted yields, the GISS and GFDL climate scenarios gave yield decreases of 18% and 26%, respectively, with respect to the irrigated baseline weather crops. The yield reduction of the GFDL scenario with respect to the GISS scenario was 10%, attributable to slightly higher temperatures (Peart et., 1989). Including the direct effects of CO_2 plus climate had little effect on the predicted yields of maize, as expected because maize is a C_4 plant.

Ritchie et al. (1989) found that higher temperature had the greatest effect of the climate change factors on predicted yields of soybean and maize in the Great Lakes and Corn Belt Area. Increases in temperature caused a decrease in the duration of the crop life cycle. Yield reductions were greatest for the GFDL climate change scenario for the southernmost locations of this region. Predicted yields increased for the northernmost stations because temperatures and growing season duration became more favorable for these crops. Overall, irrigation water requirements in this region decreased about 30% for the GISS scenario, and increased about 90% for the GFDL scenario because of low amounts of summertime rainfall.

Rosenzweig (1989) modeled corn and wheat yields in the Great Plains under GISS and GFDL climate change scenarios. She found that corn and wheat yields would be decreased particularly in the Southern Great Plains because higher temperatures shortened the life cycle of the crops. Where precipitation was predicted to decrease, irrigation requirements increased. Allen and Gichuki (1989) predicted a 15% overall increased requirement for irrigation for this region, with greater requirements for alfalfa because its growing season was increased, and lower requirements for corn and winter wheat because their growing seasons were decreased.

Direct effect of rising CO_2 offset the adverse effects of climate change at some locations, but not all, in the simulations of Ritchie et al. (1989) and Rosenzweig (1989).

Rosenberg et al. (1990) used the Penman-Monteith evapotranspiration model and GISS, GFDL, and NCAR scenarios for estimating the impact of predicted climate changes on summer evapotranspiration at Mead, NE (wheat), Konza Prairie, KS (grassland), and Oak Ridge, TN (forest). They found that inclusion of other factors (net radiation, vapor pressure, windspeed, stomatal resistance, and leaf area index) decreased the impact of temperature increases on predicted evapotranspiration. For example, a 6.3°C rise in GFDL scenario temperature for Mead, NE resulted in a 42% increase in predicted evapotranspiration, but only a 23% increase when all other climate change factors were also considered. Similarly, a GISS scenario temperature rise of 4.7°C for Konza Prairie, KS gave a 28% predicted increase in evapotranspiration. However, this increase was only 4% when all other climate change factors were included in the Penman-Monteith equation. On the basis of these computations, Rosenberg et al. (1990) caution against the presumption that evapotranspiration will increase in all scenarios of global climate change. When the decreased duration of the life cycle of plants (caused by increased temperature) is included in climate change scenarios, some crops may actually transpire less total water throughout their life cycles when grown on the Great Plains (e.g., corn and winter wheat) as shown in model studies of Allen and Gichuki (1989).

Curry et al. (1995) predicted responses of soybean for the eastern half of the USA using baseline, GISS, GFDL, and UKMO climate scenarios for an effective doubling of CO_2. Because of the anticipated contribution of other greenhouse effect gases to global warming, the actual CO_2 concentration for an effective doubling of CO_2 was estimated to be 555 μmol mol^{-1}. Yields under these conditions generally increased for the Fargo, ND simulations and decreased for the Memphis, TN simulations. These predictions support the hypothesis that crop productivity could shift to higher latitudes under anticipated climate change conditions. Except for the UKMO model, which predicted a harsher temperature regime, the results for the southeastern USA were similar to the earlier modeling studies of Peart et al. (1989) and Curry et al. (1990a; 1990b).

ADAPTATIONS AND EVAPOTRANSPIRATION REQUIREMENTS

The crop yield simulations of the previous section demonstrate two main points, namely, (a) the importance of direct CO_2 fertilization effects in the face of elevated temperatures, and (b) the adverse impact of inadequate rainfall under rising temperature scenarios on crop production.

Much of the predicted decrease in soybean yields reported by Peart et al. (1989) and Curry et al. (1990a; 1990b) was due to decreases in the length of the grain filling period under higher temperatures. Changes in standard management practices were not simulated, but they might help restore part of the yield. Planting earlier or later in the season might help offset the higher temperatures. Selection of other more-adapted cultivars could help also. In the future, it may be necessary for plant breeders to adapt combinations of temperature tolerance and photoperiod responses into new germplasms. Perhaps plant growth regulators could play a role also. Under conditions where nonstructural carbohydrates accumulate in CO_2 enriched plants, new germplasms which make better use of the photoassimilate source need to be developed.

Irrigation is not likely to be a panacea for climate change. From Peart et al.

(1989) and Curry et al. (1990a; 1990b) simulations, the irrigation requirement was increased 22% under the GISS scenario and 113% under the GFDL scenario (Table 4). However, with less summertime rainfall under the GFDL scenario, region-wide water resources would become scarce, and water may not be readily available for crops. Some areas of the USA may have to adapt by irrigating less land area, and other areas by shifting from rainfed to irrigated agriculture. Increasing temperatures and decreasing precipitation for the USA as predicted by the GFDL model (and the UKMO model) would have a serious negative impact overall on agricultural productivity and society, although producers in favorable regions may benefit from scarcity-mediated higher prices (Adams et al., 1990).

Allen (1991) analyzed the impacts of the precipitation (PPT) and temperature scenarios of the GISS and GFDL models on water availability for the Southeastern USA. Average month-by-month potential evapotranspiration (PET) was calculated for Columbia, SC, and Memphis, TN based on GISS and GFDL temperature predictions and using the methods of Doorenbos and Pruitt (1977) for estimating solar radiant energy inputs.

From these calculations, on the long-term average, monthly PPT-PET for the May through October period at Columbia would allow sufficient water for crop production for both the baseline climate and the GISS scenario (Table 5). However, PPT-PET over the May to October period was −435 mm of water for the GFDL scenario, which indicates a severe water deficit during this period. Nevertheless, during the November through April period, PPT-PET for the GFDL scenario was about 87% of the GISS scenario. Therefore, following dry summer months, there is opportunity for sizable recharge and streamflow during the winter months despite the overall annual water deficit of about 153 cm for the GFDL scenario at Columbia (Table 5).

For the Memphis example, the baseline climate showed a slight tendency toward summertime drought (Allen, 1991). The GISS scenario and the GFDL scenario had more month-to-month variation than for the Columbia example. However, the total annual PPT was very similar for all three Memphis cases (Table 5). The PPT deficit was particularly severe for the GFDL scenario during the June-July-August period. Nevertheless, the potential for recharge and subsequent streamflow was great (PPT-PET = 528 mm) during the November through April period, so the GFDL scenario has the potential of providing more streamflow water resources than the GISS scenario, although the GISS scenario provides more rainfall for crops and other vegetation during the summer months (Table 5).

These examples at two locations in a humid region show that although climate change scenarios may imply severe rainfall shortages and soil water deficits during the growing period of the year, PPT excesses over PET during the cooler parts of the year still have the potential for maintaining cool season streamflows and water storage in reservoirs.

TABLE 5. Average annual precipitation (PPT), potential evapotranspiration (PET), PPT minus PET for a 6-month period of recharge and runoff (November through April), and PPT for a 6-month period of high PET (May through October) for baseline climate (1951-1980) and for GISS and GFDL scenarios at two locations, Columbia, SC and Memphis, TN.

	BASE	GISS	GFDL
	(mm)	(mm)	(mm)
Columbia, SC			
PPT	1245	1419	1042
PET	955	1096	1195
PPT − PET (N-A)	+317	+323	+282
PPT − PET (M-O)	−27	0	−435
NET	+290	+323	−153
Memphis, TN			
PPT	1310	1314	1306
PET	895	1063	1061
PPT − PET (N-A)	+522	+361	+528
PPT − PET (M-O)	−107	−110	−283
NET	+415	+251	+245

REFERENCES

Acock, B. and L.H. Allen, Jr. (1985). Crop responses to elevated carbon dioxide concentration. pp. 53-97. In: B.R. Strain and J.D. Cure (eds.), *Direct Effects of Increasing Carbon Dioxide on Vegetation*. DOE/ER-0238. US Dep. of Energy, Carbon Dioxide Res. Div., Washington, DC.

Adams, R.M., C. Rosenzweig, R.M. Peart, J.T. Ritchie, B.A. McCarl, J.D. Glyer, R.B. Curry, J.W. Jones, K.J. Boote and L.H. Allen, Jr. (1990). Global climate change and US agriculture. *Nature* 345:219-224.

Allen, L.H., Jr. (1990). Plant responses to rising carbon dioxide and potential interactions with air pollutants. *J. Environ. Qual.* 19:15-34.

Allen, L.H., Jr. (1991). Effects of increasing carbon dioxide levels and climate change on plant growth, evapotranspiration, and water resources. In: *Managing Water Resources in the West Under Conditions of Climate Uncertainty*. Proceedings of a Colloquium, November 14-16, 1990, Scottsdale, Arizona. National Academy Press, Washington, DC.

Allen, L.H., Jr., K.J. Boote, J.W. Jones, P.H. Jones, R.R. Valle, B. Acock, H.H. Rogers and R.C. Dahlman. (1987). Response of vegetation to rising carbon dioxide: Photosynthesis, biomass, and seed yield of soybean. *Global Biogeochemical Cycles* 1:1-14.

Allen, L.H., Jr., P. Jones and J.W. Jones. (1985). Rising atmospheric CO_2 and evapotranspiration. pp. 13-27. In: *Advances in Evapotranspiration. ASAE Pub. 14-85.* American Society of Agricultural Engineers, St. Joseph, Michigan 49085-9659.

Allen, R.G. and F.N. Gichuki. (1989). Effects of projected CO_2-induced climatic changes on irrigation water requirements in the Great Plains States (Texas, Oklahoma, Kansas, and Nebraska). pp. 6-1 to 6-42. In: J.B. Smith and D.A. Tirpak (eds.), *The Potential Effects of Global Climate Change on the United States. Appendix C, Agriculture, Vol. 1.* EPA-230-05-89-053, U.S. Environmental Protection Agency, Washington, DC.

Allen, S.G., S.B. Idso, B.A. Kimball and M.G. Anderson. (1988). Interactive effects of CO_2 and environment on photosynthesis of *Azolla*. *Agric. For. Meteorol.* 42:209-217.

Allen, S.G., S.B. Idso and B.A. Kimball. (1990a). Interactive effects of CO_2 and environment on net photosynthesis of water lily. *Agric., Ecosystems, and Environ.* 30:81-88.

Allen, S.G., S.B. Idso, B.A. Kimball, J.T. Baker, L.H. Allen, Jr., J.R. Mauney, J.W. Radin and M.G. Anderson. (1990b). *Effects of Air Temperature on Atmospheric CO_2-Plant Growth Relationships–TR048.* DOE/ER-0450T, U.S. Dep. of Energy and U.S. Dep. of Agriculture, Washington, DC 20250.

Allison, F.E., E.M. Roller and W.A. Raney. (1958). Relationship between evapotranspiration and yield of crops grown in lysimeters receiving natural rainfall. *Agron. J.* 50:506-511.

Arkley, R.J. (1963). Relationship between plant growth and transpiration. *Hilgardia* 34:559-584.

Baker, D.N., L.H. Allen, Jr. and J.R. Lambert. (1982). *Effects of Increased CO_2 on Photosynthesis and Agricultural Productivity, Vol. II, Part 6 of Environmental Consequences of a Possible CO_2-Induced Climate Change, 013.* DOE/EV/10019-6, Carbon Dioxide Research Division, Office of Basic Energy Sciences, OER, U.S. Department of Energy, Washington, DC.

Baker, J.T., S.L. Albrecht, D. Pan, L.H. Allen, Jr., N.B. Pickering, and K.J. Boote. (1994). Carbon dioxide and temperature effects on rice (*Oryza sativa* L., cv. 'IR-72'). *Soil Crop Sci. Soc. Florida Proc.* 53:90-97.

Baker, J.T., L.H. Allen, Jr., K.J. Boote, P. Jones and J.W. Jones. (1989). Response of soybean to air temperature and carbon dioxide concentration. *Crop Sci.* 29:98-105.

Baker, J.T., L.H. Allen, Jr., K.J. Boote, P. Jones and J.W. Jones. (1990). Rice photosynthesis and evapotranspiration in subambient, ambient and superambient carbon dioxide concentrations. *Agron. J.* 82:834-840.

Baker, J.T. and L.H. Allen, Jr. (1993). Contrasting crop species responses to CO_2 and temperature: rice, soybean and citrus. *Vegetatio* 104/105:239-260. Also pp. 239-260. In: J. Rozema, H. Lambers, S.C. van de Geijn and M.L. Cambridge (eds.) *CO2 and Biosphere.* Kluwer Academic Publishers, Dordrecht.

Baker, J.T., L.H. Allen, Jr., K.J. Boote, and N.B. Pickering. (1997). Rice responses to drought under carbon dioxide enrichment. 2. Photosynthesis and evapotranspiration. *Global Change Biology* 3:129-138.

Barnola, J.M., D. Raynaud, Y.S. Korotkevich and C. Lorius. (1987). Vostok ice core provides 160,000-year record of atmospheric CO_2. *Nature* 329:408-414.

Barnola, J.M., D. Raynaud, C. Lorius and Y.S. Korotkevich. (1994). Historical CO_2 data from the Vostok ice core. pp. 7-10. In: T.A. Boden, D.P. Kaiser, R.J. Sepanski and F.W. Stoss (eds.), *Trends '93: A Compendium of Data on Global Climate Change*. ORNL-CDIAC-65. Carbon Dioxide Information Analysis Center, Oak Ridge National Laboratory, Oak Ridge, TN, USA.

Bennett, O.L. B.D. Doss, D.A. Ashley, V.J. Kilmer and E.C. Richardson. (1964). Effects of soil moisture regime on yield, nutrient content, and evapo-transpiration for three annual forage species. *Agron. J.* 56:195-198.

Berry, J. and O. Björkman. (1980). Photosynthetic response and adaptation to temperature in higher plants. *Ann. Rev. Plant Physiol.* 31:491-543.

Boote, K.J., N.B. Pickering and L.H. Allen, Jr. (1997). Plant Modeling: Advances and gaps in our capability to predict future crop growth and yield in response to global climate change. pp. 179-228. In: L.H. Allen, Jr., M.B. Kirkham, D.M. Olszyk and C.E. Whitman (eds.), *Advances in Carbon Dioxide Effects Research, ASA Special Publication No. 61*, ASA-CSSA-SSSA, Madison, Wisconsin.

Bowes, G. and W.L. Ogren. (1972). Oxygen inhibition and other properties of soybean ribulose-1,5-diphosphate carboxylase. *J. Biol. Chem.* 247:2171-2176.

Briggs, L.J. and H.L. Shantz. (1913a). *The Water Requirements of Plants: I. Investigations in the Great Plains in 1910 and 1911. USDA Bureau of Plant Industry Bull. 284.*

Briggs, L.J. and H.L. Shantz. (1913b). *The Water Requirements of Plants: II. A Review of the Literature. USDA Bureau of Plant Industry Bull. 285.*

Briggs, L.J. and H.L. Shantz. (1914). Relative water requirement of plants. *J. Agric. Res.* 3:1-63.

Chang, J.-H. (1968). *Climate and Agriculture, an Ecological Survey*. Aldine Publishing Co., Chicago. 304 pp.

Curry, R.B., R.M. Peart, J.W. Jones, K.J. Boote and L.H. Allen, Jr. (1990a). Simulation as a tool for analyzing crop response to climate change. *Trans. ASAE* 33:981-990.

Curry, R.B., R.M. Peart, J.W. Jones, K.J. Boote and L.H. Allen, Jr. (1990b). Response of crop yield to predicted changes in climate and atmospheric CO_2 using simulation. *Trans. ASAE* 33:1383-1390.

Curry, R.B., J.W. Jones, K.J. Boote, R.M. Peart, L.H., Allen, Jr. and N.B. Pickering. (1995). Response of soybean to predicted climate change in the USA. pp. 163-182. In: C. Rosenzweig, L.H. Allen, Jr., L.A. Harper, S.E. Hollinger and J.W. Jones (eds.), *Climate Change and Agriculture: Analysis of Potential International Impacts. ASA Special Publication No. 59*. American Society of Agronomy, Madison, Wisconsin USA.

De Bruin, H.A.R. and C.M.J. Jacobs. (1993). Impact of CO_2 enrichment on the regional evapotranspiration of agro-ecosystems, a theoretical and numerical modeling study. *Vegetatio* 104/105:307-318. Also pp. 307-318. In: J. Rozema, H. Lambers, S.C. van de Geijn and M.L. Cambridge (eds.) *CO_2 and Biosphere*. Kluwer Academic Publishers, Dordrecht.

De Wit, C.T. (1958). *Transpiration and Crop Yields*. Institut voor Biologisch en Scheikundig Onderzoek van Landouwgewassen. Wageningen Mededeling. No. 59.

Doorenbos, J. and W.O. Pruitt. (1977). *Crop Water Requirements. FAO Irrigation and Drainage Paper No. 24.* Food and Agriculture Organization of the United Nations. Rome.

Enoch, H.Z. and B.A. Kimball. (1986). *Carbon Dioxide Enrichment of Greenhouse Crops, Vol. I and II.* CRC Press, Inc., Boca Raton, Florida 33431.

Gesch, R.W., K.J. Boote, J.C.V. Vu, L.H. Allen, Jr. and G. Bowes. (1998). Changes in growth CO_2 result in rapid adjustments of ribulose-1,5-bisphosphate carboxylase/oxygenase small subunit gene expression in expanding and mature leaves of rice. *Plant Physiol.* 118:521-529.

Grotch, S.L. (1988). *Regional Intercomparisons of General Circulation Model Prediction and Historical Climate Data–TR 041.* DOE/NBB-0084, U.S. Dep. of Energy, Carbon Dioxide Res. Div., Washington, DC 20545.

Guinn, G. and J.R. Mauney. (1980). Analysis of CO_2 exchange assumptions: Feedback control. pp. 1-16. In: J.D. Hesketh and J.W. Jones (eds.), *Predicting Photosynthesis for Ecosystem Models, Vol II.* CRC Press, Inc., Boca Raton, Florida 33431.

Hanks, R.J., H.R. Gardner and R.L. Florian. (1969). Plant growth-evapotranspiration relationships for several crops in the Great Plains. *Agron. J.* 61:30-34.

Hansen, J., I. Fung, A. Lacis, S. Lebedeff, D. Rind, R. Ruedy, G. Russell and P. Stone. (1988). Global climate change as forecast by the GISS 3-D model. *J. Geophys. Res.* 93:9341-9364.

Idso, S.B., B.A. Kimball, M.G. Anderson and J.R. Mauney. (1987). Effects of atmospheric CO_2 enrichment on plant growth: The interactive role of air temperature. *Agri. Ecosystems Environ.* 20:1-10.

Jensen, M.E. (1974). *Consumptive Use of Water and Irrigation Water Requirements.* Technical Committee on Irrigation Water Requirements, Irrigation and Drainage Div., ASCE, New York. 215pp.

Jones, P., L.H. Allen, Jr., J.W. Jones, K.J. Boote and W.J. Campbell. (1984). Soybean canopy growth, photosynthesis, and transpiration responses to whole-season carbon dioxide enrichment. *Agron. J.* 76:633-637.

Jones, P., L.H. Allen, Jr. and J.W. Jones. (1985a). Responses of soybean canopy photosynthesis and transpiration to whole-day temperature changes in different CO_2 environments. *Agron. J.* 77:242-249.

Jones, P., L.H. Allen, Jr., J.W. Jones and R.R. Valle. (1985b). Photosynthesis and transpiration responses of soybean canopies to short- and long-term CO_2 treatments. *Agron. J.* 77:119-126.

Jones, P., J.W. Jones and L.H. Allen, Jr. (1985c). Seasonal canopy CO_2 exchange, water use, and yield components in soybean grown under differing CO_2 and water stress conditions. *Trans. ASAE* 28:2021-2028.

Jouzel, J., C. Lorius, J.R. Petit, C. Genthon, N.I. Barkov, V.M. Kotlyakov and V.M. Petrov. (1987). Vostok ice core: a continuous isotope temperature record over the last climatic cycle (160,000 years). *Nature* 329:403-407.

Keeling, C.D., R.B. Bacastow, A.F. Carter, S.C. Piper, T.P. Whorf, M. Heinmann, W.G. Mook and H. Roeloffzen. (1989). A three dimensional model of atmospheric CO_2 transport based on observed winds: Analysis of observational data. In: D.H. Peterson (ed.), *Aspects of Climate Variability in the Pacific and the Western*

Americas. Geophysical Monograph 55:165-235. Amer. Geophys. Union, Washington, DC.

Keeling, C.D. and T.P. Whorf. (1994). Atmospheric CO_2 records from sites in the SIO air sampling network. pp. 16-26. In: T.A. Boden, D.P. Kaiser, R.J. Sepanski and F.W. Stoss (eds.), *Trends '93: A Compendium of Data on Global Climate Change.* ORNL-CDIAC-65. Carbon Dioxide Information Analysis Center, Oak Ridge National Laboratory, Oak Ridge, TN, USA.

Kimball, B.A. (1983). Carbon dioxide and agricultural yield: An assemblage and analysis of 430 prior observations. *Agron. J.* 75:779-788.

Kimball, B.A. (1986). Influence of elevated CO_2 on crop yield. pp. 105-115. In: H.Z. Enoch and B.A. Kimball (eds.), *Carbon Dioxide Enrichment of Greenhouse Crops, Vol. II,* Physiology, Yield, and Economics. CRC Press, Boca Raton, Florida.

Kimball, B.A. and S.B. Idso. (1983). Increasing atmospheric CO_2: Effects on crop yield, water use, and climate. *Agric. Water Management* 7:55-72.

Kimball, B.A., J.R. Mauney, G. Guinn, F.S. Nakayama, P.J. Pinter, Jr., K.L. Clawson, R.J. Reginato and S.B. Idso. (1983). *Responses of Vegetation to Carbon Dioxide, Ser. 021. Effects of Increasing Atmospheric CO_2 on the Yield and Water Use of Crops.* Joint program of the U.S. Dep. of Energy and the U.S. Dep. of Agriculture, U.S. Water Conservation Lab. and U.S. Western Cotton Research Lab., USDA-ARS, Phoenix, AZ.

Kimball, B.A., R.L. LaMorte, P.J. Pinter, Jr., G.W. Wall, D.J. Hunsaker, F.J. Adamsen, S.W. Leavitt, T.L. Thompson, A.D. Matthias and T.J. Brooks. (1999). Free-air CO_2 enrichment and soil nitrogen effects on energy balance and evapotranspiration of wheat. *Water Resources Research* 35: 1179-1190.

Lorius, C., J. Jouzel, D. Raynaud, J. Hansen and H. Le Treut. (1990). The ice-core record: Climate sensitivity and future greenhouse warming. *Nature* 347:139-145.

Manabe, S. and R.T. Wetherald. (1987). Large scale changes of soil wetness induced by an increase in atmospheric carbon dioxide. *J. Atmos. Sci.* 44:1211-1235.

Mitchell, J.F.B. (1989). The "Greenhouse" effect and climate change. *Reviews of Geophysics* 27:115-139.

Morison, J.I.L. (1987). Intercellular CO_2 concentration and stomatal response to CO_2. pp. 229-252. In: E. Zeiger, G.D. Farquhar and I.R. Cowan (eds.), *Stomatal Function.* Stanford University Press, Stanford, CA.

Morison, J.I.L. and R.A. Gifford. (1984). Plant growth and water use with limited water supply in high CO_2 concentrations. I. Leaf area, water use, and transpiration. *Australian J. Plant Physiol.* 11:361-374.

Neftel, A., H. Friedli, E. Moor, H. Lötscher, H. Oeschger, U. Siegenthaler and B. Stauffer. (1994). Historical CO_2 record from the Siple Station ice core. pp. 11-14. In: T.A. Boden, D.P. Kaiser, R.J. Sepanski and F.W. Stoss (eds.), *Trends '93: A Compendium of Data on Global Climate Change.* ORNL-CDIAC-65. Carbon Dioxide Information Analysis Center, Oak Ridge National Laboratory, Oak Ridge TN, USA.

Norman, J.M. and T.J. Arkebauer. (1991). Predicting canopy photosynthesis and light-use efficiency from leaf characteristics. pp. 75-94. In: K.J. Boote and R.S. Loomis (eds.), *Modeling Crop Photosynthesis–from Biochemistry to Canopy.*

CSSA Special Publication No. 19. Crop Science Society of America and American Society of Agronomy, Madison Wisconsin.

Pearcy, R.W. and O. Björkman. (1983). Physiological effects. pp. 65-105. In: E.R. Lemon (ed.), *Carbon Dioxide and Plants: The Response of Plants to Rising Levels of Atmospheric Carbon Dioxide.* AAAS Selected Symposium 84. Westview Press, Boulder, CO.

Peart, R.M., J.W. Jones, R.B. Curry, K.J. Boote and L.H. Allen, Jr. (1989). Impact of climate change on crop yield in the Southeastern U.S.A.: A simulation study. pp. 2-1 to 2-54. In: J.B. Smith and D.A. Tirpak (eds.), *The Potential Effects of Global Climate Change on the United States. Appendix C, Agriculture, Vol. 1,* EPA-230-05-89-053, U.S. Environmental Protection Agency, Washington, DC 20460.

Penning de Vries, F.W.T., D.M. Jansen, H.F.M. ten Berge and A. Bakema. (1989). *Simulation of Ecophysiological Processes of Growth in Several Annual Crops.* Pudoc, Netherlands. 271 p.

Pickering, N.B., J.W. Jones and K.J. Boote. (1995). Adapting SOYGRO V5.42 for prediction under climate change conditions. pp. 77-98. In: C. Rosenzweig, L.H. Allen, Jr., L.A. Harper, S.E. Hollinger and J.W. Jones (eds.), *Climate Change and Agriculture: Analysis of Potential International Impacts. ASA Special Publication No. 59.* American Society of Agronomy, Madison, Wisconsin, USA.

Poorter, H. (1993). Interspecific variation in growth response of plants to an elevated ambient CO_2 concentration. *Vegetatio* 104/105:173-190. Also pp. 173-190. In: J. Rozema, H. Lambers, S.C. van de Geijn and M.L. Cambridge (eds.) *CO_2 and Biosphere.* Kluwer Academic Publ., Dordrecht.

Ritchie, J.T., B.D. Baer and T.Y. Chou. (1989). Effect of global climate change on agriculture: Great Lakes Region. pp. 1-1 to 1-42. In: J.B. Smith and D.A. Tirpak (eds.), *The Potential Effects of Global Climate Change on the United States, Appendix C, Agriculture, Vol. 1.* EPA-230-05-89-053. U.S. Environmental Protection Agency, Washington, DC 20460.

Rosenberg, N.J., B.A. Kimball, P. Martin and C.F. Cooper. (1990). From climate and CO_2 enrichment to evapotranspiration. pp. 151-175. In: P.E. Waggoner (ed.), *Climate Changes and U.S. Water Resources.* AAAS. John Wiley and Sons, New York-Chichester-Brisbane-Toronto-Singapore.

Rosenzweig, C. (1989). Potential effects of climate change on agricultural production in the Great Plains: A simulation study. pp. 3-1 to 3-43. In: J.B. Smith and D.A. Tirpak (eds.), *The Potential Effects of Global Climate Change on the United States. Appendix C, Agriculture, Vol. 1.* EPA-230-05-89-053. U.S. Environmental Protection Agency, Washington, DC.

Samarakoon, A.B. and R.M. Gifford. (1995). Soil water content under plants at high CO_2 concentration and interactions with direct CO_2 effects: A species comparison. *J. Biogeography* 22:193-202.

Schlesinger, M.E. (1984). Climate model simulation of CO_2-induced climate change. pp. 141-235. In: B. Saltzman (ed.) *Advances in Geophysics, Vol. 26.* Academic Press, New York.

Shawcraft, R.W., E.R. Lemon, L.H. Allen, Jr., D.W. Stewart and S.E. Jensen. (1974).

The soil-plant-atmosphere model and some of its predictions. *Agric. Meteorol.* 14:287-307.

Smith, J.B. and D.A. Tirpak (eds.). (1989). *The Potential Effects of Global Climate Change on the United States.* EPA-230-05-89-050. U.S. Environmental Protection Agency, Washington, DC 20460.

Stanhill, G. (1960). *The Relationship Between Climate and the Transpiration and Growth of Pastures.* In: Proceedings, Eighth International Grasslands Congress.

Tolbert, N.E. and I. Zelitch. (1983). Carbon metabolism. pp. 21-64. In: E.R. Lemon (ed.). *CO_2 and Plants: The Response of Plants to Rising Levels of Atmospheric Carbon Dioxide.* AAAS Selected Symposium 84. Westview Press, Boulder, CO.

Waggoner, P.E. (1990). *Climate Change and U.S. Water Resources. AAAS.* John Wiley and Sons, New York-Chichester-Brisbane-Toronto-Singapore. 496 p.

Warrick, R.A., R.M. Gifford and M.L. Parry. (1986). CO_2, climate change, and agriculture. pp. 393-473. In: B. Bolin, B.R. Doos, J. Jager and R.A. Warrick (eds.), *The Greenhouse Effect, Climate Change, and Ecosystems (SCOPE 29).* John Wiley & Sons, Chichester-New York-Brisbane-Toronto-Singapore.

Washington, W.M. and G.A. Meehl. (1984). Seasonal cycle experiment on the climate sensitivity due to a doubling of CO_2 with an atmospheric general circulation model coupled to a simple mixed layer ocean model. *J. Geophys. Res.* 89:9475-9503.

Washington, W.M. and G.A. Meehl. (1986). General circulation model CO_2 sensitivity experiments: Snow-sea ice albedo parameterizations and globally averaged surface air temperature. *Clim. Change* 8:231-241.

Wilson, C.A. and J.F.B. Mitchell. (1987). A doubled CO_2 climate sensitivity experiment with a GCM including a simple ocean. *J. Geophys. Res.* 92:13,315-13,343.

Crop Responses to Water Shortages

P. D. Jamieson

SUMMARY. This paper examines the effects that water shortages have on crop production through influences on transpiration, growth rate, phenology, light interception and biomass partitioning. It begins by considering the physics of transpiration and water transport from soil to atmosphere through plants, and examines biological adaptation and responses. It starts with some simple models that link growth and water use, and develops these through to more complex simulation models that allow the balance of mechanisms to water response to be examined. *[Article copies available for a fee from The Haworth Document Delivery Service: 1-800-342-9678. E-mail address: getinfo@haworthpressinc.com <Website: http://www.haworthpressinc.com>]*

KEYWORDS. Potential deficit, simulation models, drought response, transpiration efficiency

INTRODUCTION

Crop growth and water use are inextricably intertwined. The ingress of CO_2 and the egress of water from leaves share much of the same pathway, and the cooling effect of evaporating water is necessary to keep plants below lethal temperatures by shedding the heat load from solar radiation. If the supply of water is insufficient to meet the demand of the energy supplied,

---

P. D. Jamieson is Agricultural Meteorologist, New Zealand Institute for Crop & Food Research Ltd., Private Bag 4704, Christchurch, New Zealand (E-mail: jamiesonp @crop.cri.nz).

[Haworth co-indexing entry note]: "Crop Responses to Water Shortages." Jamieson, P. D. Co-published simultaneously in *Journal of Crop Production* (Food Products Press, an imprint of The Haworth Press, Inc.) Vol. 2, No. 2 (#4), 1999, pp. 71-83; and: *Water Use in Crop Production* (ed: M. B. Kirkham) Food Products Press, an imprint of The Haworth Press, Inc., 1999, pp. 71-83. Single or multiple copies of this article are available for a fee from The Haworth Document Delivery Service [1-800-342-9678, 9:00 a.m. - 5:00 p.m. (EST). E-mail address: getinfo@haworthpressinc.com].

then plant communities use a variety of strategies to limit water use. An inevitable consequence of this is a reduction in the growth rate, which, under most circumstances, leads to a reduction in crop yield. In this paper the effects that water shortages have on transpiration, growth, partitioning, final biomass and yield are explored.

EFFECTS ON TRANSPIRATION

Transpiration involves the extraction of soil water by roots, its transport through plant tissues to stomata where it is evaporated and carried away by diffusion and turbulent eddies into the atmosphere. The process is dominated by demand and supply. The evaporative demand depends on the rate at which energy can be supplied to the sites of evaporation. Supply depends on the rate at which the system can supply water to those sites. The energy required to evaporate water is orders of magnitude more than the energy required to transport water from the soil to the leaves. Hence plants are adapted to control evaporation at the stomata by increasing stomatal resistance so that plant tissue does not dehydrate.

One of the simplest mechanistic descriptions of the transpiration process is that due to Monteith (1965):

$$\lambda E = \frac{sR_n + \rho C_p(e^* - e)/r_a}{s + \gamma(1 + r_c/r_a)} + \lambda E_s \qquad (1)$$

The particular formulation comes from Black et al. (1970), and separates the transpiration rate (E) from the soil evaporation rate E_s. Other symbols have their usual meaning and the result has the dimensions of energy per unit area per unit time. The canopy resistance r_c depends both on stomatal resistance (r_s) and leaf area index (LAI) as:

$$r_c = r_s/(2\text{LAI}) \qquad (2)$$

for amphistomatous leaves (Monteith, 1965).

Equations 1 and 2 provide two mechanisms by which transpiration can be reduced. The energy load can be reduced by reducing the effective LAI, by reducing leaf expansion, accelerating senescence, or by curling leaves so that they intercept less radiant energy. In addition, the resistance to vapour flow can be reduced by partial or full stomatal closure. The balance between these mechanisms will vary with species. Jamieson et al. (1995a) showed that, in barley, stomatal effects were more important than variations in LAI in reducing transpiration losses under drought, and Jamieson et al. (1998a) obtained similar results for wheat crops. In semi-arid environments crops are often grown at low density to minimise the energy load and to share available water amongst fewer plants (Stewart and Steiner, 1990).

Whatever the actual balance between these mechanisms, equations 1 and 2 describe the physics of the transpiration process at the plant-atmospheric interface. The biological control mechanisms at this level are a plant response to the physics of the soil-root system. The rate at which the system can supply water is approximated by:

$$\theta_i - \theta_{i+1} = \theta_i[1 - \exp(-0.5D)] \tag{3}$$

where the indices refer to successive days and the root length density D is in cm cm^{-3} (Jamieson and Ewert, 1998). If the soil-root system cannot supply water to the plant at the rate imposed by the atmospheric energy environment, then the plant must impose controls or dehydrate and die. Dead plants do not reproduce well, so this adaptation to drought has developed. The physics is inescapable, and the biology adapts to it, regardless of the precise detail of the mechanisms coupling flow of water through soils, roots, xylem, stomata to the atmosphere.

Jamieson and Ewert (1999) recently investigated the magnitude and effect of restrictions on water uptake by the roots of wheat plants experiencing drought, by comparing water uptake measurements with calculations from a simulation model incorporating a range of hypotheses about the influence of root density on uptake. They concluded that the effects are far larger than allowed in most simulation models, but nevertheless most models provide reasonably accurate estimates of transpiration under moderate drought to well watered conditions (Jamieson et al., 1998a) because they underestimate atmospheric demand. However, these models overestimate water uptake in more severe drought conditions.

ASSOCIATIONS BETWEEN TRANSPIRATION AND GROWTH RATES

Growth and transpiration both involve the absorption of radiant energy by the canopy and, as noted above, water and CO_2 share much of the same pathway. A major difference between growth and transpiration is that for transpiration there is another source of energy provided by sensible heat in the air. The result of the transfer of this energy is cooling of the air and a decrease in the saturation deficit.

Beirhuizen and Slatyer (1965), through a consideration of the flux equations for water and CO_2 transport, concluded that the transpiration efficiency, the ratio of production to transpiration, should be inversely proportional to the saturation deficit, at least at the timescale (hours) of the processes involved:

$$C/E = k/(e^* - e) \qquad (4)$$

and that k would be constant for a species. Their analysis of some published data showed that the rule may also be applied on seasonal basis to a range of crops. Tanner and Sinclair (1983) extended the analysis by considering the energy supply for transpiration and photosynthesis in sunlit and shaded leaves, and argued the relationship would hold for short time scales in daylight. A survey of literature suggested that seasonal values of k would be stable for given crops, provided that the saturation deficit was taken as the mean over daylight hours.

Some assumptions were made in deriving equation 4. The most important ones were that, at least during daylight hours when stomata are open, the ratio of internal (substomatal) to external CO_2 concentration (c_i/c_a) is constant (0.7 for C_3 crops, 0.3 for C_4 crops). This means that the chief control of stomatal resistance is substomatal CO_2 concentration, and that any effects of water deficit are likely to be on assimilation first and, through these effects, stomatal resistance increases to maintain the gradient. The other assumption was that the saturation deficit provides a good estimate of the saturation gradient from the substomatal cavity to the air, because air within the cavity is saturated with water vapour. We need to examine these assumptions further.

Measurements made by Zhang and Nobel (1996) and published work reviewed by them suggests that c_i/c_a is itself affected directly by the saturation deficit, and reduces as the saturation deficit increases. This will have the result that the product of transpiration efficiency and the saturation deficit will decrease as the saturation deficit increases.

The assumption that the saturation deficit is equal to the vapour pressure gradient across the leaf is good only when leaf and air are at the same temperature. This is seldom so. In temperate climes (air temperature less than about 33°C) for crops not short of water, the energy use in transpiration is usually less than that supplied by the incoming radiation. There is a positive upward sensible heat flux, and the canopy is warmer than the air. In these circumstances the saturation deficit underestimates the vapour pressure gradient. Conversely, in arid environments well watered crops can be substantially cooler than the overlying air, and the saturation deficit overestimates the gradient. The reduction in transpiration associated with water shortage will also cause the canopy to be warmer than the air, once more causing the saturation deficit to overestimate the gradient.

The combination of these two effects would cause the transpiration efficiency to decrease with stress, and also for the product of the transpiration efficiency and the saturation deficit to decrease with stress. This contrasts with experimental findings of Jamieson et al. (1998a) in wheat. They found that both the transpiration efficiency and the product of transpiration efficiency and saturation deficit increased with stress. Hence the stomatal response to stress which limited water vapour egress in this experiment did not limit CO_2 ingress. For this to be so, the flux of CO_2 could have been maintained by increasing the gradient, i.e., by decreasing the concentration within the stomata.

Although the arguments above suggest that the relationship between transpiration efficiency and saturation deficit given by Tanner and Sinclair (1983) is not exactly true (and no such claim was made by them), its approximate truth led them to suggest that the likelihood of major improvements in water use efficiency through genetic manipulation is small.

RELATIONSHIP BETWEEN YIELD LOSS AND POTENTIAL DEFICIT

The argument that there is a more or less predictable relationship between water use and biomass production suggests that conditions that decrease transpiration below its potential rate will also reduce biomass production below its potential rate. Hence one might expect a reasonably consistent relationship between yield and potential soil moisture deficit (D_p), defined as the accumulated difference between demand for water (potential evapotranspiration) and its supply from precipitation and irrigation. This is the basis of the model first suggested by Penman (1971) and expanded on by French and Legg (1979). D_p is an index of drought that is determined mostly by meteorology and is nearly independent of crop type, and completely independent of soil type. Although D_p is independent of soil type, the relationship between yield and D_p is likely to be influenced by soil type, particularly the available water holding capacity (AWC) in the crop rootzone (French and Legg, 1979). D_p cannot be completely crop independent because some account must be made in potential evapotranspiration estimates of the reduction in transpiration and variations in soil evaporation during the early life of the crop when ground cover is substantially less than 100% (French and Legg, 1979, Jamieson et al., 1995b).

The yield response model suggested by Penman (1971) was that there is a decline in yield (Y) proportional to the amount by which the maximum value of D_p exceeds some threshold value (D_c).

$$Y = Y_o[1 - a(D_{pmax} - D_c)] \tag{5}$$

except that $Y = Y_o$ when $D_{pmax} < D_c$. Note that a requirement of the model is some estimate of the potential yield Y_o.

French and Legg (1979) showed this model provided a good description of the response to drought of various crops in the UK including wheat, potatoes and pasture leys. The model was successfully calibrated in New Zealand for peas (Jamieson et al., 1984; Wilson et al., 1985; Martin and Jamieson, 1996), potatoes (Jamieson, 1985; Martin et al., 1992) and for wheat, barley and maize (Jamieson et al., 1995b). With the exception of maize, the values of D_c and a were independent of the time of occurrence of D_{pmax}. In the case of maize, D_c increased as the season progressed and the volume of soil explored by the roots increased. However there was no evidence that a changed with drought timing. Values of a found in the above studies ranged from 0.1 to 0.3% of potential yield for each mm that D_{pmax} exceeded D_c. Interesting, Stone et al. (1997) found a very similar value of a for sweet corn (0.1% mm^{-1}) to that reported by Jamieson et al. (1995b) for maize, but found no evidence of a shift in D_c with drought timing. In other crops, French and Legg (1979) found that the D_c varied among crops in a single soil, but for a single crop the ratio of the values of D_c between sites was approximately the same as the ratio of available water holding capacities of the soils in the crop rootzone.

In the analyses reported above, there were no substantial departures from the bilinear model even though the timing and duration of drought was varied substantially. Hence, in the environments tested, there was no evidence that any stage of growth was more sensitive to drought than any other, at least when drought was quantified is a sensible way. However, in harsher environments, there is substantial evidence that drought at some times can have a much larger effect than at others. For instance, Retta and Hanks (1980) showed that drought at around silking in maize in a mid-west US environment caused substantially larger decreases in yield than drought at other times. This environment is much warmer than the New Zealand environment where Jamieson et al. (1995b) did their work. It is likely that elevated temperature, associated with water stress, rather than water shortage *per se,* was the cause.

The above potential soil moisture deficit response model is one of the more useful models for operational use. It can be calibrated for a location with reasonably simple experiments. D_p is simply calculated with readily available data, and calculations of the value of water can be made on the back of an envelope. Hence a farmer growing wheat with irrigation can determine that a delay in starting irrigation leads to an excess in D_p over D_c of, say, 50 mm. He or she can calculate the cost in a crop, with a yield potential of 8 t ha^{-1}, as $50 \times .003 \times 8$ or about 1.2 t ha^{-1}. Hence he or she can make decisions about the value of the water based on the value of the crop, and use the information to choose among crops when there is a limited supply of water.

MECHANISMS OF GROWTH RESPONSES

A deficiency of the models described above is that they are based on rates alone or on accumulations over unspecified amounts of time. Timing is at least implicit within any analysis, because processes occurring before and after a crop exists are of limited relevance. However, timing can be made explicit by including the dates of crop emergence and maturity.

A simple model of production can be written:

$$\text{biomass} = \int_{\text{emergence}}^{\text{maturity}} C\, dt \tag{6}$$

and

$$\text{yield} = \int_{\text{emergence}}^{\text{maturity}} \xi\, C\, dt \tag{7}$$

where C is the daily growth rate and ξ is a partitioning coefficient. From this it is apparent that yield can be affected when stress influences the growth rate, the growth duration, or the manner in which material is partitioned to the economically important portion of the crop. From these equations the simplest response may be a reduction in growth duration that leads to smaller final biomass and yield. ξ may become larger or smaller in response to stress to offset or increase the effect. Effects on the growth rate are more complicated, because it is itself, in the simplest description, the combination of two processes.

$$C = A\, Q \tag{8}$$

where A is the light use efficiency (photosynthetically active radiation energy per unit biomass) and Q is the amount of light intercepted by the crop (energy per unit area). Q in turn depends on incident radiation and the LAI:

$$Q = Q_o[1 - \exp(-\kappa\, \text{LAI})] \tag{9}$$

where κ is an extinction coefficient. From equations 8 and 9, C can vary either because of variations in LAI or A. The availability of gas exchange measuring apparatus in recent years has meant that there has been a large concentration of effort in measuring stress effects on gas exchange properties per unit leaf area. However, analysis of experimental data for barley (Jamieson et al., 1995c) and wheat (Jamieson et al., 1998a) showed that drought caused substantial variations in leaf area, through reduced leaf expansion rates and/or accelerated senescence. The resulting change in light interception was the major cause of reductions in growth rates. In barley (Jamieson et al. 1995c), there was some evidence that A was decreased by drought early in the season, but the effects on biomass accumulation of drought occurring later in the season could be explained almost entirely by premature leaf senescence reducing light interception. In wheat (Jamieson et al., 1998a), changes in LAI dominated the variation in biomass accumulation regardless of when drought occurred.

PARTITIONING TO GRAIN

In addition to effects on biomass accumulation, drought may influence yield in grain crops by affecting the proportion of biomass partitioned to grain during grain filling. The simplest hypothesis is that the harvest index (HI, the ratio of grain to total biomass) is constant, and that changes in yield simply reflect changes in total biomass. However, Jamieson et al. (1984) showed that for field peas, variation in yield with drought was dominated by changes in HI; there was much less variation in total biomass than in grain yield. Moot et al. (1996) also showed that final HI varied substantially among wheat crops. They investigated an hypothesis suggested by Spaeth and Sinclair (1985), Muchow (1988) and Amir and Sinclair (1991) that partitioning of biomass to grain proceeds so that the rate of change of HI with time is constant. Using the data for wheat reported by Jamieson et al. (1995b) and Jamieson et al. (1998a), Moot et al. (1996) found that plots of HI with time all fell on a common line, but that the duration of grain filling varied so that final HI differed among treatments. Hence final yield depended both on final biomass and the duration of grain filling.

There is a common observation that variations in grain yield in small grain cereals are much more strongly associated with changes in grain number per unit area than grain size (Gallagher et al., 1975; Daniels et al., 1982). There is also often very little variation in the rate of individual kernel growth (Blacklow and Incoll, 1981). Hence grain yield ultimately will depend on how many grains are set and the duration of grain growth. Does this mean that partitioning to grain is almost entirely controlled by the size of the carbohydrate sink? Or is the size of the sink set so that demands are most likely to be met? In contrast, the preceding section could be interpreted to mean that the whole process depended on source strength–the available assimilate during grain growth, affected most by the collapse of the canopy during water stress.

Grain number in wheat is set during the period of rapid ear growth before anthesis (Fischer, 1985). Jamieson et al. (1996, 1998a) showed that, alone, the variation in the accumulation of biomass during this time was insufficient to explain the variation in grain number. Jamieson et al. (1998b) showed water stress also modified the partitioning of biomass to the developing ears, in addition to its direct effect on biomass accumulation.

AFRCWHEAT2 (Porter, 1993) is a wheat simulation model that uses a calculation of grain number (sink size) as an intermediate step in the calculation of grain growth and yield. Sirius (Jamieson et al., 1998b) is a model that transfers biomass to grain biomass without reference to grain number (therefore source limited only). These two models were equally successful in predicting the response of wheat grain yield to periods of drought that varied in both timing and duration (Jamieson et al., 1998a). From this it is obviously impossible to say whether sink size or source strength dominates the response of grain yield to drought. It appears that the two are closely dependent and only an unnaturally fast transition from a non-stressed to a stressed condition is a difference likely to appear. In natural situations, the onset of water-stress is gradual, so that substantial stress during the grain-filling period is necessarily preceded by stress during rapid ear growth. In most situations where stress occurs during pre-anthesis ear growth, it will be followed by stress during grain growth. Limiting grain number during this period appears to be a good survival strategy for maximising the survival of seeds. At the same time, stress accelerates the senescence of leaves and the continuing stress means that sink size and source strength are likely to remain in balance.

SIMULATION MODELS

Transpiration and growth are dynamic processes that change with the state of the plant and its environment. Although they are useful in illustrating some processes, and in perhaps in assisting in operational crop management, the simple models described heretofore provide little physiological insight into the dynamics of crop growth processes. In the past 15 years or so, many physiologically based simulation models of varying detail have been created. Because they seek to mimic the dynamics of processes that lead to production and yield, and how these change with environmental variation, they offer a better prospect of predicting the effects of a variety of environmental variations, of which water stress is but one. Crop simulation models are formalised collections of testable hypotheses about how environmental variations affect plant processes. Therefore they also offer a good prospect of increasing our understanding of these effects.

Jamieson et al. (1998a) compared five wheat simulation models of varying complexity that contained a range of hypotheses about environmental effects on crop processes. The comparison was made in terms of the detailed predictions of such things as biomass accumulation, LAI gain and loss, partitioning to the grain, water extraction from the soil, and final grain yield and shoot biomass. Importantly, the simulations were also compared with measurements from an experiment in a rainshelter where a range of treatments produced yields varying threefold. The models were broadly similar in that their main timestep was a day, and they used small variations on the Penman equation to calculate transpiration. Processes within the models are modified in response to a stress index, and all the models used as a stress index the ratio between the rate at which the system could supply water and the rate at which it was demanded. The major differences among the models were:

- the detail of simulation of leaf area development, from simulation of the size and appearance of cohorts of leaves on mainstem and tillers, to a simple thermal time model;
- the rate at which roots extended downwards and the influence of root length density on water extraction;
- the prediction of grain number as an intermediate in grain yield determination, or not;
- the calculation of biomass accumulation from photosynthesis equations at the sub-canopy level or light use efficiency at the canopy level; and
- the calculation of the influence of water stress on LAI and photosynthesis or light use efficiency.

Despite these differences, four of the five models predicted both grain yield and its response to drought very well. The remaining model underestimated the response to drought. Three of the models predicted total seasonal evapotranspiration close to that observed, except they overestimated water extraction in severe drought conditions. The success of the predictions of final biomass and its response to drought was much more variable. The dynamic performance was also different, mostly in predicting the time courses of LAI, biomass accumulation and evapotranspiration. The study identified differences in the calculation of the water supply at any time to be a major cause of the differences in performance. The models universally identified variations in light use efficiency and photosynthesis per unit leaf area as the major causes of variation in biomass accumulation, whereas analysis of the experimental data showed they were minor contributors, and most of the variation was associated with changes in light interception. Nevertheless, the models integrated the seasonal influences that determine the variation of grain yield very well, despite their varying performance in the calculation of the intermediate state of the system, and despite substantial differences in the level of detail of the processes simulated. One advantage of such a comparison is that it identified the strengths and weaknesses of the approaches used, so that better overall descriptions can be made, and the importance of particular processes in responding to environmental variation. This is not to say that we are likely to end up with just one universal wheat simulation model. There are advantages in diversity, and simulation models have many purposes.

CONCLUSION

The responses of major arable crops to water shortages are in the main well enough understood that operational decisions and policy can be made from present knowledge. The balance between the rates at which water is demanded and can be supplied by the system determine the manner of response in limiting water loss, and this inevitably leads to responses in production. Our knowledge specifies what experiments and measurements are needed to determine irrigation needs with particular crops, and can be used to determine whether environments can be matched to crops.

REFERENCES

Amir, J. and T.R. Sinclair. (1991). A model of the temperature and solar-radiation effects on spring wheat growth and yield. *Field Crops Research* 28: 47-58.

Bierhuizen, J.F. and R.O. Slatyer. (1965). Effect of atmospheric concentration of water vapour and CO_2 in determining transpiration-photosynthesis relations in cotton leaves. *Agricultural Meteorology* 2: 259-270.

Black, T.A., C.B. Tanner, and W.R. Gardner. (1970). Evapotranspiration from a snap bean crop. *Agronomy Journal* 62: 66-69.

Blacklow, W.M. and L.D. Incoll. (1981). Nitrogen stress of winter wheat changed the determinants of yield and the distribution of nitrogen and total dry matter during grain filling. *Australian Journal of Plant Physiology* 8: 191-200.

Daniels, R.W., M.B. Alcock, and D.H. Scarisbrick. (1982). A reappraisal of stem reserve contribution to grain yield in spring barley. *Journal of Agricultural Science, Cambridge* 98: 347-355.

Fischer, R.A. (1985). Number of kernels in wheat crops and the influence of solar radiation and temperature. *Journal of Agricultural Science, Cambridge* 105:447-461.

French, B.K. and B.J. Legg. (1979). Rothamsted irrigation 1964-1976. *Journal of Agricultural Science, Cambridge* 92: 15-37.

Gallagher, J.N., P.V. Biscoe, and R.K., Scott. (1975). Barley and its environment. V. Stability of grain weight. *Journal of Applied Ecology* 12: 319-336.

Jamieson, P.D., D.R. Wilson, and R. Hanson. (1984). Analysis of responses of field peas to irrigation and sowing date 2. Models of growth and water use. *Proceedings of the Agronomy Society of New Zealand* 14: 75-81.

Jamieson, P.D. (1985). Irrigation response of potatoes. In *Potato Growing: A Changing Scene*, eds. G.D. Hill and G. Wratt. Agronomy Society of New Zealand Special Publication No 3: 17-20.

Jamieson, P.D., G.S. Francis, D.R. Wilson, and R.J. Martin. (1995a). Effects of water deficits on evapotranspiration from barley. *Agricultural and Forest Meteorology* 76: 41-58.

Jamieson, P.D., R.J. Martin, and G.S. Francis. (1995b). Drought influences on grain yield of barley, wheat and maize. *New Zealand Journal of Crop and Horticultural Science* 23: 55-66.

Jamieson, P.D., R.J. Martin, G.S. Francis, and D.R. Wilson. (1995c). Drought effects on biomass production and radiation use efficiency in barley. *Field Crops Research* 43: 77-86

Jamieson, P.D., R.J. Martin, G.S. Francis, and J.R. Porter. (1996). Analysing wheat biomass and grain yield response to drought using AFRCWHEAT2. *Proceedings of the 8th Australian Agronomy Conference, Brisbane*, pp. 669.

Jamieson, P.D. and F. Ewert. (1999). The role of roots in controlling soil water extraction during drought: an analysis by simulation. *Field Crops Research* 60: 267-277.

Jamieson, P.D., J.R. Porter, J. Goudriaan, J.T. Ritchie, H. van Keulen, and W. Stol. (1998a). A comparison of the models AFRCWHEAT2, CERES-Wheat, Sirius, SUCROS2 and SWHEAT with measurements from wheat grown under drought. *Field Crops Research* 55: 23-44.

Jamieson, P.D., M.A. Semenov, I.R. Brooking, and G.S. Francis. (1998b). *Sirius*: a mechanistic model of wheat response to environmental variation. *European Journal of Agronomy* 8: 161-179.

Martin, R.J., P.D. Jamieson, D.R. Wilson, and G.S. Francis. (1992). Effects of soil moisture deficits on yield and quality of 'Russet Burbank' potatoes. *New Zealand Journal of Crop and Horticultural Science* 20:1-9.

Martin, R.J. and P.D. Jamieson. (1996). The effect of timing and intensity of drought on the growth and yield of field peas. *New Zealand Journal of Crop and Horticultural Science* 24: 174-176.

Monteith, J.L. (1965). Evaporation and environment. *Symposia of the Society for Experimental Biology* 19: 205-234.

Moot, D.J., P.D. Jamieson, A.L. Henderson, M.A. Ford, and J.R. Porter. (1996). Rate of change of harvest index during grain filling in wheat. *Journal of Agricultural Science, Cambridge* 126: 387-395.

Muchow, R.C. (1988). Effect of nitrogen supply on the comparative productivity of maize and sorghum in a semi-arid tropical environment. *Field Crops Research* 18: 31-43.

Penman, H.L. (1971). Irrigation at Woburn VII. *Report Rothamsted Experimental Station for 1970, Part 2.* pp. 147-170.

Porter J.R. (1993). AFRCWHEAT2: A model of the growth and development of wheat incorporating responses to water and nitrogen. *European Journal of Agronomy* 2: 69-82

Retta, A. and R.J. Hanks. (1980) Corn and alfalfa production as influenced by limited irrigation. *Irrigation Science* 1: 135-147.

Spaeth, S.C. and T.R Sinclair. (1985). Linear increase in soybean harvest index during seed-filling. *Agronomy Journal* 77: 207-211.

Stewart, B.A. and J.L. Steiner. (1990). Water-use efficiency. *Advances in Soil Science* 13: 151-173.

Stone, P.J., D.R. Wilson, and R.N. Gillespie. (1997). Water deficit effects on growth, water use and yield of sweet corn. *Proceedings Agronomy Society of New Zealand* 27: 45-50.

Tanner, C.B. and T.R. Sinclair. (1983). Efficient water use in crop production: research or research? In *Limitations to Efficient Water Use in Crop Production*, eds. H.M. Taylor, W.R. Jordan, and T.R. Sinclair. American Society of Agronomy, Crop Science Society of America, Soil Science Society of America, Madison, Wisconsin, pp. 1-27.

Wilson, D.R., P.D. Jamieson, W.A. Jermyn, and R. Hanson. (1985). Models of growth and water use of field peas (*Pisum sativum* L.). In *The Pea Crop: A Basis for Improvement*, eds. P.P.D. Hebblethwaite, M.C. Heath, and T.C.K. Dawkins. Proceedings of the University of Nottingham 40th Easter School in Agricultural Science, April 1984. Butterworths, London, pp. 139-151.

Zhang, H. and P.S. Nobel. (1996). Dependency of c_i/c_a and leaf transpiration efficiency on the vapour pressure deficit. *Australian Journal of Plant Physiology* 23: 561-568.

Effect of Irrigation Water Quality Under Supplementary Irrigation on Soil Chemical and Physical Properties in the "Southern Humid Pampas" of Argentina

José Luis Costa

SUMMARY. Salinity and sodicity were quantified at five locations irrigated with central pivots and different water qualities in the southeast of Buenos Aires Province. Changes in electrical conductivity (EC_e) and sodium adsorption ratio (SAR_e) of saturated soil extract were determined from soil cores collected to a depth of 0.6 m between 1996 and 1998. Unsaturated hydraulic conductivity was measured with disk infiltrometers. Soil from one location, Balcarce, was treated with synthetic waters to obtain variable levels of SAR_e. Laboratory studies on optical transmission (OT), aggregate percentage (AP), and saturated hydraulic conductivity (Ks) were performed.

Infiltration of rainfall caused a substantial dilution of salt concentration. Changes in sodicity were persistent, but tended to stabilize between $SAR_e = 5$ to $SAR_e = 7$. When irrigation was discontinued for a year and 1129 mm of rainfall leached the soil, the SAR_e value decreased from 5.3 to 1.8.

Ks showed a threshold at exchangeable sodium percentage (ESP) of

José Luis Costa is Ingeniero Agrónomo, (Agricalatural Engineer) Facultad de Agronomía, Universidad Nacional de Buenos Aires, Argentina and is affiliated with INTA (Instituto Nacional de Tecnología Agropecuaria), Dto Agronomía, EEA INTA Balcarce, CC 276 (7620), Balcarce, Argentina.

[Haworth co-indexing entry note]: "Effect of Irrigation Water Quality Under Supplementary Irrigation on Soil Chemical and Physical Properties in the 'Southern Humid Pampas' of Argentina." Costa, José Luis. Co-published simultaneously in *Journal of Crop Production* (Food Products Press, an imprint of The Haworth Press, Inc.) Vol. 2, No. 2 (#4), 1999, pp. 85-99; and: *Water Use in Crop Production* (ed: M. B. Kirkham) Food Products Press, an imprint of The Haworth Press, Inc., 1999, pp. 85-99. Single or multiple copies of this article are available for a fee from The Haworth Document Delivery Service [1-800-342-9678, 9:00 a.m. - 5:00 p.m. (EST). E-mail address: getinfo@haworthpressinc.com].

5. The results from the Ks laboratory experiment were non-consistent with field estimation of Ks using disk infiltrometers. *[Article copies available for a fee from The Haworth Document Delivery Service: 1-800-342-9678. E-mail address: getinfo@haworthpressinc.com <Website: http://www.haworthpressinc.com>]*

KEYWORDS. Supplementary irrigation, water quality, salinity, sodicity

INTRODUCTION

Supplementary irrigation of corn, wheat, and soybean in the "Humid Pampas" of Buenos Aires Province, Argentina, is relatively new. The most common source of water for irrigation in the southeast of Buenos Aires Province is from ground water. Since 1800, there have been standards for irrigated water quality (Rhoades, 1972). The prevailing criteria for irrigation water quality and their associated potential hazard to crop growth are: (i) salinity, (ii) sodicity, and (iii) toxicity (Rhoades, 1972). The effect of water quality on soils of arid conditions is well known (U.S. Salinity Laboratory Staff, 1954).

Supplementary irrigation is a management strategy that can be used by producers to achieve crop-moisture needs during periods of below-normal precipitation under humid or sub-humid conditions. Irrigation and rainfall interact closely with soil physical properties (Oster, 1994). The effect of the combination of sodic water and rainwater on the chemical and physical properties of irrigated soil of Buenos Aires Province is not well established (Génova, 1993). Uncertainty about the effects that irrigation will have on physical and chemical properties of these soils is a reason for producers to be unwilling to invest in an irrigation system.

Most of the laboratories in the southeast of Buenos Aires Province use the diagram of the U.S. Salinity Laboratory Staff (1954) for classifying irrigation water. The recommendations generated by this diagram could restrict the expansion of irrigation in the region (Costa, 1995).

Eaton (1950) proposed the concept of residual sodium carbonate for rating the quality of irrigation waters high in bicarbonate. Wilcox et al. (1954) showed that, when the value of bicarbonate increased in water, precipitation of $CaCO_3$ increased. The increment of the ESP was linearly related to the $CaCO_3$ precipitation. Ayers and Wescot (1976) introduced the concept of adjusted SAR, originally proposed by Rhoades (1972). Ayers and Wescot (1989), based on work of Oster and Schroer (1979) and Suarez (1981), suggested that the adjusted SAR overestimated the sodium hazard. They recommended the new procedure proposed by Suarez (1981).

Most of the ground water in the humid Pampas is high in sodium and bicarbonate (Costa, 1995). Waters of this type have a tendency for CO_3^{-2} and HCO_3^- to precipitate, which thereby increases the resultant SAR as do evapotranspiration processes (Rhoades, 1972).

Under the humid conditions of the southeast of Buenos Aires Province the average amount of water applied is 150 mm. The excess over the annual water-balance range (between 100 and 200 mm) seems to be enough for salt to be leached out of the root zone, and salinization does not occur. For this reason, we can use poor water quality for irrigation. Nevertheless, annual monitoring of soil salinity must be carried out (Costa, 1995).

Two simulation models that describe the water regimen and the chemistry and transport of solutes in unsaturated or partially saturated soils have been formulated to simulate transient movement of inorganic ions: LEACHM (Wagenet and Hutson, 1989) and UNSATCHEM (Simunek et al., 1997). These models could be used to study the effect of irrigation water quality under field conditions, particularly with supplementary irrigation.

Dispersion and swelling of clays within the soil matrix are interrelated phenomena and either can reduce soil Ks. Swelling reduces soil pore size and dispersion clogs soil pores by enabling migration of small particles. Swelling is not generally appreciable unless ESP exceeds about 2 or 3 (Shainberg and Caiserman 1971). But dispersion can occur at an ESP level as low as 1 to 2 if the electrolyte level is less than 1 meq/liter (Felhendler et al., 1974). Yousaf et al. (1987) concluded that soil Ks decreases correspondingly with clay dispersion as electrolyte concentration is decreased and SAR is increased over the ranges often encountered in irrigated sodic soils.

Shainberg et al. (1981), working with a fine-loam soil, found that when the salt concentration is 3 meq/liter clay dispersion decreased if the ESP value exceeded 12. Conversely in distilled water clay dispersion decreased at ESP values as low as 2. Frenkel et al. (1978) present data that show that the plugging of pores by dispersed clay particles is the major cause of reduced Ks.

Under supplementary irrigation even low values of ESP could induce soil crust formation and reduce infiltration during rainfall events. Wienhold and Trooien (1995) mention that supplementary irrigation of alfalfa is viable in the Northern Great Plains with waters having an EC = 3.4 dS m^{-1} if enough irrigation water can be applied to leach salts. However, sodicity did increase when the SAR of the water was 16, producing in the soil a SAR = 11 after 7 years of irrigation. The results of Admasen (1992) indicate that sodic water can be used in supplementary irrigation to irrigate corn in the mid-Atlantic coastal plain.

Soils of the north of Buenos Aires Province with 3 to 7 years of supplementary irrigation have ESP values ranging between 2 to 5 with no increase in salinity (Irurtia and Mon, 1998). Peinemann et al. (1998) showed data of 11 Hapludolls and 7 Argiudolls irrigated with C3S1 and C3S3 waters. The initial range of values for ESP, 0.5 to 1.0, increased to a range of 2.6 to 8.7 because of the irrigation. Andriulo et al. (1998) found that after 11 years of supplementary irrigation with C3S3 water in a typic Argiudoll (Pergamino series) the ESP increased from 2 to 12.2.

The main objective of this work is to evaluate the effect of water quality on soil chemical and physical properties under supplementary irrigation to assess the suitability of water for irrigation and suggest new directions of approach and research needs.

MATERIALS AND METHODS

Five locations irrigated with different water qualities (Table 1) in the southeast of Buenos Aires Province were selected for this study (Figure 1). Table 2 reports classification and some physico-chemical characteristics of the fine-silty illitic mixed thermic Typic Argiudolls. Those sites were irrigated with central pivots. The main crops were corn, wheat, and soybean except for the site irrigated with high saline water. That one was planted with potatoes. Each site was divided into four areas for sampling. A composite soil sample of six sub-samples was taken at each location, one in the irrigated part and the other in the non-irrigated part during the wintertime. The sampling depths were 0-10, 10-20, 20-40, and 40-60 cm.

Unconfined field infiltration rates at 60 and 160 mm water tensions were taken using 20-cm diameter tension infiltrometers (Ankeny et al., 1991). Two infiltrometers were installed in each of the four areas of sampling in the irrigated and non-irrigated Balcarce soil. Bulk density was measured by the cylinder method, and 6 soil cores were taken at each of the four areas of sampling for irrigated and non-irrigated soil (Blake and Hartge, 1986).

In the soil saturation extract of the soil samples from the six locations the

TABLE 1. Composition of waters for the five different locations: ionic content, sodium adsorption ratio (SAR), electrical conductivity (EC_e) and pH.

	Location Energía	EC_e	pH	Ca	Mg	Na	K	CO_3^{-2}	HCO_3^-	Cl^-	SAR
1	Alvarado Iraizos	1.3	8.5	0.8	0.6	13.0	0.5	0.6	9.6	2.5	15.5
2	Lobería	1.6	7.7	0.9	0.8	16.0	0.3	0.0	9.6	4.6	17.4
3	Necochea Energía	2.8	7.5	1.7	2.0	26.0	0.4	0.0	10.0	11.1	19.1
4	Balcarce	1.1	7.9	0.3	0.3	11.0	0.2	0.0	9.1	1.9	20.1
5	Necochea Orense	4.0	7.8	2.4	1.6	41.3	0.5	0.0	9.4	34.0	29.1

FIGURE 1. Location of the sampled sites.

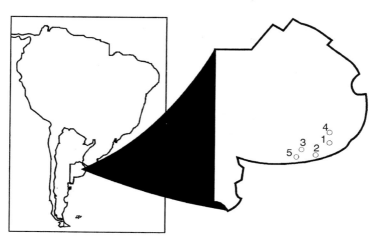

TABLE 2. Report classification and some physico-chemical characteristics of the soils.

	Balcarce	Alvarado-Iraizos	Lobería	Necochea-Energía	Necochea-Orense
Classification	Typic Argiudoll	Typic Argiudoll	Typic Argiudoll	Typic Argiudoll	Typic Argiudoll
pH	5.9	6.0	6.1	6.3	6.5
C %	4.0	3.4	3.2	3.2	3.7
Clay %	28.7	24.4	21.0	27.0	20.3
Silt %	35.9	36.6	25.6	29.3	28.9
Sand %	35.4	39.0	53.2	43.6	50.8
Texture	Clay loam	Clay loam	Sandy loam	Clay loam	Sandy loam

following chemical analysis were determined: EC_e, pH, and soluble ions (Rhoades, 1982). Cation exchange capacity (CEC) (Chapman, 1965) and exchangeable cations were measured by saturation with ammonium acetate (pH = 7) and displaced with Na-acetate (Thomas, 1982). Ca^{++} and Mg^{++} were read with an atomic absorption spectrophotometer and Na^+ and K^+ with a flame photometer.

Surface Balcarce soil (0-20 cm depth) was air dried and crushed to pass through a 2 cm sieve and packed in columns 40 cm high and 20 cm diameter. The columns were leached with 13057 mm of synthetic waters with SAR = 6, 12, 18, 26, and 35 to obtain different ESP values. After the treatment with synthetic waters, the columns were leached with 1600 mm of distilled water. The soil columns were air dried and divided through their vertical axes. Half of the sample was passed through an 8 mm mesh for OT and AP measurement. The rest was passed through a 2 mm mesh for saturated Ks measurements and chemical analysis.

For the soil samples treated with synthetic waters, the following was measured: (i) soil Ks in disturbed soil samples using distilled water (Klute and Dirksen, 1986), (ii) AP using a modification of the U.S. Salinity Laboratory Staff (1954) proposed by Costa et al. (1991), (iii) optical transmission at a wavelength of 410 nm in 1:5 soil water ratio samples, which were shaken for 30 minutes and centrifuged for one minute (Suarez et al., 1984), and (iv) chemical analysis similar to those done on the irrigated soil samples from the farm fields.

RESULTS AND DISCUSSION

Laboratory Studies

The sodium adsorption ratio was closely related to another quantity, the exchangeable sodium ratio (ESR). The Gapon coefficient for the range of conditions represented by U. S. Salinity Laboratory Staff (1954) is 0.01465 $mmol_c \, l^{-1}$. The relationship between ESR and SAR_e for the 190 soil samples from irrigated farm fields yielded a Gapon coefficient of 0.0115 $mmol_c \, l^{-1}$ with an $R^2 = 0.67$. The same coefficient was obtained for the soil samples treated with synthetic waters, i.e., $R^2 = 0.94$ (Figure 2).

Beside the differences founded in soil texture, the chemical behavior of these soils with respect to Na/Ca interchange seems to be similar.

Optical transmission was closely related to the ESP value (Figure 3).

FIGURE 2. Exchangeable sodium ratio (ESR) as related to sodium adsorption ratio (SAR) of the saturation extract.

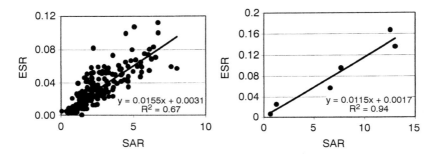

FIGURE 3. Optical transmission (OT) and aggregate percentage (AP) related to exchangeable sodium percent.

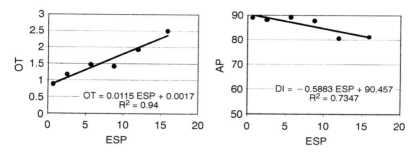

Those values could be associated with soil organic matter (OM) dispersion. Coloration due to OM dispersion was also observed in the saturation extract from irrigated soils with ESP > 4. The AP values for the soil samples treated with synthetic waters were less related to ESP than to the OT (Figure 3). Optical transmission and AP were linearly related to ESP, and, in the range of data, no critical ESP threshold was evident.

It could be expected that there would be some loss of OM from the surface horizon when ESP values are less than 15. But this result would be different for clay. The AP was maintained at values greater than 80% for ESP values lower than 15. Similar results were reported by Costa et al. (1991).

Soil samples from the Balcarce site treated with synthetic waters with SAR = 0, 6.1, 12.2, 18, 26.1, and 35 yielded ESP = 0.7, 2.5, 5.7, 8.7, 11.9, and 15.9, respectively. For the Ks studies, soil samples of ESP = 0.7, 5.7, 11.9, and 15.9 were selected. When the soil samples were leached with distilled water, Ks decreased markedly. The Ks data were adjusted to the volume leached using an exponential equation. The regression equation gave R^2 = 0.75, 0.9, 0.82, and 0.88 for ESP values of 0.7, 5.7, 11.9, and 15. 9, respectively. Using the exponential equations, values of Ks for each ESP for different volumes of leaching were calculated to compare the effect of ESP and distilled water on Ks (Figure 4).

The OT and the AP showed a linear relationship with ESP and no critical ESP threshold could be established. Crescimanno et al. (1995) found no critical ESP value for Sicilian soils and a linear relationship between soil properties and ESP at low cationic concentrations. In our study, the Ks was reduced by 50% after 50 mm of distilled water passed through the column and the ESP increased from 0.7 to 5.7. Further increase in ESP resulted in a negative exponential relationship. The Ks was reduced as more distilled water passed through the soil. Frenkel et al. (1978) found a 60% reduction in Ks after 100 mm of distilled water passed through a soil column with ESP =

FIGURE 4. Saturated hydraulic conductivity (Ks) as related to exchangeable sodium percentage (ESP) for different volumes of leaching.

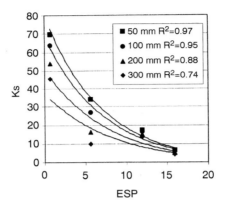

10. From these data we could expect a marked reduction in soil Ks during rainstorms even at low ESP values. This reduction would be still greater when the amount of rainwater passing through the soil profile is increased (Figure 4).

Salinity and Sodicity Under Field Conditions

Figure 5 shows the salinity profile on the farm at the Necochea location where potatoes were irrigated with water with an $EC_e = 4$ dS m^{-1} and the profile was recharged with rainwater. After 284 mm of irrigation (from October through January) the EC_e reached a maximum value of 3.2 dS m^{-1}. In February, 120 mm rainfall reduced the EC_e value in the top 10 cm of the soil from 3.2 to 2 dS m^{-1}. After 290 mm of accumulated rainfall (from February to August), the EC_e decreased to 0.4 dS m^{-1}, a value similar to a non-irrigated soil.

At the other five locations, where waters in a range of 1 to 2.8 dS m^{-1} were used, there were no differences between the EC_e of irrigated and non-irrigated crops when soil samples were taken during wintertime. The average EC_e values remained close to 0.4 dS m^{-1} (Figure 6). These data suggest that rainwater was sufficient to remove salt at a rate exceeding salt precipitation.

Three sites (Lobería, Alvarado-Iraizos, and Necochea Energía) were sampled in two consecutive years at the end of winter to analyze changes in the soil chemical properties. Table 3 shows the amount of irrigated water and rainwater that each location received.

The average SAR_e value for the top 20 cm of soil at Lobería increased from 2 to 5.5 after being irrigated with 527 mm of SAR = 17.4 water (Figure 7).

FIGURE 5. Saturated paste extract electrical conductivity (EC_e) as a function of the soil depth during January to August of 1998. Error bars indicate a ± 1 standard deviation.

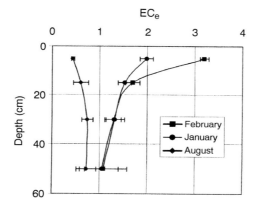

FIGURE 6. Saturated paste extract electrical conductivity (EC_e) as a function of the soil depth for irrigated (I) and non-irrigated (NI) soil.

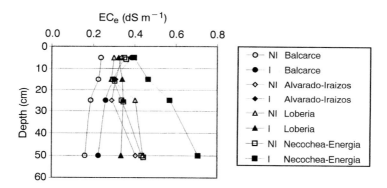

The average SAR_e for the top 20 cm of soil at the Necochea Energía site was 5 when the soil was irrigated with 380 mm of SAR = 19.1 water. At both sites no differences were observed in SAR_e of the irrigated soils in two consecutive years. The SAR_e of the irrigated part of these two sites always was greater than the non-irrigated part.

The soil irrigated with 341 mm of SAR_e = 16.3 water at Alvarado-Iraizos reached in 1997 an average SAR_e = 5.3 in the top 20 cm of soil. During July 1997 through June 1998, the soil received 1192 mm of rainfall and no irriga-

TABLE 3. Irrigation, precipitation, and crop for each grown period (July to June) for three locations.

| Grown period | Location | Data from July to June | |
		Irrigation	Precipitation
Alvarado-Iraizos		Mm	
1993-94	Corn	45	985
1994-95	Wheat	48	871
1995-96	Corn	183	588
1996-97	Wheat	65	1028
1997-98	Soybean	0	1129
	Total	341	4601
Lobería			
1992-93	Corn	100	1184
1993-94	Wheat	75	897
1994-95	Corn	202	637
1995-96	Wheat	20	804
1996-97	Corn	130	970
	Total	527	4492
Necochea-Energía			
1993-94	Wheat	40	897
1994-95	Corn	190	637
1995-96	Wheat	30	804
1996-97	Soybean	120	970
	Total	380	3308

tion. In 1998 the SAR_e value decreased to 1.8. The explanation for the decreased SAR_e, and the associated ESP, through the effect of rainwater is not yet clear. One reason could be that part of $CaCO_3$ precipitated during the irrigation period was redissolved by rainwater. This soil could mineralize in the top 40 cm up to 330 kg ha^{-1} of N-NO$_3$ (Echeverria and Bergonzi 1995). The CO_2 produced by soil respiration plus the H^+ released by nitrogen mineralization could provide the necessary acidity to redissolve this $CaCO_3$ bringing enough Ca to the soil solution to displace the Na of the exchangeable site.

Bulk density measured by the cylinder method did not show significant differences between irrigated and non-irrigated soil at Balcarce with values of 1.3 and 1.33 Mg m^{-3}, respectively. Results of the field infiltration measurements at the Balcarce site taken using tension infiltrometers are shown in Figure 8. The laboratory studies showed a reduction in Ks when the ESP of the soil increased from 0.7 to 5.2% (Figure 4). Field estimation of hydraulic conductivity at saturated, -80 mm, and -160 mm tension with disk infiltrometers (Ankeny et al., 1991) showed no differences between irrigated and non-irrigated soil when the ESP for this soil was 4% and 0.5% for the irrigated and non-irrigated soil, respectively (Figure 9).

This result suggested that under laboratory conditions, where the soil sample is crushed and sieved and loses part of the natural soil structure, the

FIGURE 7. Saturated paste extract sodium adsorption ratio (SAR$_e$) of Alvarado-Iraizos, Lobería, and Necochea-Energía for irrigated (I) and non-irrigated (NI) soil. Error bars indicate a ± 1 standard deviation.

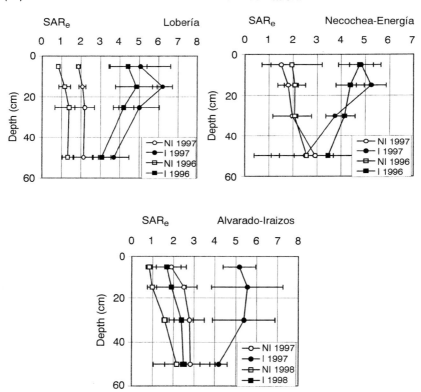

soil has more tendency to disperse reducing its Ks than under field conditions. The N mineralization also maintains a solute concentration in the soil solution, which prevents soil dispersion when rainwater infiltrates into the soil. The average EC$_e$ for the top 20 cm of the non-irrigated soil was 0.37 dS m^{-1}. Echeverria and Ferrari (1993) reported an EC$_e$ = 0.56 dS m^{-1} for 417 observations in the topsoil at Balcarce, Lobería, and Necochea.

CONCLUSION

We have presented data, which show that supplementary irrigation under the southeast Buenos Aires Province climate conditions allows the utilization

FIGURE 8. Hydraulic conductivity for irrigated and non-irrigated soil measured at 80 and 160 mm water tension and saturated hydraulic conductivity estimated by the method of Ankeny et al. (1991).

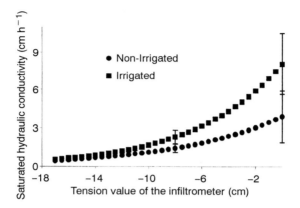

FIGURE 9. Exchangeable sodium percent (ESP) at the Balcarce site. Error bars indicate a ±1 standard deviation.

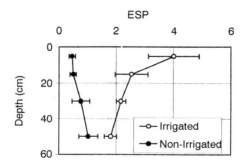

of waters with salt concentration EC > 4 with no risk of salt accumulation in the soil profile. The drainage capacity of these soils and the rainfall are sufficient to ensure leaching of salts. The leaching potential of precipitation was demonstrated during the fall of 1998 in the Necochea-Orense site.

Sodicity increases are more persistent. Fall and winter rainfall seems to be adequate in preventing a long-term increase in soil-Na. SAR did increase but appeared to stabilize between SAR = 5 and SAR = 6. Only a year of rainfall infiltration (1129 mm) with no irrigation caused a substantial reduction of sodium concentration, from SAR = 5.3 to SAR = 1.8.

An ESP value around 5, which is associated with low concentration values in the leaching water, caused Ks to decrease by 50% under laboratory conditions. This result could not be reproduced under field conditions. Using disk infiltrometers and demineralized water, no differences were obtained between ESP 4 and 0.46. Information on the field dispersion characteristics of the soils in Buenos Aires Province would be an asset in deciding what level of water sodicity is acceptable for irrigation, but such information is not available. More study is needed to define water-quality parameters to predict the effect of irrigation water quality under supplementary irrigation

REFERENCES

Admasen, F. J. (1992). Irrigation methods and water quality effects on corn yield in the Mid-Atlantic Coastal Plain. *Agron. J.* 84:837-843.

Andriulo, A., M. L. Galetto, C. Ferreyra, G. Cordone, C. Sasal, F. Abrego, J. Galina Y F. Rimatori. (1998). Efecto de 11 años de riego complementario sobre algunas propiedades del suelo. I: Propiedades físico-químicas. *Actas XVI Congreso Argentino de la Ciencia del suelo, Villa Carlos Paz*, 247-258.

Ankeny, M. D., M. Ahmed, T. C. Kaspar, and R. Horton. (1991). Simple field method for determining unsaturated hydraulic conductivity. *Soil Sci. Soc. Am. J.* 55:467-470.

Ayers, R. S. and D. W. Westcot. (1976). Water quality for agriculture. *FAO, Irrigation and drainage*. Paper N° 29, Rome.

Ayers, R. S. and D. W. Westcot. (1989). La calidad de agua de la agricultura., *Estudio FAO, Riego y drenaje*. N° 29 rev., Roma.

Blake G. R. and K. H. Hartge. (1986). Bulk density. Pp. 363-375. In A. Klute (ed.) *Methods of soil analysis*, Part 1: 2nd ed. Agron. Monogr. 9. Am. Soc. Agron. and Soil Sci. Soc. Am., Madison, WI.

Chapman, H. D. (1965). Cation-exchange capacity. Pp. 891-901. In C. A. Black, D. D. Evans, L. E. Ensminger, J. L. White, and F. E. Clark (eds.). *Methods of soil analysis.*, Part 2: 1st ed. Agron. Monogr. 9. Am. Soc. of Agron., Madison, WI.

Costa José Luis, (1995). Calidad de agua. In H. Cetrángolo et al. (ed.). *Manual de riego del productor pampeano*. SAPyA. ISBN Nro 987-95327-4-0. Bs As.

Costa, J. L., L. Prunty, B. R. Mongomery, J. L. Richardson, and R. L. Alessi. (1991). Water quality effects on soil and alfalfa: II. Soil physical and chemical properties. *Soil Sci. Soc. Am. J.* 55: 203-209.

Crescimanno G., M. Iovino, and G. Provenzano. (1995). Influence of salinity and sodicity on soil structural hydraulic characteristics. *Soil Sci. Soc. Am. J.* 59: 1701-1708

Eaton, M. F. (1950). Significance of carbonates in irrigation waters. *Soil Sci.* 69: 123-133.

Echeverria, H. E. and J. Ferrari. (1993). Relevamiento de algunas caracteristicas de los suelos agricolas del sudeste bonaerense. *Boletin Tecnico Nro* 112 ISSN 0522-0548. E.E.A. INTA Balcarce.

Echeverria, H. E. and R. Bergonzi. (1995). Estimacion de la mineralización de nitrogeno en suelos del sudeste bonaerense. *Boletin Tecnico Nro* 135 ISSN 0548-0548. E.E.A. INTA Balcarce.

Felhendler R., I. Shainberg, and H. Frenkel. (1974). Dispersion and hydraulic conductivity of soil in mixed solution. *Int. Congr. Soil Sci. Trans.* 1:103-112.

Frenkel H., J. O. Goertzen, and J. D. Rhoades. (1978). Effect of clay type and content exchangeable sodium percentage and electrolyte concentration of clay dispersion and soil hydraulic conductivity. *Soil Sci. Soc. Am. J.* 42:32-39.

Genova, L. J. (1993). Estudio de la degradación de suelos bajo riego complementario de cultivos extensivos con aguas subterráneas del acuífero pampeano en el norte de Buenos Aires. *XIV Congreso Argentino de la Ciencia del Suelo*, Mendoza 25 al 29 de octubre de 1993.

Irurtia, C. B. and R. Mon. (1998). Cambios en las propiedades físicas y químicas de los suelos de la región pampeana después de 5 años de riego suplementario. *Actas XVI Congreso Argentino de la Ciencia del suelo*. Villa Carlos Paz, Mayo 1998: 241-242.

Klute A. and C. Dirksen. (1986). Hydraulic conductivity and diffusivity: laboratory methods. Pp. 687-734. In A. Klute (ed.) *Methods of soil analysis*, Part 1. 2nd ed. Agron. Monogr. Am. Soc. Agron. and Soil Sci. Soc. Am., Madison, WI.

Oster, J. D. (1994). Irrigation with poor water quality. *Agric. Water Manage.* 25:271-297.

Oster, J. D. and F. W. Schroer. (1979). Infiltration as influenced by irrigation water quality. *Soil Sci. Am. J.* 43: 444-447.

Peinemann, N., M. Diaz Zorita, M. B. Villamil, H. Lusarreta, Y. D. Grunewald. (1998). Consecuencias del riego complementario sobre propiedades edáficas en la llanura pampeana. *Actas XVI Congreso Argentino de la Ciencia del suelo*, Villa carlos Paz, mayo de 1998: 7-8.

Rhoades, J. D. (1972). Quality of water for irrigation. *Soil Sci.* 113: 277-284.

Rhoades, J. D. (1982). Soluble salts. Pp. 167-179. In A. L. Page, R. H. Miller, and D. R. Keeney (ed.) *Methods of soil analysis*, Part 2. 2nd ed. Agron. Monogr. 9. Am. Soc. Agron. and Soil Sci. Soc. Am., Madison, WI.

Shainberg I. and A. Caiserman. (1971). Studies of Na/Ca montmorillonite systems II. The hydraulic conductivity. *Soil Sci.* 111:271-281.

Shainberg I., J. D. Rhoades, D. L. Suarez, and R. J. Prather. (1981). Effect of mineral weathering on clay dispersion and hydraulic conductivity of sodic soils. *Soil Sci. Soc. Am. J.* 45:287-291.

Simunek J., D. L. Suarez, and M. Sejna. (1997). The UNSATCHEM software package for simulating one-dimensional variably saturated water flow, heat transport, carbon dioxide production and transport, and solute transport with major ion equilibrium and kinetic chemistry. *USSL Research Report* No. 141. U. S. Salinity Laboratory, Riverside, CA. 186p.

Suarez, D. L. (1981). Relation between pHc and sodium adsorption ratio (SAR) and an alternative method of estimating SAR of soil or drainage waters. *Soil Sci. Soc. Am. J.* 45: 469-475.

Suarez, D. L., J. D. Rhoades, R. Lavado, and C. M. Grieve. (1984). Effect of pH on saturated hydraulic conductivity and soil dispersion. *Soil Sci. Soc. Am. J.* 48:50-55.

Thomas, G. W. (1982). Exchangeable cations. Pp. 159-165. In A. L. Page, R. H. Miller, and D. R. Keeney (eds.) *Methods of soil analysis*, Part 2. 2nd ed. Agron. Monogr. 9. Am. Soc. Agron. and Soil Sci. Soc. Am., Madison, WI.

U. S. Salinity Laboratory Staff. (1954). *Diagnosis and improvement of saline and alkali soils*. USDA Agric. Handb. 60. U. S. Gov. Print. Office, Washington, DC.

Wagenet, R. J. and J. L. Hutson. (1989). A process-based model of water and solute movement, transformations, plant uptake and chemical reactions in the unsaturated zone. *Continuum V 2, Water Res. Inst. Cornell University*. Ithaca, New York.

Wienhold, B. J. and T. P. Trooien. (1995). Salinity and sodicity changes under irrigated alfalfa in the Northern Great Plains. *Soil Sci. Soc. Am. J*. 59:1709-1714.

Wilcox, L. V., G. Y. Blair, and C. A. Bower. (1954). Effect of bicarbonate on suitability of water for irrigation. *Soil Sci*. 77: 259-266.

Yousaf M., O. M. Ali, and J. D. Rhoades. (1987). Clay dispersion and hydraulic conductivity of some salt-affected arid land soils. *Soil Sci. Soc. Am. J*. 51:905-907.

Water Use by the Olive Tree

J. E. Fernández
F. Moreno

SUMMARY. This is a review of water use by the olive tree, in which the most-relevant knowledge from the literature is combined with key results from experiments just finished or currently in progress. We describe the plant characteristics and mechanisms conferring drought tolerance on the olive tree. The root system functionality, hydraulic characteristics of the conductive system, leaf water relations, and transpiration behavior are considered. We explain the most-advanced techniques for optimizing irrigation, based on a more accurate calculation of the crop water needs. The crop responses to deficit irrigation strategies and to the use of wastewater for irrigation are also included. *[Article copies available for a fee from The Haworth Document Delivery Service: 1-800-342-9678. E-mail address: getinfo@haworthpressinc.com <Website: http://www.haworthpressinc.com>]*

KEYWORDS. Olive, irrigation, evapotranspiration, drought, roots, sap flow, hydraulic characteristics, leaf water relations, orchard management

J. E. (Enrique) Fernández is Senior Researcher in Agricultural Engineering, and F. (Félix) Moreno is Senior Researcher in Chemistry, Spanish Council for Scientific Research (CSIC), Instituto de Recursos Naturales y Agrobiología, Avda. de Reina Mercedes 10, Apartado 1052. 41080-Sevilla (Spain) (E-mail: Enrique: jefer@irnase.csic.es Félix: fmoreno@irnase.csic.es).

The authors wish to thank M. J. Palomo, A. Díaz-Espejo, and I. F. Girón for their help with the experiments, and Drs. B. E. Clothier and S. R. Green for their valuable contribution to the sap flow studies. Thanks are also due to the Consejo Superior de Investigaciones Científicas, to the Comisión Interministerial de Ciencia y Tecnología (Plan Nacional de I + D) and to the Junta de Andalucía (Research Group AGR151) for funding the experiments.

[Haworth co-indexing entry note]: "Water Use by the Olive Tree." Fernández, J. E. and F. Moreno. Co-published simultaneously in *Journal of Crop Production* (Food Products Press, an imprint of The Haworth Press, Inc.) Vol. 2, No. 2 (#4), 1999, pp. 101-162; and: *Water Use in Crop Production* (ed: M. B. Kirkham) Food Products Press, an imprint of The Haworth Press, Inc., 1999, pp. 101-162. Single or multiple copies of this article are available for a fee from The Haworth Document Delivery Service [1-800-342-9678, 9:00 a.m. - 5:00 p.m. (EST). E-mail address: getinfo@haworthpressinc.com].

WHY THE OLIVE TREE?

There is an increasing interest in the olive tree and its products in many areas of the world. The reasons for this popularity of the species are not only agronomic, and therefore economic, but also related to the environment and human health. The tolerance of the tree to drought, and its capacity to grow in shallow, poor quality soils, make the species among the most interesting for cultivation in arid and semiarid areas. This agronomic interest of the olive is enhanced by the fact that, despite its tough character, the tree shows a remarkable response to any improvement in the cropping conditions. Reduced water supplies by irrigation, for instance, produce substantial increases in yield. This, together with its salt-tolerant character, makes the olive one of the very few profitable crops in many of the extensive areas of the world with high salinity levels and scarce water for irrigation. Environmentally speaking, the olive is one of the most rewarding cultivated species. It is enough to see the potential conditions for erosion and desertification in many areas where the tree is cultivated, to realize that it plays a mayor role both in minimizing soil losses with its roots and in reducing the air dryness with its transpiration. Other aspects we cannot forget are the crucial contribution of the olive tree to the typical landscape in many areas of Mediterranean climate, and the deep influences of this crop on the culture and tradition of the people living in those latitudes. Finally, but of utmost importance in the recent and growing popularity of the olive as a fruit tree, is the increasing demand of its products, both the oil and the fruits. This favorable market trend is partly due to rigorous dietetic studies proving the advantages of the regular consumption of olive oil for human health.

In contrary to the norm for scientific papers, we have just made a stack of statements about the nature of the olive tree and its importance as a crop without referring to the published work. This is not fair to the reader, since any rational and productive use of the plant must be based on the rigorous application of existing knowledge. We will refer to specific published works in the following sections, mentioning here only the most comprehensive works. Thus, books such as those by Loussert and Brousse (1978), Ferguson, Sibbett and Martin (1994), COI (1996), Guerrero (1997), and Barranco, Fernández-Escobar and Rallo (1998) provide the reader with information on a variety of aspects of the olive, as a plant and as a crop. Excellent reviews about more concrete aspects are those by Lavee (1985, 1986) on flowering, Bongi and Palliotti (1994) on the response to the environment, Xiloyannis et al. (1996) on drought tolerance, and Gucci and Tattini (1997) on salinity tolerance.

Water management in the orchard is one of the issues where farmers, agronomists, and environmentalists demand more information. This is not surprising, taking into account the need for water saving in the areas where the olive is usually grown, as well as the significant improvement in crop performance when the trees are irrigated. Echoing this interest, the objectives of this work were (a) to describe the characteristics and mechanisms of the cultivated olive tree regulating the water use by the plant, and (b) to analyze the most widely used and advanced irrigation techniques designed for optimizing irrigation practice.

Many of the aspects considered here are illustrated with data obtained from the different research projects we and other members of our Group have carried out from the early 70s at the experimental farm *La Hampa*, of the *Instituto de Recursos Naturales y Agrobiología*, Seville, southwest Spain. The farm is located in the heart of *El Aljarafe* county, considered the most representative area in the world for the cultivation of 'Manzanilla de Sevilla', a cultivar considered by many as the best for table consumption, and which we will refer to here as 'Manzanilla'. Three experimental 'Manzanilla' olive orchards were planted in *La Hampa* in 1968, with the trees at 5×5 m^2, 5×7 m^2, and 7×7 m^2 apart. The farm is at $37° \ 17'$ N, $6° \ 3'$ W, and 30 m above sea level. The climate is typically Mediterranean: a wet, mild season from October to March, with an average rainfall of 500 mm (period 1971-1995), and the rest of the year being dry and hot. More details on the orchard characteristics are given in Moreno, Vachaud and Martín-Aranda (1983), Moreno et al. (1988) and Fernández et al. (1991).

THE SPECIES AND ITS HABITAT

The olive tree, *Olea europaea* L., is a subtropical evergreen plant of great longevity, probably the most cold-hardy of the subtropical fruit trees (Denney and McEachern, 1985). It is a sclerophyllous and glycophytic species, being more salt- and drought-tolerant than other temperate fruit trees. Some authors include the olive within the category of desert shrubs (Schwabe and Lionakis, 1996). The olive tree is the only species with edible fruits in the family Oleaceae. Although there are different systems for the botanical classification of the species, it is generally accepted that the commercial cultivars are included in the subspecies *sativa* and the wild types belong to the subspecies *sylvestris* (Lavee, 1985, 1996). There is evidence of olive cultivation at around 4800 BC in Cyprus (Loukas and Krimbas, 1983). The origin of the species is not clear, though it seems to be somewhere in the eastern part of the Mediterranean basin or in Asia Minor. It appears that olive plants were sent from Spain and Italy to Central America, South Africa and Australia (Yáñez and Lachica, 1971; Denney and McEachern, 1985).

Although growth is possible in other latitudes, the area for olive cultivation is between $45°$-$30°$ north and south latitude, or lower if the altitude is higher (Hartmann, 1953; Yáñez and Lachica, 1971). The olive can be commercially grown in a wide variety of soils, even in shallow and low-quality soils. Only very compact, poorly drained soils are a limiting factor for the crop, due to the sensitivity of the plant to hypoxia (Martín-Aranda, Arrúe and Muriel, 1975; Denney and McEachern, 1985). Optimum values of pH are between 7 and 8, though the olive can grow in soils with pH from 5.5 to 8.5 (Denney and McEachern, 1985). The plant does not respond to photoperiod (Hackett and Hartmann, 1964). There are abundant references in the literature, however, to the importance of temperature for growth and production. For some cultivars, cold requirements in winter may be more than 1000 h of a temperature below 7°C (Hartmann, 1953). Denney and McEachern (1985) referred to several published papers mentioning that the olive sets flower buds in the late winter, approximately eight to ten weeks before full bloom, in response to the progress of winter temperatures. They also mentioned the work of Badr and Hartmann (1971), who showed that exposure of trees to a constant temperature of 12.5°C resulted in significant production of flowers. This temperature, called the "compensation point," is believed to be cold enough to effect vernalization but also warm enough to allow for necessary concomitant cell division. Bongi and Palliotti (1984) reported that the species requires at least 10 weeks below 12.2 to 13.3°C for full expression of flowering, this being best induced when temperature fluctuates between 2 and 15°C for 70-80 days. Sensitivity to temperature is cultivar-dependent (Bongi et al., 1987). The effect of extreme temperatures must also be taken into account. The plant may suffer severe damage with minimum temperatures below − 12°C, or at higher minimum temperature if exposure is longer. High temperatures before and during bloom may reduce fruit productivity substantially (Hartmann and Opitz, 1980).

THE BIENNIAL CYCLE OF GROWTH AND PRODUCTION

The different growth stages of the olive tree must be taken into account for a correct water management in the orchard, especially when deficit irrigation strategies are used. The most complete diagram we have found for the biennial cycle of the olive tree is that of Rallo (1995, 1998), shown in Figure 1.

For the conditions of *La Hampa*, shoot growth takes place from February to August, as in most areas of the Mediterranean basin. Growth late in autumn has been reported in some cases (Abdel-Rahman and Sharkawi, 1974). Cimato, Cantini and Sani (1990) found that shoot elongation was correlated with the average monthly temperature. Leaves became fully expanded in about three to four weeks, depending on environmental conditions. The olive tree is day-neutral, temperature being the driving factor for flowering and fruiting (Denney and McEachern, 1983). Flowering takes place on 1-year-old wood. High yields are produced in the "on" years, followed by "off" years with

FIGURE 1. Biennial cycle of the olive tree (adapted from Rallo, 1998).

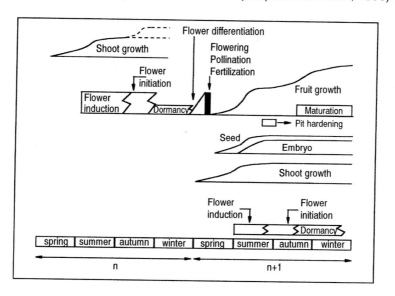

yields being rarely commercial. Alternate bearing in the olive is governed by both endogenous and external factors. We will describe management practices for reducing alternate bearing. The induction of flower buds takes place in summer, around the time of endocarp sclerification (Fernández-Escobar et al., 1992). The number of flowers able to set fruit and to remain until harvest is much smaller in the "off" than in the "on" years. The olive flowers are hermaphroditic and grouped in dichotomous panicles with up to about 40 flowers, but generally set only one fruit per inflorescence. A good crop can be obtained with 1% only of the original flowers setting and fruit remaining until harvest (Griggs et al., 1975). Cross-pollination may increase yield, since most olive cultivars are partially incompatible (Lavee 1986; Fernández-Escobar and Gómez-Valledor, 1985). Flower and fruit abscission takes place in the five to six weeks following full bloom, and is responsible for the small percentage of fruit retained to maturity (Rallo and Fernández-Escobar, 1985). The peak of fruit abscission in 'Manzanilla' has been reported to be when the fruits are 3-4 mm in diameter (Troncoso, Prieto and Liñán, 1978). The final number of developing fruits is reached about six to seven weeks after full bloom.

Climatic conditions are critical for fruit set. Hard rain, dry winds, and high temperatures during flowering may affect fruit setting markedly. Lack of light also reduces fruit set, as do thinning practices required to obtain a marketable size in table cultivars (Suárez, Fernández-Escobar and Rallo, 1984). The olive fruit is a drupe with the seed enclosed in a hard endocarp forming the pit or stone, which solidifies after four to six weeks after fruit set and hardens gradually until about three months after fruit set. The mesocarp continues to grow throughout the whole season. Fruit growth follows a double sigmoid curve, as in most drupe fruits. The seed reaches maturity just before the fruit changes color, in the stage called green maturation. Fruit growth stops when the fruit begins to change color. After the green maturation stage, there is a decrease in chlorophyll content and an increase in anthocyanin accumulation, responsible for the black color of the fruits. In "on" years, when flowering and fruit set are abundant, ripening is delayed, mean fruit size reduced, and oil accumulation slow. Fruit size, oil accumulation, and ripening are highly dependent on the relative load per tree (Lavee, 1986). Sánchez-Raya (1988) and Lavee and Wodner (1991) gave curves for fruit growth and oil accumulation for different cultivars, showing the marked influence of environmental conditions. Fruit yield can be up to 22,000 kg ha^{-1} or more. Fruit weight varies between less than 1 g and more than 12 g, depending on the cultivar. The percent of oil content is also a function of the cultivar, varying from about 12 to 28%.

TOLERANCE TO DROUGHT

The Root System

The root system of the olive tree seems to be designed for absorbing the water of the light and intermittent rainfall usual in its habitat, rather than for taking up water from deep layers. Most of the main roots seem to grow more or less parallel to the soil surface, without a dominant tap root. Olive roots are rather sensitive to hypoxia, and poorly drained soils are inadvisable for the crop. A high portion of the root length is of small diameter, which favors the absorption capacity. Absorption by olive roots is also enhanced by high potential gradients between roots and soil caused by osmotic adjustment. The highest root density is found close to the trunk, although the volume explored by the roots can easily extend beyond the canopy projection. Apart from genetic disposition, the distribution of the roots can be markedly influenced by the soil conditions, by neighboring trees, and by the irrigation practice. Olive roots are able to react quickly after a long period of drought, absorbing water immediately that is finally available in the soil. Root growth dynamics are also markedly affected by water in the soil. Irrigation makes roots grow during the dry season, preventing root shrinking and increasing the period of activity of each root. There is evidence from anatomical studies suggesting that the conductive capacity of olive roots is not reduced by drought, which is another feature of the high adaptability of the species to water stress. All those aspects are described with detail in the following sections. A detailed description of the mentioned variables and techniques can be found in Böhm (1979) and in Smit et al. (1999).

Root Distribution and Activity

The earliest studies on root distribution showed that the main development of the olive root system occurs in the most superficial soil layers. Abd-El-Rahman, Shalaby and Balegh (1966) measured the root length density (L_v, cm cm^{-3}) of soil samples taken around 7-year-old olive trees growing in a desert area in Egypt, with only 150 mm mean annual rainfall. They found maximum L_v values in the layer of 0.15 to 0.30 m in depth and up to 0.30 m from the trunk. Michelakis and Vougioucalou (1988) used the trench method to study the root distribution of 5-year-old 'Kalamon' trees in Crete, observing the highest number of roots in the upper 0.4 m of soil. Pisanu and Corrias (1971) observed a very shallow root system in the roots of 7-year-old olive trees in Sardinia. Their photographs and drawings of the excavated root system clearly illustrate the horizontal development of the roots, and the fact that the roots of contiguous trees avoid competition by developing outwards from the tree row. Núñez-Aguilar et al. (1980) studied the distribution of the root system of rainfed, 12-year-old 'Manzanilla' olive trees in *La Hampa*, using the trench method and the cylinder method based on sampling by auger. They separated the roots in groups of different diameter, finding that most of the roots were of a diameter smaller than 0.5 mm. They observed the highest L_v values, of about 0.7 cm cm^{-3}, at 0.45 m from the trunk and in the top layers of the soil, though high L_v values were also found at the depth of 1.4-1.6 m and at 1 m from the trunk. They found high water depletion in soil volumes with high root density–the first attempt we have found in the literature to relate root distribution with root activity. Later, Fernández et al. (1991) carried out in the same orchard, and also in a neighboring orchard, a more detailed study on the distribution and activity of the root system of 20-year-old 'Manzanilla' trees, planted at 7×7 m^2 and under different water regimes (see legend of Figure 2 for details). As they said, "works dealing only with distribution offer little information on root activity. Activity may be higher in zones of low root density than in zones of high density as a compensation mechanism." They used the cylinder and the trench methods to determine root distribution, and labeling with ^{32}P to determine root activity. Figure 2 shows some of their results on root distribution. Much higher values of root density were found in tree 1 in the direction of the emitters than in that of the perpendicular from the trunk, where the soil was not affected by irrigation. Some roots were found at 2 m, the maximum sampling depth, though the

FIGURE 2. Root length density (L_v) measured in two 20-year-old 'Manzanilla' olive trees at 7 × 7 m² and irrigated by a single lateral drip line placed on the soil surface in each tree row, with four 4 L h⁻¹ emitters per tree, 1 m apart in tree 1 and 1.2 m in tree 2. Irrigation doses were calculated with a class A evaporation pan, with a reduction coefficient of 0.4 for tree 1 and of 0.7 for tree 2. Both trees had a single trunk and a canopy of about 4.6 m diameter. Tree 1 was in a sandy loam deep soil, while tree 2 was in a sandy clay loam soil with a hard pan at a depth of about 0.8 m. Root density measurements were taken at different depths and at different distances from the trunk, shown in boxes, following the dripper line (gray bars) and its perpendicular from the trunk (blank bars). Values of L_v were determined in 0.2 m long soil samples of about 1 L volume, taken by auger at each location (adapted from Fernández et al., 1991).

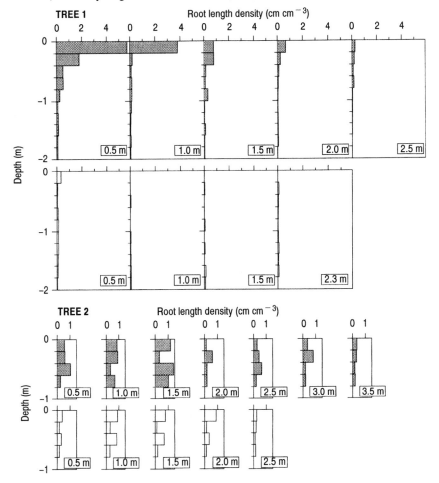

highest root densities were measured in the top layers of soil and close to the trunk. In tree 2, the impedance of the hard calcareous layer at the depth of 0.8 m restricted water and root penetration. This, together with the fact that this tree received larger volumes of water than tree 1, favored the existence of wider bulbs of wetted soil in tree 2 than in tree 1. As a result, the soil in the direction perpendicular to the drip line of tree 2 was affected by irrigation and, consequently, relatively high root densities were found also in that direction (Figure 2). This is a clear example of how water and soil conditions may affect root growth and distribution. Fernández et al. (1991) reported maximum L_v values similar to those found by Núñez-Aguilar et al. (1980), and their results also agree with the most-abundant roots having a diameter < 0.5 mm. Greater values of L_v of up to 1.167 cm cm^{-3} were observed by Fernández et al. (1987), also in *La Hampa*, at 0.5 m from the trunk and 0.2-0.4 m depth. At other depths, however, root density was generally below 0.360 cm cm^{-3}. With the ^{32}P-labeling technique, Fernández et al. (1987, 1991) found a maximum root activity at 0.5-0.6 m from the trunk and between 0.5 and 1 m depth. They observed that during the dry season root activity was high in the part of the root system well watered by the localized irrigation, and significantly lower in the soil volumes under increasing soil water depletion. Arambarri and Madrid (1974) used the ^{32}P-labeling technique to study the root activity of rainfed 'Manzanilla' olive trees in *La Hampa*. In the non-irrigated trees, the greatest root activity was found at 0.5 m from the trunk and at 0.6-1 m depth. In an earlier study carried out in Tunisia also with mature rainfed trees, Scharpenseel, Essafi and Bouguerra (1966) recommend applying the fertilizers at 0.5 m depth and at two distances from the trunk, either close to it or at 3-4 m from the trunk, a distance slightly larger that the radius of the canopy.

The most-recent studies on root activity in olive are based on sap velocity measurements in main roots. Moreno et al. (1996) used for the first time in olive the compensation heat-pulse technique, as described by Green and Clothier (1988), to determine the uptake strategy employed by the roots. They worked in *La Hampa*, with 25-year-old 'Manzanilla' trees under different water regimes. Moreno et al. (1996) observed that the non-irrigated tree switched from an extraction by deep roots beyond the dry surface of the soil, to one of near surface following irrigation. Fernández et al. (1996) gave additional data from the same experiment, corroborating the existence of an immediate response in water absorption by the roots of the olive tree after a long period of drought. They monitored the sap velocity profiles at different depths below the cambium of a root belonging to a non-irrigated olive tree, detecting no sap flow in the outer annuli, normally the region of highest flow in irrigated trees. Both Moreno et al. (1996) and Fernández et al. (1996) stated that the outer xylem vessels of the root must have cavitated after being in dry soil for a long period. Using the same technique for measuring sap flow, we carried out in *La Hampa* further studies on the hydraulic functioning of roots belonging to a semi-irrigated tree, in which water was applied in either the north or the south half of the soil explored by the roots (Fernández et al., 1998a; Díaz-Espejo et al., 1998). We observed an immediate absorption of water by the roots when this was finally available in the soil. In addition, we were able to see how the absorption rate of a root decreased when water was available for other roots growing in different soil areas finally irrigated, as some kind of compensation within the root system (Figure 3). Root absorption at night was also detected in our sap flow measurements, accounting for the recovery of the water stored in the tree when there is no transpiration.

Another feature showing the adaptation of the olive tree to dry areas is that the absorption capacity of its roots seems to be higher than that in other fruit trees. In early studies, Abd-El-Rahman, Shalaby and Balegh (1966) found a high osmotic potential in the sap of 7-year-old olive trees growing in an area of Egypt with only 150 mm per year of average rainfall. They stated that this

FIGURE 3. Sap flows monitored in two main roots—one in the north side (N root, 14.8 mm in radius) and the other in the south side (S root, 15.2 mm in radius)—of a 29-year-old 'Manzanilla' olive tree at *La Hampa*. Sap flows were determined with the compensation heat-pulse technique. A 2.5 m radius pond was built around the tree, divided into two sides, north and south, by a small earthen dyke, in order to semi-irrigate the tree by applying water in one side at a time. The arrow represents an irrigation of 70 mm of water applied to the north side on day of year 287, at 10.25 GMT (adapted from Fernández et al., 1998a).

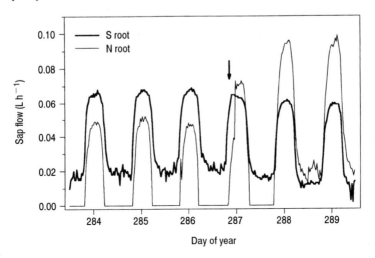

enables the roots to extract water from the soil at low soil water potentials. Xiloyannis et al. (1996) mentioned that osmotic adjustment in olive roots allowed the plant to maintain turgor and so to prevent or delay the separation of the root from the surrounding soil. In roots of olive plants under high water stress, with predawn leaf water potential of −5.2 MPa, Xiloyannis et al. (1996) measured an osmotic adjustment of 1.67 MPa in roots of 4-5 mm in diameter, 1.42 MPa in roots between 4 and 1 mm, and 0.2 MPa in roots of diameter smaller than 1 mm. They mentioned that the olive tree is able to extract water from the soil up to soil water potentials of −2.5 MPa, due to the high potential gradients between leaves, root, and soil created by lowering the water content and water potential of its tissues. Dichio, Nuzzo and Xiloyannis (1997) found in young 'Coratina' olive plants that both the osmotic potential at full turgor and the elastic modulus of tissues increased with water stress, which lead to high water potential gradients between leaf and soil. It seems, therefore, that for the olive tree, the wilting point is reached when the soil is much drier than for other fruit tree species, which are thought to be unable to extract water when the soil water potential is not much lower than −1.5 MPa.

The efficiency of the olive root system is also due to a root/canopy ratio that is usually bigger in non-irrigated trees than in irrigated ones, contributing to the drought tolerance of the species. This has been observed by Nuzzo, Dichio and Xiloyannis (1995), Celano et al. (1997) and Nuzzo et al. (1997), with young 'Coratina' olive trees in southern Italy. In *La Hampa*, Fernández et al. (1991, 1992) observed that the volume of soil explored by the roots of mature 'Manzanilla' olive trees under localized irrigation was smaller than in the case of rainfed trees, in which higher root densities were measured deeper and farther from the trunk. This is a factor to take into account when choosing the irrigation system of the orchard. The fact that trees irrigated by localized irrigation explore a reduced volume of soil may represent a risk for water stress if irrigation is interrupted. The tree may consume the water of the bulbs in very few days, suffering from water stress if the irrigation is not then re-established. This problem is minimized with irrigation systems that wet larger volumes of soil. An incorrect application of water may also affect the activity of olive roots due to their sensitivity to hypoxia. This may not be a problem in soils with hydraulic characteristics favoring drainage, but it should be taken into account if the soil is heavy, with a low hydraulic conductivity. In the rather sandy soils of *La Hampa*, Fernández et al. (1991) did not observe a diminution of root density near the drippers compared with other zones in which aeration conditions were supposed to be more favorable. Michelakis (1986), however, working with drip-irrigated 'Kalamon' olive trees in Crete, reported lower root densities in a small volume of soil under the drippers than in the rest of the wet bulb. He did not give the soil characteristics. Other aspects to consider when choosing the irrigation system are discussed later.

Root Dynamics

When and where the roots grow is crucial for understanding the functioning of the root system and its relation with the aerial part, and for optimizing irrigation and fertilization. There are, however, very few studies on root dynamics of the olive tree. We have found just the work by Fernández, Moreno and Martín-Aranda (1990) and Fernández et al. (1992) in *La Hampa*, and the work by Celano et al. (1998) in southern Italy. Both of these used mini-rhizotrons made with transparent tubes buried in the soil explored by the roots.

Fernández, Moreno and Martín-Aranda (1990) and Fernández et al. (1992) made their studies of root dynamics in the same orchards where they studied root distribution and activity, to get a comprehensive view of the behavior of the root system of mature olive trees under field conditions and different water regimes. They observed that the period of emergence and growth of new roots was limited to the spring in rainfed trees, while it extended to the whole summer period and autumn in the trees irrigated throughout the dry season. When analyzing the root appearance, they observed that root color changed from the typical pale color of the newly formed roots to a dark brown color in about one month in the case of the rainfed tree, and up to three months in the irrigated trees. Root shrinking was observed in some roots of the rainfed tree when the soil dried out, with partial separation from the surrounding soil. Celano et al. (1998) also observed that root dynamics is very much affected by irrigation. They studied the growth dynamics of the roots together with the growth dynamics of different aerial organs, concluding that the growth pattern is determined by the sink-source competition established between the aerial and subterranean parts throughout the year.

Root Anatomy

The only work we have found in olive relating root anatomy with root functioning is that of Fernández et al. (1994). They examined the root development of 2-year-old 'Manzanilla' olive trees grown for four months in 0.8 m^3 containers under wet and dry water regimes. They studied the transition to secondary growth and how the water treatment affected the radius of the root, the central cylinder, and the xylem vessels. Their most relevant results are shown in Figure 4. They found complete transition to secondary growth closer to the apex in the roots grown in dry soil (at about 90 mm) than in the roots grown in watered soil (at about 120 mm). Up to about 50 mm from the

FIGURE 4. Stage of development from primary to secondary growth scored according to a maturity index indicating the degree of vascular cambium development (Figure 4a). Observations were made along roots taken from 2-year-old 'Manzanilla' olive trees grown in 0.8 m^3 containers with sandy-loam soil and subjected to two water regimens. In the irrigation treatment, the soil was kept about field capacity throughout the four months experimental period. No water was supplied in the dry treatment, with only one irrigation immediately after planting in the containers. Measurements were made in six sequential 5 mm samples taken at 30 mm intervals along the root axis. Each point represents the mean of the six measurements. Vertical bars indicate twice the standard error. Measurements of total root and central cylinder (stele) radius are shown in Figure 4b. The inner wall of the endodermal cells was considered the external limit of the central cylinder. The radial width of each zone was determined by averaging the horizontal and vertical radii. Each point represents the average of eight to 87 measurements for each stage. Vertical bars indicate twice the standard error. The central cylinder was only considered through stage 3 (adapted from Fernández et al., 1994).

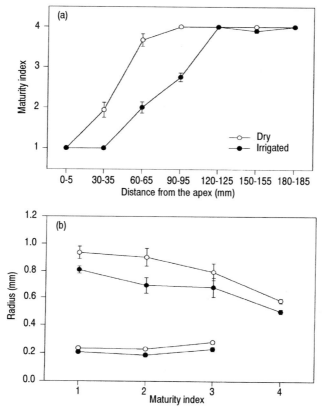

root tip, the cortical tissue was wider in the roots grown in dry soil than in watered soil, though the authors warned about the possibility of mechanical stress in the dry soil influencing those results. They did not find significant differences between treatments in the diameter of the metaxylem vessels, which were a mean 10.21 $\pm$ 0.34 μm in roots of the irrigation treatment and 9.86 $\pm$ 0.35 μm in those of the dry treatment. In addition, the radius of the central cylinder was slightly greater in the dry than in the irrigated treatment. They concluded that "although further studies are necessary, certain speculation is possible as to the significance of these features in the adaptability of the olive root system to water stress. The more rapid maturation found closer to the apex in the dry treatment may allow increased water movement through the root system due to the increased water flux associated with secondary vascular development . . . The lack of reduction in metaxylem vessel and central cylinder size under drought suggests that resistance to water flow is not reduced in these tissues."

Root Signaling

One of the most interesting mechanisms affecting water consumption has its origin in the root system, though it affects stomata behavior. Evidence has been found in different species, including a fruit tree such as apple (Gowing, Davies and Jones, 1990), suggesting stomatal control by signals coming from the roots. It seems that the roots are able to sense the dryness of the soil and "order" the stomata to close, thereby reducing water losses and preventing excessive water stress. One of the most relevant papers we have found in this line is that by Tardieu and Davies (1993). They suggested an integrated chemical and hydraulic signaling mechanism controlling leaf water relationships. The only reference we have found to a root-to-leaf signaling mechanism in olive is an experiment in pots mentioned by Bongi and Palliotti (1994). The root system of young 'Frantoio' olive plants was split into two parts, leaving a small portion in drought (-1.1 MPa) and the rest in optimal water conditions (-0.2 MPa). They observed stomatal closure in those plants, which appeared to be mediated by a translocated signal coming from the stressed roots. We have recently carried out experiments in *La Hampa* with 30-year-old 'Manzanilla' olive trees, whose results have not yet been published. Figure 5 shows some preliminary results suggesting that the root signaling mechanism is also present in mature trees under field conditions.

Hydraulic Characteristics of the Wood

The olive is a diffuse-porous tree having a dense wood with abundant fibers and little parenchyma. We have measured a mean wood density of

FIGURE 5. Daily evolution of the transpiration ratio calculated by dividing the daily transpiration of an olive tree in which the total volume of soil explored by its roots was wetted by irrigation (T1) by the daily transpiration of a tree in which part of its roots was affected by irrigation and the rest remained in dry soil (T2). The daily transpired amounts were determined from sap flow measurements in the trunk of the trees. Measurements were made in *La Hampa*, with 30-year-old 'Manzanilla' olive trees irrigated throughout the dry season of 1998 (treatment I, open circles), and with trees under dry farming conditions until the beginning of this experiment on August 23 (treatment D, closed circles). Each T1 tree was pond-irrigated, with a pond big enough for wetting the total volume of soil explored by the roots. The T2 trees were irrigated by the a localized irrigation system consisting on a single lateral drip line placed on the soil surface in each tree row, with five 3 L h^{-1} emitters per tree 1 m apart. The arrow close to the open circle indicates the day in which irrigation was applied to the trees T1 and T2 of treatment D. The arrow close to the closed circle indicates the day on which the T1 tree of treatment I was irrigated by pond; before that day, the T1 and T2 trees of that treatment were irrigated by the described localized irrigation.

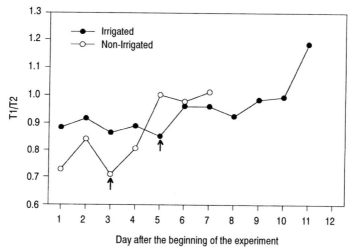

Day after the beginning of the experiment

0.623 kg L^{-1} in samples taken from the trunk of 'Manzanilla' olive trees. The large amount of fibers, which makes olive wood so hard, accounts for the low vessel lumina of the species in comparison with other diffuse-porous Mediterranean plants. Salleo, Lo Gullo and Oliveri (1985) observed that the vessel lumina, expressed as a percentage of the total xylem cross-sectional area, was about 8% in olive, being 17% in *Vitis vinifera* and *Populus deltoides*. The xylem vessels are thick-walled and generally grouped in radial multiples of two to four vessels, rarely solitary. Growth ring boundaries are often indistinct. The analysis of the radial sections shows heterogeneous rays, numerous though small intervessel and rayvessel pits, and simple perforation plates (Schweingruber, 1990). In our calibration experiments of the compensation heat-pulse technique for measuring sap flow in olive, from which some preliminary results are given in Fernández et al. (1997), we stained the sapwood of the trunk of a 12-year-old 'Manzanilla' olive tree with safranin. At about 0.74 m above the ground, the average radius below the cambium was 68.2 mm, the area of the sapwood being 112.3 cm^2, and that of the heartwood 33.9 cm^2. The maximum depth of the sapwood was highly heterogeneous, varying between about 12 and 34 mm, with an average value of 26.6 mm. Scaramella and Ricci (1988) observed the presence of interxylary phloem in olive, which contributes to the sapwood heterogeneity.

The diameter of the xylem vessels varies depending on the conductive organ. In roots of 2-year-old 'Manzanilla' olive trees, Fernández et al. (1994) measured a mean metaxylem vessel diameter of 10.03 μm. In wood samples taken from the trunk of 29-year-old 'Manzanilla' olive trees, we have found that most of the xylem vessels have a diameter between 33 and 39 μm, with maximum values rarely greater than 50 μm. These values are low in comparison with other diffuse-porous species. The laurel, for instance, has xylem conduits between 50 and 80 μm in diameter (Lo Gullo and Salleo, 1988), and the kiwifruit between 100 and 500 μm (Green and Clothier, 1988). Salleo, Lo Gullo and Oliveri (1985) measured the mean vessel radius in internodes of 1-year-old twigs, finding values of about 11 μm in the proximal internodes, 10 μm in the middle, and 8 μm in the distal internodes. There is little information about the effect of those characteristics on the hydraulic functioning of the conductive tissues of the olive tree. The small diameter of the vessels may account for a low hydraulic conductance, and may also be a protection against cavitation. Tyree and Sperry (1989), however, claimed that the vulnerability to cavitation is determined by the diameter of the intervessel pit membrane pore, rather than by that of the xylem vessel. Measurements of the sap velocity profiles made by Fernández et al. (1996) in roots, and our measurements in the trunk, have shown no flow in the outer annuli of the sapwood of water-stressed 'Manzanilla' olive trees. We believe that this is a consequence of air emboli forming in the xylem vessels of that area.

Thompson et al. (1983) studied the hydraulic architecture of young olive trees. They measured the leaf specific conductivity (LSC, kg s^{-1} m^{-2} MPa^{-1}), which relates the flow rate with the amount of transpiring surface area by the following equation:

$$LSC = \frac{V}{P_x a} = \frac{E}{P_x} \tag{1}$$

where V is the mass flow rate (kg s^{-1}) through a section of stem supplying water to leaves of green surface area a (m^2, one side only), and P_x (MPa

m^{-1}) is the pressure gradient over the length of the section. The value of V/a is the average evaporative flux E (kg s^{-1} m^{-2}) for all the leaves fed by the stem segment. Thompson et al. (1983) found for olive larger values of E (2.6 × 10^{-5} kg s^{-1} m^{-2} against 1.1 to 1.6 × 10^{-5} kg s^{-1} m^{-2}) and smaller values of LSC (4 × 10^{-4} kg s^{-1} m^{-2} MPa^{-1} against 28 to 250 × 10^{-4} kg s^{-1} m^{-2} MPa^{-1}) than those reported by Zimmermann (1978) for maple, white birch, and poplar, all of them diffuse-porous trees. Salleo, Lo Gullo and Oliveri (1985) mentioned a number of reasons supporting the utility of the LSC measurements for plant physiologists, despite LSC not allowing for quantitative measurements of the stem hydraulic conductivity. They stated that calculating the hydraulic conductivity of plant stems on the basis of the whole xylem cross-sectional area can cause substantial underestimates of this parameter and that, on the other hand, measurements of the hydraulic conductivity of vessel lumina ignore the permeability to water of fibers and wood parenchyma cells.

Water flow through the soil-plant-atmosphere system is generally assumed to be well described by a model similar to Ohm's law. Larsen, Higgins and Al-Wir (1989) followed this approach to estimate the liquid pathway resistance to water flow (R, MPa µg^{-1} cm^2 s) in apple, apricot, grape, peach, and olive plants. They calculated the difference between measured values of predawn xylem pressure potential, assumed to be equivalent to the soil water potential, and diurnal xylem pressure potential values measured at the petiole. The value of R was calculated by dividing that difference by the estimated transpirational flux density. The resulting R values were 0.511 for apricot, 0.465 for olive, 0.329 for grape, 0.319 for peach, and 0.182 for apple. They stated that the high value of R in olive, together with stomatal closing, may account for the low transpiration losses per unit leaf area measured in this species. Bongi and Pallioti (1994) mentioned that the large water potential differences between leaves and roots usually found in olive might reflect a strong resistance to water movement. In our field experiments with well-irrigated trees, in which the soil water potential can be assumed to be similar to that of the absorbing roots, we have observed that the drop in water potential from leaves to roots is easily greater than 2 MPa, with a maximum difference of about 4 MPa.

Leaf and Canopy Characteristics

Olive leaves are well-adapted to avoid excessive water loss under the highly demanding conditions of the areas where the tree usually grows. They show not only several sclerophyllous characteristics, but also active mechanisms controlling water loss. The olive tree can have leaves of up to three years of age. Leaf aging significantly modifies the leaf characteristics and leaf response to the environment, as we will discuss in this and other sections.

In the 'Manzanilla' olive trees of *La Hampa*, we have measured a mean leaf surface of 359 mm^2 one side only, and a mean specific leaf weight (SLW), i.e., the leaf dry weight divided by the leaf surface, of 203 g m^{-2} for the current year leaves and 319 g m^{-2} for 1-year-old leaves. In samples taken in April, we found that the water content, expressed as percentage of the fresh leaf weight, was 59.9% in the current year leaves and 47.6% in 1-year-old leaves. Details on leaf growth and development are given in a former section.

The high content of cuticular wax (Leon and Bukovac, 1978) prevents water diffusion through the cuticular membrane, transpiration taking place only through the stomata. Another characteristic contributing to reducing water loss under stress is the dense packing of the mesophyll layer in the olive, which produces a low proportion of cell walls exposed to the air (Bongi, Mencuccini and Fontanazza, 1987). The stomata are present only on the abaxial surface. Stomatal density has been reported to range from about 250 mm^{-2} to more than 700 mm^{-2}, depending on cultivar and nutrient status (Leva, 1977; Bartolini, Roselli and Di Milla, 1979; Bongi, Mencuccini and Fontanazza, 1987). Cultivars resistant to cold have a lower stomatal density than those that are susceptible (Roselli, Benelli and Morelli, 1989). The values of stomatal density found in olive are similar to those of rapidly transpiring plants (Bongi, Mencuccini and Fontanazza, 1987). Stomatal pore widths have been measured by Schwabe and Lionakis (1996) and Vitagliano et al. (1997), among others. The presence of cuticular ledges in the stomata, described by Durán-Grande (1977) and Leon and Bukovac (1978), may make it difficult to measure the stomatal opening. In addition, the olive leaf is subject to patchy stomatal closure over the total leaf surface (Loreto and Sharkey, 1990; Natali, Bignami and Fusari, 1991). Stomatal closure is an active mechanism for preventing excessive water stress under conditions of high atmospheric demand, described with detail in the section *Gas Exchange*. Peltate trichomes of about 130 μm in diameter are present on both faces of the leaf, but their number is about eight times greater on the abaxial than on the adaxial surface. Leon and Bukovac (1978) identified specialized cells at the base of the peltate stalk, reported by Bongi, Mencuccini and Fontanazza (1987) as being effective in limiting water loss. Palliotti, Bongi and Rocchi (1994) observed that the trichomes are a barrier to the diffusion of CO_2 and H_2O, lowering the boundary layer conductance in the air surrounding the stomata. After removing the trichomes from leaves of 'Manzanilla' olive trees, they found that the total boundary layer resistance was reduced more than 5-fold. Stomatal conductance and leaf transpiration were significantly higher (21.2% and 20.5%, respectively) in trichome-free leaves than in intact leaves. Schwabe and Lionakis (1996), however, questioned the efficiency of the trichomes in reducing water loss, apart from increasing the reflection of radiation. The trichomes have flavonoid constituents that absorb ultraviolet-B radiation, protecting the leaf against the negative effects of this radiation on growth and development (Karabourniotis, Kyparissis and Manetas, 1993; Grammatikopoulos, Karabourniotis and Kyparissis, 1994).

The optical properties of the olive leaf play an important role in controlling water consumption. Baldy, Lhotel and Hanocq (1985) found that the adaxial surface of the olive leaf absorbs more photosynthetically active radiation (PAR) than the abaxial surface. The abaxial surface reflected 20 to 40% of the PAR. The olive tree is able to reduce the amount of intercepted radiation during drought by increasing leaf rolling and reducing the angle of the leaf with the stem (Schwabe and Lionakis, 1996). The capacity of the olive leaf for leaf rolling, for the upward movement of the leaf, and for the control of stomatal opening–the three main mechanisms influencing water consumption–is lower in the older leaves than in the young ones. In addition, both the reflectance and the transmittance are greater in young than in older olive leaves (Trigui, 1984). Another mechanism that helps the plant to withstand drought is the water intake by the leaves (Spiegel, 1955; Natali, Bignami and Fusari, 1991), though not enough information exists to evaluate the effect of this phenomenon in the reduction of plant water stress.

The tree size and canopy architecture depend very much on pruning practices and plant density. One of the most widely accepted tree shapes for an intensive olive orchard is that of a single trunk with two or three main branches at 0.8-1.5 m from the ground surface, and a round canopy. In any case, pruning practices must favor light interception, since the dry matter production of the olive tree is directly proportional to intercepted PAR. The radiation use efficiency (RUE) is the ratio between accumulated dry matter and intercepted PAR. Mariscal, Orgaz and Villalobos (1998) measured an RUE value of 1.25 g $(MJ\ PAR)^{-1}$ in young 'Picual' olive trees. The RUE seems to be lower in winter and summer than in spring and autumn. The height of the tree must also be controlled by pruning, to minimize harvesting costs, especially in the cultivars for table consumption which are harvested manually. This point will be further discussed in the last section. Villalobos, Orgaz and Mateos (1995) used a plant canopy analyzer and a simulation model to determine leaf area in the olive. They found that the leaf area index (LAI) exceeds 90% of the plant area index (PAI), so the surface area of green leaves alone is more than 90% of the surface corresponding to leaves and stems. We have made rough estimations of LAI in 29-year-old 'Manzanilla' olive trees of *La Hampa*, planted at 7×5 m^2, by counting the number of leaves of a fraction of the canopy and extrapolating to the rest. Once the total number of leaves was estimated, we measured the leaf area of a representative sample of leaves with a leaf area meter and calculated the total leaf area of the tree. Results showed that the maximum leaf area of a tree of average size was around 60 m^2, measured at the end of the summer after the growing period, yielding an LAI of 1.71. The usual pruning practices in olive make inadvisable the methods for estimating LAI based on diameter measurements of branches and twigs.

Leaf Water Relations

Leaf measurements of water potential (Ψ_l, MPa), stomatal conductance to H_2O (g, mm s^{-1} or mmol H_2O m^{-2} s^{-1}), net photosynthesis rate (P_N), also called CO_2 assimilation rate (A, μmol CO_2 m^{-2} s^{-1}), and the evaporative flux from the leaves (E, kg m^{-2} s^{-1} or mmol H_2O m^{-2} s^{-1}) are usually made in studies of plant water relations. Most of the variables can be expressed in different units, to be coherent with those of related variables and with the purpose of the study. Also, E_t can be used instead of E when the transpiration of the whole canopy is considered. The value of E in young plants growing in pots can be monitored by the use of instruments measuring gas exchange between leaf and air, or by weighing, in which case the units are kg m^{-2} per unit of time, normally day or hour. In mature plants in the field, the value of E can be monitored with techniques for determining the sap flow in the trunk or in the main branches, and the units are L m^{-2}, also per day or per hour. In trees, the wide range of leaf conditions in the canopy make it difficult to evaluate the meaning of the measurements at leaf level in the water behavior of the tree. Despite this, a good deal of information can be obtained by the measurement of those variables, which, in addition, are relatively easy to make even in fully established orchards with mature trees under field conditions. We are going to summarize here the most-relevant results obtained from the measurement of the variables used when studying leaf water relations in the olive tree. However, it must borne in mind that, as Tardieu and Davies (1993) noted, "Stomatal conductance, leaf and root water potential, water flux, and xylem [ABA] have multiple interrelations which cannot be summarized by a relationship between any of these variables." For those interested in this sort of study, an excellent collection of review articles on ecophysiology was published in 1997 in the *Journal of Experimental Botany* (Environmental Perspectives 1996/7). We analyze in the section *Determining crop evapotranspiration* other techniques designed to study water use at the tree and orchard levels, which avoid the mentioned limitations of the measurements at leaf level.

Water Status

Leaf water potential is probably the most widely measured variable for knowing the water status of the plant. Daily and seasonal changes in Ψ_l measured in different olive cultivars and conditions are shown elsewhere

(Agabbio, Dettori and Azzena, 1983; Xiloyannis et al., 1996; Fernández, Moreno and Martín-Aranda, 1993; Fernández et al., 1997). The predawn leaf water potential (Ψ_{pd}) is well-correlated with soil water content (Natali, Xiloyannis and Angelini, 1985). The value of Ψ_{pd} can be used as an indicator of the degree of water recovery of the tree at night. A value of Ψ_{pd} below -0.5 MPa is considered to be a threshold for satisfactory recovery (Dettori, 1987). Fernández et al. (1997) plotted the Ψ_{pd} values measured on mature 'Manzanilla' olive trees under different water treatments against relative extractable water (REW) in the soil (Figure 6). Fernández et al. (1997) found a mean value of $\Psi_{pd} = -0.46$ MPa for REW ≥ 0.4, and assumed this value of REW to be a threshold for soil water deficit. At midday, the values of Ψ_l can be quite negative even in well-irrigated trees, if the atmospheric demand is high (Fernández, Moreno and Martín-Aranda, 1993; Moreno et al., 1997). Bongi and Palliotti (1994) mentioned that the large midday Ψ_l drop in well-irrigated trees might reflect a strong resistance to water movement. They stated that it

FIGURE 6. Relationship between relative extractable water and predawn leaf water potential measured on 26-year-old 'Manzanilla' olive trees at the experimental farm *La Hampa*. Trees were under different water regimes (● weekly pond irrigation to cover the crop water demand; ▽ the same as in ●, but with about 1/3 of the water applied; ○ non-irrigated trees, with rainfall as the only water supply). Data from the experiment by Fernández (1989) carried out in the same orchard (△ non-irrigated trees) have been used. Each point represents the average of six measurements of predawn leaf water potential per treatment. The mean value of predawn leaf water potential is -0.46 MPa for a relative extractable water ≥ 0.4 (adapted from Fernández et al., 1997).

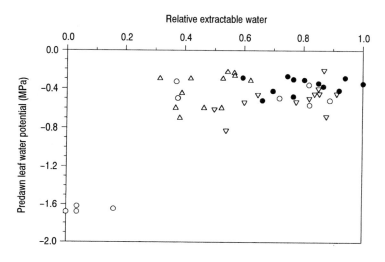

is difficult to establish a critical ψ_l for growth or physiological processes, and referred to a paper by Rhizopoulos, Meletiou-Christou and Diamantagiou (1991) in which it is reported that olive leaves can tolerate Ψ_l values near -9 or -10 MPa without losing rehydration capacity. After reaching a minimum value during the central hours of the day, Ψ_l became less negative in the afternoon, its value at sunset indicating the degree of recovery from water stress.

The water content at saturation in the olive leaf is lower than in other fruit tree species. Abd-El-Rahman, Shalaby and Balegh (1966) measured 1.59 g water g^{-1} dry weight in olive, 5.77 g g^{-1} in fig, and 5.85 g g^{-1} in grape. Fernández et al. (1997) measured in 'Manzanilla' 1.49 g g^{-1} in current year leaves and 0.89 g g^{-1} in 1-year-old leaves. That means that the olive tissues can reach turgidity with a lower water uptake than other plants, and are able to be at full turgidity after a limited amount of rainfall. Xiloyannis et al. (1996) found a clear linear correlation ($r^2 = 0.96$) between the relative water content (RWC) of olive leaves and their counterpart Ψ_{pd} values, from 0 to -7.0 MPa. They found that RWC was about 40% when Ψ_{pd} was as low as -7.0 MPa. Under field conditions, RWC depends not only on the water conditions in the soil and the atmosphere, but also on other factors such as the time of the year and the leaf area per plant (Abdel-Rahman and El-Sharkawi, 1974).

Olive leaves have a high volumetric modulus of elasticity (Bongi and Palliotti, 1994), also called elastic modulus ($\in$, MPa). This is a variable inversely proportional to tissue elasticity, representing an applied pressure divided by a fractional change in volume size:

$$\in = \frac{\Delta P}{\Delta V / V} \qquad (2)$$

where P is turgor pressure and V is cell volume. In the olive, $\in$ tends to increase with drought. Bongi and Palliotti (1994) observed that at 87.5% of maximal cell volume, $\in$ was 7 MPa in hazelnut and 22.5 MPa in olive. Loss of turgor of the more rigid cells in olive occurred at 80% of maximal cell volume, while positive turgor was maintained in hazelnut leaves at 66% of maximal volume. Plants with less rigid cells, such as *Agave deserti*, displayed a decrease in $\in$ and tended to retain higher turgor pressure (Ψ_P) under drought conditions (Schulte, 1992). The olive tree does not maintain a low $\in$ in response to drought, but reduces the osmotic potential (Ψ_p). The reduction in Ψ_p causes the reduction in Ψ_l responsible for the high capacity of the olive tree for water absorption already mentioned. This reduction in Ψ_p is responsible for the drop in water potential between the soil and the leaves. Such behavior has been observed by Dichio, Nuzzo and Xiloyannis (1997) in

2-year-old 'Coratina' olive trees under different water stress. They reported that the high osmotic adjustment and the rigidity of the cell wall induced high potential gradients between leaves and roots. They calculated the maximum $\in$ at full turgor in plants at three water stress levels. In the control plants (Ψ_{pd} = -0.45 MPa) and in those of a first level of stress (Ψ_{pd} = -1.6 MPa), the maximum $\in$ caluated at full turgor was 11.6 MPa. In more-stressed plants (Ψ_{pd} from -3.3 MPa to -5.2 MPa) the elastic modulus was 18.6 MPa. The Ψ_p value at saturation was -2.06 MPa in the control plants and -2.81 MPa in the most-stressed plants, indicating a total active osmotic adjustment of 0.75 MPa. At incipient plasmolysis, Ψ_p varied from -3.07 in controls to -3.85 in the most-stressed plants, with an RWC of 77.8% for the control plants and 74.5% for the more-severely stressed plants. When the olive plants are under stress for a long time, for instance during the long dry season of the Mediterranean areas, maximum $\in$ can be higher than during the wet season (Dichio, personal communication).

Monthly variations of Ψ_p in mature olive trees under desert conditions in Egypt were measured by Abd-El-Rahman, Shalaby and Balegh (1966). They found minimum values of about -6.8 MPa in September. They also observed that the values of Ψ_p in olive were lower than in other xerophytes. They reported mean annual Ψ_p values of about -4.8 MPa for olive, -2.4 MPa for almond, -1.5 MPa for fig, and -1.3 MPa for grape, among other species. With drought, olive leaves tend to overcome water deficit by solubilizing sugar from the starch reserve, and so Ψ_p becomes more negative (Tombesi, Proietti and Nottiani, 1986). Starch depletion in conjunction with the rise in soluble carbohydrates and mannitol during the summer has been observed in olive by Drossopoulos and Niavis (1988). Mannitol is a sugar alcohol which in olive represents from 1/2 to 2/3 of the total soluble sugars in leaves and bark (Bongi and Palliotti, 1994). Xiloyannis et al. (1996) also outlined the significant role of the high osmoregulation capacity of the olive tree in its tolerance to drought. They differentiated between the passive osmotic adjustment, due to the loss of water by the tissues, and the active adjustment due to the synthesis of osmolytes.

Gas Exchange

We have mentioned in previous sections that stomatal closure is a mechanism used by the olive tree for restricting water loss on days of high atmospheric water demand. Figure 7, taken from Fernández et al. (1997), is a good example of this behavior. The figure shows diurnal time courses of g in olive trees under different conditions of water in the soil, recorded in *La Hampa* on two summer days with different atmospheric water demand. On July 25, a relatively dry, clear-sky day, the stomata opened as soon there was light, and g increased very quickly during the first hours of the morning. Maximum

FIGURE 7. Diurnal time course of leaf water potential and stomatal conductance measured in 16-year-old 'Manzanilla' olive trees at the experimental farm
La Hampa. Measurements were made in non-irrigated trees (○), and in trees
with weekly irrigation to cover the crop water demand (●) and 1/3 of it (∇). Each
point represents the average of six values per treatment. Vertical bars indicate
twice the standard error. Values of photon flux density and vapor pressure
deficit of the air recorded on the measurement days are also plotted. The 25th
of July was a somewhat fresh and partially cloudy day, whereas the 27th was
a clear, hot, very dry day. Relative extractable water in the soil was, for the 25th
and the 27th of July, respectively, 0.04 and 0.04 for the non-irrigated trees, 0.65
and 0.61 for the medium watered trees, and 0.87 and 0.77 for the most irrigated
trees (adapted from Fernández et al., 1997).

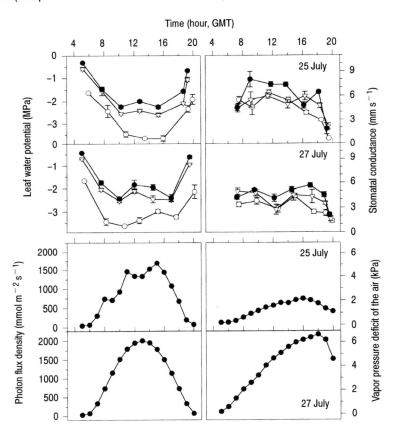

values of g were recorded early in the morning, before the atmospheric water demand was high enough to cause stomatal closure. For the conditions of *La Hampa*, we have usually measured maximum values of g before 10.00 GMT, recording values of up to 12 mm s^{-1}, though they are rarely over 10 mm s^{-1}. The minimum values of Ψ_l occurred later in the day, usually between 14.00 and 16.00 h GMT. On July 27, the values of g remained fairly constant throughout most of the day, and they were lower than on the 25th for all treatments, despite low or negligible variations in soil water content for all treatments between the measuring days. This was a consequence of an earlier and more marked stomatal closure on July 27, a day when both radiation and atmospheric demand were greater than on July 25 (Figure 7). The earliest references we have found reporting stomatal control in the olive are those of Migahid and Abd-El-Rahman (1953) and Hammouda (1954), cited by Abd-El-Rahman, Shalaby and Balegh (1966). Apart from the root signaling phenomenon already described, the meteorological driving variables for stomatal opening are light intensity and vapor pressure deficit of the air (Abdel-Rahman and El-Sharkawi, 1974; Fernández et al., 1997). Wind speed seems to have low influence. Upper-bound relationships between g and D_a and the photon flux density (I_P, μmol m^{-2} s^{-1}) were obtained by Fernández et al. (1997) for 'Manzanilla' olive trees in *La Hampa*. They found maximum values of g at relatively low levels of I_P, from about 500 μmol m^{-2} s^{-1}, and a proportional decrease in g with increasing D_a for values up to approximately 3.5 kPa. The stomata remained partially opened at higher D_a. Higher values of g were observed in the morning, during the opening phase, than in the afternoon for a similar level of D_a and I_P. This behavior could be explained by the fact that maximum values of I_P occur earlier in the day than the maximum values of D_a (Jarvis, 1976). A similar behavior has been observed in other fruit tree species, such as oak (Hinckley et al., 1975), apple (Jarvis, 1976), and peach (Punthakey, McFarland and Worthington, 1984). The fact that D_a is the main driving variable for midday stomatal closure seems to be true except for winter time. At that time of the year, soil temperature (Gimenez et al., 1996), or perhaps other factors related to root functioning (Fereres, Moriana and Ruz, 1998), could cause low Ψ_l and stomatal closure despite high soil water potential and relatively low atmospheric demand.

Some authors have found certain correlation between Ψ_l and g values (Sorrentino, Giorio and d'Andria, 1998), but at other times, such correlation has not been found (Fernández, Moreno and Martín-Aranda, 1993; Fernández et al., 1997). The influence of environmental factors such as D_a and I_P (Jarvis, 1976) and the possibility that the water potential of the stomatal apparatus is different to the bulk leaf water potential (Castel and Fereres,

1982) probably account for the large scatter sometimes found when plotting Ψ_l against g.

Maximum A values are usually measured early in the morning before stomatal closure, the same as for g. Maximum values of A of up to 22 μmol CO_2 m^{-2} s^{-1} have been reported by Angelopoulos, Dichio and Xiloyannis (1996) and Díaz-Espejo et al., (1998). This value is rather low in comparison with that of other fruit trees, though the olive tree is able to maintain relatively high photosynthetic rates over a long period of drought. This low photosynthetic capacity has been related to certain leaf characteristics, such as dense, thick cell walls and the presence of trichomes, as well as the low density of photosynthetic reaction centers (Bongi and Palliotti, 1994). On a typically Mediterranean summer day there is a continuous decrease of A after the peak value reached in the morning, due to stomatal closure and to other effects caused by water stress. Larcher, Moraes and Bauer (1981) observed in 'Leccino' that A began to decrease when Ψ_l fell to -1.3 MPa, and that the photosynthetic capacity was reduced by 50% when Ψ_l reached -2.2 MPa. With 3-year-old olive plants of the same cultivar, Tombesi, Proietti and Nottiani (1986) reported a reduction of 50% in A when the available water in the soil was 40% of that at field capacity. If the stomatal closure due to water stress is accompanied by a high light intensity, A is also reduced by photoinhibition. This phenomenon has been observed in the olive by Bongi and Palliotti (1994) and Angelopoulos, Dichio and Xiloyannis (1996), among others. Basically, it consists of an imbalance in the photosynthetic apparatus caused by a lack of CO_2 accompanied by high temperature and light fluence. It has been stated that, in moderately stressed plants, the decline of A after the peak values reached early in the morning is due to the limited CO_2 supply to the chloroplast caused by stomatal closure. In severely stressed plants, however, the reduction in A is also due to the inactivation of photosynthetic activity (Angelopoulos, Dichio and Xiloyannis, 1996; Xiloyannis et al., 1996). This may explain why a certain correlation has been found between g and A for low or moderately stressed olive plants, but for more-severely stressed plants the two variables are no longer correlated (Natali, Bignami and Fusari, 1991; Angelopoulos, Dichio and Xiloyannis, 1996).

The Ψ_l threshold for A seems to be between -4.2 and -6.0 MPa, depending on the stress conditions and plant acclimation (Larcher, Moraes and Bauer 1981; Jorba, Tapia and Sant, 1985; Tombesi, Proietti and Nottiani, 1986). Light saturation for A occurs from 1000-1200 μmol m^{-2} s^{-1} of photosynthetic photon flux density (PPFD) (Baldy, Lhotel and Hanocq, 1985; Natali, Bignami and Fusari, 1991). Fluorescence measurements can be used to evaluate the damage caused by photoinhibition (Bongi, Rocchi and Palliotti, 1994). The same technique was used by Bongi and Lupattelli (1986) to assess the limit of salt tolerance in photosynthesis.

Transpiration and Water Use Efficiency

The olive tree is considered a parsimonious consumer of water. We have already mentioned the marked stomatal control on transpiration, making the maximum stomatal conductance be achieved early in the morning on the days of high atmospheric demand. This does not mean, however, that the maximum transpiration rate is achieved at the same time, as we will see below. Stomatal control, together with the high hydraulic resistance, may be responsible for the transpiration rates in the hottest and driest months of the year being lower than before and after the dry season. This explains why the crop coefficients for the olive tree in the Mediterranean basin are lower in the summer than in spring or autumn. Such behavior was observed in the earliest studies on transpiration. For instance, Abd-El-Rahman, Shalaby and Balegh (1966) cited the works by Evenari and Richter (1937) and Rouschal (1938) in which the limitation of transpiration during the dry summer conditions was outlined. Larsen, Higgins and Al-Wir (1989) compared the transpiration rates of different fruit species. Apple had the highest transpiration rate (100%), followed by peach (57%), grape (39%), apricot (34%), and olive (34%). The water consumption per unit of leaf area of young olive plants in pots has been found to be between $1 \text{ L m}^{-2} \text{ d}^{-1}$ (Cruz-Conde and Fuentes-Cabanas, 1986) and $1.7 \text{ L m}^{-2} \text{ d}^{-1}$ (Natali, Bignami and Fusari, 1991). Water consumption was calculated by weighing the pots. Despite differences between the conditions of the pot experiments and those of mature trees in the field, the amounts are not very different from what we have measured in *La Hampa* with 'Manzanilla' olive trees from 25 to 30 years-of-age and with leaf area ranging from about 55 to 60 m². Some of our measurements have already been reported by Moreno et al. (1996). We calculated transpiration from sap flow velocities measured with the compensation heat-pulse technique (see section *Sap flow measurements*) in the trunk of trees under different water regimes. The maximum daily water consumption we found in a well-irrigated tree was $1.65 \text{ L m}^{-2} \text{ d}^{-1}$, on a day with maximum values of global solar radiation and air vapor pressure deficit of 850 W m^{-2} and 3 kPa, respectively. The highest E values, however, were rarely higher than $1.20 \text{ L m}^{-2} \text{ d}^{-1}$ for well-irrigated trees. Maximum sap flow rates were measured in the trunk between 13.00 and 14.00 GMT, despite the porometer measurements' showing that stomatal closure occurred much earlier in the day, at about 10.00 GMT. Even assuming a certain delay between the sap flow in the trunk and water loss by transpiration, it is clear that the maximum transpiration rates occurred later than the maximum stomatal aperture. We have already mentioned that the stomata do not fully close, but remain partially open. Between 13.00 and 14.00 GMT, there was probably the best balance between the degree of stomatal opening and the environmental conditions for enhancing transpiration. The average maximum value of transpiration rate we have recorded on well-irrigated trees was 3.05×10^{-5} L m^{-2} s^{-1}. There is no agreement among the different authors reporting on the maximum E for olive, probably due to the different cultivars and experimental conditions in which the measurements were made. Jorba, Tapia and Sant (1985) worked with 1-year-old 'Arbequina', 'Manzanilla' and 'Sevillana' olive plants in pots. They reported a maximum E of 8×10^{-5} kg m^{-2} s^{-1}, without mentioning differences between cultivars. Thompson et al. (1983) reported maximum E values of about 2.6×10^{-5} kg m^{-2} s^{-1} for 4-year-old plants of 'Coratina' and 'Nocellara' grown in pots. It has been reported that the transpiration rate per unit of leaf surface may be lower in irrigated than in non-irrigated olive trees, though the total amount of water lost by transpiration is greater in the irrigated trees due to their larger leaf area (Abdel-Rahman and El-Sharkawi, 1974). Xiloyannis et al. (1996) mentioned that a significant part of the water lost by transpiration comes from the water stored in the tissues of the olive tree during the afternoon and night, which ensures a certain level of leaf functionality in drought conditions. As mentioned before, they observed that olive leaves can lose about 60% of the water stored in their tissues under severe water stress ($\Psi_l = -7.0$ MPa). Using the compensation heat-pulse technique in the 'Manzanilla' trees of *La Hampa*, we have been able to record sap flow at night in both the trunk and main roots, accounting for the nocturnal water recovery.

The olive tree uses water more efficiently than other fruit tree species. Bongi and Palliotti (1994) calculated that for the southern Mediterranean area, the number of grams of dry fruit matter per kilogram of water consumed was 3.17 for olive, 2.46 for *Citrus*, and 1.78 for *Prunus*. Xiloyannis et al. (1996) showed the water use efficiency (WUE) values given by different authors for various fruit species. He mentioned WUE values between 5.5 and 9.6 g CO_2 kg H_2O^{-1} for olive, between 3.2 and 4.4 g CO_2 kg H_2O^{-1} for grape, and between 2.3 and 3.5 g CO_2 kg H_2O^{-1} for peach, among other species. Natali, Bignami and Fusari (1991) determined the diurnal time course of WUE for 4-year-old 'Frantoio' olive plants. They found the highest WUE values early in the morning (2.28 g CO_2 kg H_2O^{-1} at 08.30 h), which later decreased (1.43 g CO_2 kg H_2O^{-1} at 18.00 h). They mentioned that the decrease of WUE in the afternoon could be due to photoinhibition and high transpiration rates in the central hours of the day. Bongi and Palliotti (1994) estimated WUE values of 2.16 and 3.48 g CO_2 kg H_2O^{-1} for the cultivars 'Ascolana' and 'Moraiolo', respectively.

Influence of Leaf Aging

During aging, there are various changes in the olive leaf affecting water use by the tree. Leaf thickness increases with age, reducing light transmittance and photosynthetic capacity of the leaves in the inner parts of the

FIGURE 8. Water loss of detached, bench-dried 1-year-old leaves and the current season leaves, taken from 27-year-old 'Manzanilla' olive trees irrigated weekly to cover the crop water demand. A group of 10 leaves was weighed in each case. Fresh weight of the 10 old leaves: 2.1706 g; fresh weight of the 10 young leaves: 1.8190 g. Water content of the 10 old leaves: 1.0256 g; water content of the 10 young leaves: 1.0898 g (adapted from Fernández et al., 1997).

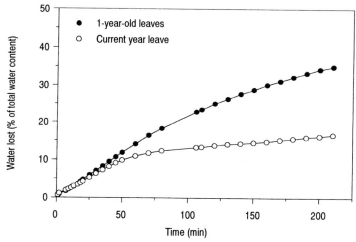

canopy (Bongi and Palliotti, 1994). Tissue elasticity is also reduced with leaf aging. Bongi and Palliotti (1994) calculated that in olive leaves at 87.5% of maximal cell volume, $\in$ was 22.5 MPa in mature leaves and 8.4 MPa in young leaves. The leaf movements and stomatal control already described as mechanisms controlling the water loss in olive are also reduced with leaf aging. Schwabe and Lionakis (1996) observed steeper angles with the stem in very young leaves than in older leaves. Fernández et al. (1997) found that during a water stress period, the stomata remained more open in 1-year-old leaves than in the current year leaves. This loss of stomatal control with aging is illustrated by Figure 8 and by the average values of Ψ_l (-1.11 MPa in young leaves; -1.35 MPa in old leaves), g (5.2 mm s^{-1} in young leaves; 6.3 mm s^{-1} in old leaves) and A (14.57 μmol CO$_2$ m^{-2} s^{-1} in young leaves; 20.66 μmol CO$_2$ m^{-2} s^{-1} in old leaves) measured by Fernández et al. (1997) in leaves of the current year and in 1-year-old leaves. Bongi, Mencuccini and Fontanazza (1987) also reported variations of A with leaf aging, with a increase of A in the first two months, a plateau from month 2 to month 11-13, and a decrease of 50% or more when the leaf is about 2 years old or even older, just before dying.

Recovery After Drought

The Ψ_l of stressed olive trees recovers quickly after rewatering. It takes longer for the values of g and A to recover, the delay being related to the level of water stress previously reached. Natali, Xiloyannis and Angelini (1985) reported the recovery of Ψ_l the day after irrigating stressed 3-year-old plants of 'Moraiolo', 'Leccino' and 'Plantoio'. Fernández, Moreno and Martín-Aranda (1993) observed that Ψ_l of water-stressed 20-year-old 'Manzanillo' olive trees recovered significantly just 12 h after rewatering, reaching similar values to those on control plants in just three days. The day before rewatering, Ψ_{pd} was -0.22 MPa for the control trees and -0.51 MPa for the stressed trees. The minimum values of Ψ_l reached that day were -2.19 MPa for the control trees and -2.83 MPa for the stressed trees. In later studies with more-severely water-stressed 'Manzanilla' trees ($\Psi_{pd} = -0.28$ MPa for control trees and -1.65 MPa for stressed trees; Ψ_l at midday $= -2.38$ MPa for control trees and -3.62 MPa for stressed trees), Fernández et al. (1997) reported very little difference in Ψ_{pd} just two days after rewatering, and no difference at all six days after rewatering. From the second day after rewatering, the formerly stressed plants showed greater values of Ψ_l at midday than the control plants. This phenomenon was also observed by Jorba, Tapia and Sant (1985). They mentioned that this behavior "could be due to the after-effects of stress (Fischer, 1967) perhaps mediated by ABA (Aspinall, 1980) and its control of stomatal conductance and transpiration." A full recovery of g was also observed in just two days after rewatering. It seems that if the trees reach more-severe levels of stress, the recovery of Ψ_l, g and A takes longer. Fereres et al. (1996) studied the recovery of 22-year-old 'Picual' after a period of severe water stress in which minimum Ψ_l values were close to -8.0 MPa. They found that Ψ_l recovered in about four days, but it took several weeks for g to recover. The recovery of the photosynthetic capacity was studied by Angelopoulos, Dichio and Xiloyannis (1996) in 2-year-old 'Coratina' olive plants. Their severely water-stressed plants (-6.5 MPa) had not completely recovered five days after rewatering, but the authors observed that the olive has a strong capacity for repairing the inactivation of the primary photochemistry associated with photosystem II (PSII) after long-term photoinhibition and water stress. The authors stated that this is another feature showing the olive tree's tolerance to drought.

We have already mentioned that our measurements of sap flow in roots indicated a quick water absorption by roots of olive trees after a long period of drought, when the water was finally available in the soil. This explains the quick recovery of Ψ_l mentioned above. Despite this quick response, the tree may not fully recover if the water supply after the drought period is scarce. In our experiments (Fernández et al., 1996, and Moreno et al., 1996), we measured sap velocities at different depths below the cambium in main roots of a

25-year-old 'Manzanilla' olive tree in *La Hampa*, which was without any water supply from the end of May until September 12. Results showed that, despite a significant increase in water uptake, the conductance capacity of the roots did not recover after rewatering, since no flow was detected in the outer sapwood annuli, probably because the vessels remained cavitated. Measurements of Ψ_l and water consumption by the tree also showed a partial recovery, indicating that the irrigation on September 12 was not enough for a full recovery of the tree.

IRRIGATING THE ORCHARD

The olive tree's outstanding adaptation to drought enables it to grow and to produce commercial yields under rainfed conditions in areas where the average rainfall is not much more than 500 mm, and where the dry season can last for five or six months. There are, however, two main reasons for irrigating the olive orchard. On one hand, the plant has a marked response to additional water supplies, even if only small doses of water are applied. On the other hand, in the new intensive orchards, where the maximum crop productivity is pursued, plant densities range from 250 to 400 trees ha^{-1} or more, which means a significant increase in leaf surface per unit of soil surface compared with traditional, rainfed orchards. Under those conditions, rainfall is not enough and irrigation becomes a necessity. Water, however, is scarce and sometimes of low quality in the areas where the olive is cultivated. There is, therefore, an increasing interest in new techniques designed for a more accurate estimation of the irrigation doses. Deficit irrigation and irrigation with both saline and wastewater are also major subjects for current research on water management in olive orchards.

Crop Response to Irrigation

Most of the critical biological processes for growth and production of the olive tree (Figure 1) occur during the dry season in most areas where the olive is cultivated. Water supplied by irrigation minimizes the negative effects of water stress on crop performance, summarized in Table 1. When water stress is present, shoot growth is reduced, though less than root growth. Here the olive shows one of its mechanisms of adaptation to drought–the root/canopy ratio is usually greater in non-irrigated than in irrigated olive trees (Xiloyannis et al. 1996; Celano et al., 1999). This enhances root water uptake under rainfed conditions. Flowering and fruiting are negatively affected by water stress. Lavee (1985) cited the work by Fahmi (1958) showing that an adequate water supply during differentiation is critical for normal inflorescence

TABLE 1. Effects on the growth and production of the olive tree of water stress in different periods of the annual cycle (adapted from Beede and Goldhamer, 1994, and Fereres, 1995).

Phenological event	Period of the year	Effect of water stress
Shoot growth	Mainly from late winter to the beginning of summer, and autumn	Reduced shoot growth
Flower bud development	February to April	Reduced flower formation
Bloom	April to May	Incomplete flower
Fruit set	May to June	Poor fruit set, increased alternate bearing
Fruit growth due to cell division	June to July	Reduced fruit size due to the decreased cell division
Fruit growth due to cell enlargement	August to harvest	Reduced fruit size due to the decreased cell expansion
Oil accumulation	September to harvest	Reduced fruit oil content

development. It has been reported that the occurrence of water stress two or three weeks after fruit set, before pit hardening, reduces pit size (Lavee, 1986). When the stress occurred later, the pericarp/pit ratio was smaller in the fruits from stressed trees than in fruits from the non-stressed ones, though this was due in most cases to reduced pericarp weight, pit size not being affected. Appropriate irrigation management prevents significant water stress and causes marked positive effects on the olive performance. Irrigation increases trunk diameter and shoot growth, as well as the number and size of the fruits (Michelakis 1990; d'Andria et al., 1998; Magliulo et al., 1998). Agabbio, Dettori and Azzena (1983) reported that irrigation increased the pulp/stone ratio and reduced the percentage of dry matter in the fruits. Manrique (1996) observed in 'Picual' olive trees that fruit growth did not stop during the dry season in irrigated trees, resulting in a greater oil production in the irrigated trees than in the non-irrigated ones. Lavee et al. (1990) also observed an increase in total fruit and oil production with irrigation, though the relative oil content based on fresh weight was considerably higher in the fruits of non-irrigated trees than in those with irrigation. Irrigation, together with an adequate orchard management, reduces alternate bearing (see last section). The contrary can be obtained if irrigation is applied in a badly managed orchard, since alternate bearing depends on a variety of factors (Poli 1986a, 1986b; Lavee and Wodner, 1991). Pastor, Castro and Vega (1997) reported that irrigation increases the canopy volume, the number and size of the fruits, and the total oil production. Beltrán, Jiménez and Uceda (1995) and Pastor et al. (1995) reported a decrease with irrigation in the polyphenol content and in the specific absorptions K225, K232, and K270. Patumi et al. (1998) also reported a lower polyphenol content in irrigated trees compared with non-irrigated trees, though this did not affect the organoleptic and storage characteristics of the oil.

Determining Crop Evapotranspiration

The amount of water consumed by the crop is called crop evapotranspiration (ET_c, mm). Dettori (1987) reported ET_c values for the olive crop of 620 and 560 mm for areas of 1200 and 1000 mm of annual potential evapotranspiration (ET_p, mm), respectively. Fereres (1995) and Villalobos et al. (1998) estimated an ET_c of 700-800 mm for olive orchards in southern Spain, where ET_p is about 1400 mm. Goldhamer, Dunai and Ferguson (1993, 1994) made an economic analysis of olive crop revenues following use of increasing irrigation doses in a mature 'Manzanilla' olive orchard in California. They found an increasing response in the crop gross revenue with irrigation up to about 950 mm. It can be concluded that the ET_c value varies for each orchard, depending on environmental conditions, crop characteristics, and orchard management. The optimum value of ET_c must be known for calculating the irrigation requirements (IR, mm). These are given by the equation

$$IR = ET_c - P_e \tag{3}$$

where P_e (mm) is the effective precipitation. This is calculated as a percentage of the total rainfall, normally 70%, though this value can be lower in sloping orchards with restricted infiltration and greater in flat orchards with soils of high infiltration rate. The resulting value of IR can be increased for inefficiencies of the application system and leaching requirements. In the following sections, different approaches for estimating ET_c are described.

The Soil Water Balance Approach

The value of ET_c can be derived from the water balance equation applied at orchard level. The method has a low temporal resolution, a few days at best. It is laborious and time consuming, requiring detailed monitoring of the equation components. In addition, the hydraulic properties of the soil profile, such as the hydraulic conductivity-soil water content relationship, must be accurately determined (Moreno, Vachaud and Martín-Aranda, 1983; Moreno et al., 1998; Vanderlinden, Gabriels and Giráldez, 1998). The spatial variability of these properties has to be taken into account in order to establish to what degree local measurements can be extrapolated to the plot scale. Temporal variability can also be significant. For given climatic conditions and soil type, tillage methods and irrigation practices are the main factors affecting the soil structure of the top layers, and consequently their hydraulic properties (Messing and Jarvis, 1993; Moreno et al., 1997, 1998). Despite the difficulties, studies of soil hydraulic properties are needed for running models simulating water and solute transport in the soil and water use by the crop. Moreno et al. (1988) evaluated the representativeness of soil water content measurements in an olive orchard. They took into account the spatial variability of soil texture and the space-time series of water content measurements in the soil. Caution is needed, however, when estimating ET_c from the water balance equation applied to extensive field areas, because of errors due to the mentioned spatial variability of parameters such as soil water storage and hydraulic conductivity. If such variability is high, aerodynamic and empirical approaches, although based mainly on atmospheric observations, are a better choice for estimating ET_c (Villagra et al., 1995).

Examples on the use of the water balance approach for determining the amount of water used by mature olive trees under different water regimes can be found in the literature. In most cases, the water balance equation was applied for the whole irrigation period, taking into account just the total amount of water supplied by irrigation and the soil water content at the beginning and end of the experimental period (Michelakis and Vougioucalou, 1988; Goldhamer, Dunai and Ferguson, 1993). In other cases, the changes in water storage within the root zone were frequently monitored as a way of determining the root water extraction patterns (Martín-Aranda et al., 1982). At *La Hampa*, Moreno et al. (1988) estimated ET_c from the water balance equation applied to an olive orchard of 16-year-old 'Manzanilla' olive trees at a spacing of 7×7 m^2 and under drip irrigation. The irrigation system consisted of a lateral drip line per row of trees with four 4 L h^{-1} drippers per tree. A deficit irrigation was applied, calculated from total weekly evaporation of a class A pan, adjusted by a reduction coefficient of 0.4. With this reduction coefficient, the trees used during the dry season part of the water stored in the soil at the end of the rainy season. On average, the trees depleted 30% of the total available water in the root zone. The estimation of the water balance components was carried out using a model that separated the soil areas affected by irrigation from the non-affected ones. The model included a weighting method for changes of water storage in the mentioned areas. The affected and non-affected areas were established according to differences in the soil water content profiles at different distances from the tree trunk (Figure 9). The same weighting approach was followed for the drainage calculations. Drainage was estimated by calculating the amount of water lost below

FIGURE 9. Water profiles at different distances from the trunk of an olive tree irrigated by a single lateral drip line placed on the soil surface in each tree row, with four 4 L h^{-1} drippers per tree, 1 m apart. The ground area wetted by each emitter was a circle of about 0.7 m in diameter, and the maximum width of the wetted bulb was about 1.1 m (adapted from Moreno et al., 1988).

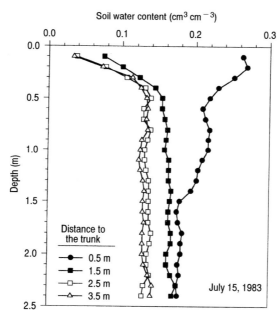

the maximum depth reached by the roots. In the calculations, the hydraulic conductivity-soil water content relationship, the hydraulic gradient, and the changes of water storage in each period between two consecutive measurements were taken into account. Results from the hydrological year 1982-83 are shown in Figure 10. The values of the water balance components in the orchard for six consecutive years, determined by Moreno et al. (1988) and Fernández (1989), are shown in Table 2. The greatest water losses by drainage were recorded in years with high rainfall and low ET_p. In years with heavy rainfall, greater drainage was recorded during the first part of the irrigation period, due to the high soil hydraulic conductivity below the root zone (Moreno, Vachaud and Martín-Aranda, 1983). Water losses by runoff (R, mm) were significant only in periods of heavy rainfall. Evaporative demand was low in those periods, allowing R to be calculated from the water balance equation after assuming $ET_c = ET_p$. The values of R obtained by Moreno et al. (1988) using this indirect approach agreed with previous results

FIGURE 10. Cumulative value of the water balance components during the hydrological year 1982-1983, calculated for an olive orchard with 12-year-old 'Manzanilla' olive trees at 7 × 7 m² spacing. The trees were irrigated daily during the dry season, from April to September, with a drip-irrigation system. A deficit irrigation approach was followed, designed for the trees to use 30% of the available water in the soil. The unknown component in the equation was ET_c. P = precipitation, I = irrigation; ΔS = variation of the water stored in the soil; D = drainage; ET_c = crop evapotranspiration (adapted from Moreno et al., 1988).

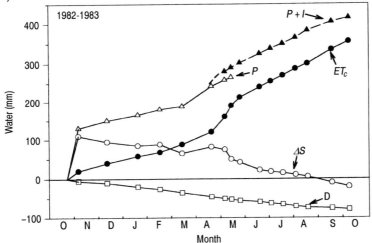

obtained experimentally by Moreno (1986). The increasing ET_c throughout the years was due to the increase in tree size. Table 2 also shows the value of the water balance components for the hydrological year 1997-98, in which the orchard was irrigated with enough water to cover the crop demand. This was a year of heavy rainfall (the average calculated for the period 1971-95 in the area is 500 mm), with rains of high intensity in winter, which caused a large runoff.

The total value of ET_c determined for an olive orchard can be split into its two components, provided soil evaporation (E_s, mm) or tree transpiration (E_t, mm) is determined separately. Moreno, Vachaud and Martín-Aranda (1983) estimated E_s in a drip-irrigated olive orchard at *La Hampa* as a function of the water content of the top soil layers. Recently, Palomo et al. (1998a,b) conducted some experiments in the same orchard to validate the results obtained by Moreno, Vachaud and Martín-Aranda (1983), and obtained an empirical equation for estimating E_s for the orchard conditions of *La Hampa* (Figure 11).

TABLE 2. Annual values of the components of the water balance equation determined in an olive orchard at *La Hampa* when the trees were irrigated (1) for six consecutive years with a deficit irrigation designed to use up to 30% of the available water content in the soil, and (2) for one year in which irrigation was enough to replace the crop water demand. Crop evapotranspiration was the unknown component of the equation. P = precipitation; I = irrigation; ΔS = difference between the soil water stored in the soil at the beginning and at the end of the experimental period; D = drainage; R = water runoff; ET_c = crop evapotranspiration; ET_p = potential evapotranspiration calculated by the FAO-Penman equation (adapted from Moreno et al., 1988, Fernández, 1989, and unpublished data).

Irrigation mode	Hydrological year	P (mm)	I (mm)	ΔS (mm)	D (mm)	R (mm)	ET_c (mm)	ET_p (mm)
(1)	1982-1983	258	133	− 20	73	0	338	1486
	1983-1984	542	65	+45	157	82	323	1239
	1984-1985	584	87	+2	238	69	362	1381
	1985-1986	438	101	− 37	87	95	394	1410
	1986-1987	544	95	+35	61	77	466	1492
	1987-1988	707	87	− 34	167	225	436	1570
(2)	1997-1998	717	387	1	118	345	640	1406

Evaporation Pans

Evaporation pans are cheap and easy to use, and give reasonably good estimates of ET_c provided siting and management are correct. The most common type is the class A pan of the U.S. Weather Service. The standard installation and management are described in Doorenbos and Pruitt (1977). The evaporation of a reference crop (ET_r, mm) is obtained by multiplying the pan evaporation (E_{pan}, mm) by a pan coefficient (K_p). The ET_r is the ET_p of an identifiable crop, such as alfalfa or grass, covering the whole ground surface and well-supplied with water. The ET_c for a particular crop can be calculated by multiplying ET_r by the corresponding crop coefficient K_c. This is the approach followed by Milella and Dettori (1987) and Dettori (1987) for olive. They determined the crop water requirements in a young orchard of 'Olia manna' olive plants in Sardinia, concluding that the optimum K_c was 0.5 from May to September, and 0.55 in April and October. They justified the use of greater K_c, of about 0.6, only in the case of the biggest fruits having a high commercial value. Evaporation pans have been used sometimes in a different way than that described by Doorenbos and Pruitt (1977). Michelakis, Vouyoukalou and Clapaki (1996), for instance, estimated ET_c by multi-

FIGURE 11. Relationship between the evaporation from the soil surface of an olive orchard and the volumetric soil water content of the top 80 mm soil layer. The experiments were conducted in an olive orchard at *La Hampa* on typically hot, dry summer days (average ET_p = 8 mm d^{-1}). The orchard was planted with manure 'Manzanilla' olive trees at 7 × 7 m². Data from two experiments were considered. In experiment 1, evaporation from the soil surface was empirically determined with microlysimeters. In experiment 2, data obtained during the hydraulic characterization of the orchard soil by Moreno et al. (1988) were taken into account (adapted from Palomo et al., 1998).

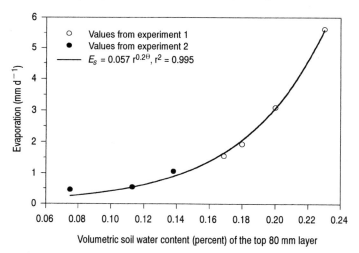

Volumetric soil water content (percent) of the top 80 mm layer

plying E_{pan} by a coefficient Kc that they called crop coefficient, which must not be confused with the standard definition of K_c. They established different irrigation regimes for 12-year-old 'Kalamon' olive trees in Crete by using different irrigation systems and water doses for keeping the soil water potential around the roots in the ranges from -0.02 to -0.06 MPa (treatment ψ0.2) and from -1.0 to -1.5 MPa (treatment ψ15). They found that Kc increased from May to September from 0.4 to 0.65 for the ψ0.2 treatment, that it was about 0.3-0.4 for the ψ15 treatment, and that it decreased from 0.2 to 0.05 for the non-irrigated treatment. Another example of a simplified use of the evaporation pan is that of Moreno et al. (1988) and Fernández (1989). They wanted the olive trees at *La Hampa* to use part of the available water in the soil. They calculated the irrigation doses from weekly E_{pan} measurements adjusted by a reduction coefficient of 0.4 and referred to the projected area of the canopy, which was 12 m². This way of using the evaporation pan is easier than the standard procedure, and it can be recommended if the applied reduction coefficient has been previously validated as appropriate for the purpose.

It is difficult, however, to extrapolate the same reduction coefficient to orchards with different crop characteristics and environmental conditions. A rough estimate of ET_p can be obtained by dividing E_{pan} by 1.24 (Beede and Goldhamer, 1994).

Combination Methods

When climatic data recorded close to the orchard are available, ET_r can be calculated by combination methods. These are described elsewhere (Jensen, Burman and Allen, 1990; Burman and Pochop, 1994). They enable accurate estimates of ET_r provided the combination equation used has previously been validated as appropriate for the area. Any other combination equation can lead to significant errors in the estimation of ET_r. In *La Hampa*, for instance, we use the FAO-Penman equation, which Mantovani (1994) evaluated as the best for the area. The mean annual ET_r calculated by this equation for the period 1987-96 was 1499 mm. However, the value of ET_r calculated by the Kimberly Penman equation for the same period was only 1185 mm (79% of the FAO-Penman ET_r). In the case of olive orchards, ET_c can be calculated by multiplying ET_r by a crop coefficient K_c accounting for crop differences between the olive and the reference crop, and by a reduction coefficient K_r accounting for the percentage of ground surface covered by the crop:

$$ET_c = K_r\, K_c\, ET_r. \tag{4}$$

One of the main disadvantages of this approach is that information on K_r and K_c is scarce. The value of K_r for the olive tree is unknown. K_r values obtained for fruit tree species living in similar conditions to those of the olive tree are used, such as the K_r values obtained by Fereres et al. (1981) for almond. Monthly values of K_c have been determined for the conditions of southern Spain (Table 3). Pastor and Orgaz (1994) determined K_c for the conditions of Córdoba, assuming negligible soil evaporation. Orgaz and Fereres (1997) took evaporation into account, which explains their higher K_c values. Fernández et al. (1998b) experimented in *La Hampa* with the K_c values given by Pastor and Orgaz (1994) for the irrigation season, from March to October. They reported that the values must be increased by 0.05 for getting good results. The olive orchard in which Fernández et al. (1998b) validated K_c was planted with mature trees covering 34% of the ground surface, irrigated daily by a drip line lateral placed on the soil surface in each tree row, with five 3 L h^{-1} drippers per tree 1 m apart. Using the increased K_c values, it was found that the soil matric potential in the wet bulb was greater than -0.02 MPa throughout the irrigation season, and water losses by drainage were 13% of the total amount of water supplied by irrigation. Values of

TABLE 3. Crop coefficients (K_c) obtained for olive orchards in south Spain.

Month	Pastor and Orgaz (1994)	Orgaz and Fereres (1997)
January	0.50	0.65
February	0.50	0.65
March	0.65	0.65
April	0.60	0.60
May	0.55	0.55
June	0.50	0.55
July	0.45	0.50
August	0.45	0.50
September	0.55	0.55
October	0.60	0.60
November	0.65	0.65
December	0.50	0.65

K_c shown in Table 3, and those reported by Fernández et al. (1998b), are applicable to orchards of the same characteristics and conditions as those for which they have been validated. Apart from the lack of information on K_c and K_r, another disadvantage of the method is that no less than a week of weather records is needed for an accurate estimate of ET_r. Then, irrigation doses for the week are calculated based on the ET_r of the previous week. This lack of temporal accuracy can lead to either excess or lack of water in the soil, especially if the latter has a low water retention capacity.

This approach is, however, perhaps the most widely used nowadays in the olive orchards where irrigation is controlled. It was the one used by Pastor and Orgaz (1994) in a 'Picual' olive orchard in Córdoba, and by Fernández et al. (1997, 1998b) at *La Hampa*. Table 4 shows the average values of *IR* calculated by Fernández et al. (1998b) in *La Hampa* for an olive orchard of 286 mature trees ha^{-1}, with a maximum LAI of 1.7 and 34% of ground cover. Following the approach described here, the calculated *IR* for satisfying the crop water demand was about 3,800 m^3 ha^{-1} yr^{-1}. Some authors claim that a localized irrigation system must wet a minimum of 25-30% of the corresponding ground surface of each tree. We believe, however, that the volume of soil affected by irrigation is more critical than the wetted ground surface. This is supported by the root signaling phenomena occurring in the olive tree, already described. With the irrigation system of Fernández et al. (1997, 1998b), consisting of five 3 L h^{-1} drippers per tree 1 m apart, the average ground surface wetted by the drippers was 3.9 m^2 per tree, and the volume of wetted soil was 28% of the total soil volume corresponding to each tree.

TABLE 4. Daily potential evapotranspiration (ET_p) and effective precipitation (P_e) calculated for every week of the irrigation season of an olive orchard at the experimental farm *La Hampa*. Data are the average of a 25-year-data set measured from 1971 to 1995 at the weather station of the farm. Details of the values of the crop coefficient (K_c) and the reduction coefficient (K_r) are given in the text. These have been used for calculating the crop evapotranspiration (ET_c) and the irrigation requirements (IR), using Equations 3 and 4 mentioned in the text. Data are from a mature olive orchard with mature trees planted at 7×5 m^2, covering 34% of the ground surface, and drip irrigated to satisfy the crop water demand (adapted from Fernández et al. 1998b).

Month	Week	K_r	K_c	ET_p (mm d^{-1})	ET_c (mm d^{-1})	P_e (mm d^{-1})	IR (mm d^{-1})
March	1	0.70	0.70	2.5	1.22	0.76	0.46
	2	"	"	2.9	1.42	1.00	0.42
	3	"	"	3.2	1.57	0.81	0.76
	4	"	"	3.7	1.81	0.88	0.93
April	1	"	0.65	3.7	1.68	1.02	0.66
	2	"	"	4.2	1.91	0.77	1.14
	3	"	"	4.3	1.96	0.79	1.17
	4	"	"	4.5	2.05	1.20	0.85
May	1	"	0.60	5.2	2.18	1.23	0.95
	2	"	"	5.6	2.35	0.50	1.85
	3	"	"	5.6	2.35	0.43	1.92
	4	"	"	5.8	2.44	0.22	2.22
June	1	"	0.55	6.1	2.35	0.22	2.13
	2	"	"	6.2	2.39	0.39	2.00
	3	"	"	6.4	2.46	0.50	1.96
	4	"	"	6.5	2.50	0.00	2.50
July	1	"	0.50	6.9	2.42	0.00	2.42
	2	"	"	7.2	2.52	0.00	2.52
	3	"	"	7.4	2.59	0.00	2.59
	4	"	"	7.2	2.52	0.00	2.52
August	1	"	0.50	7.1	2.49	0.00	2.49
	2	"	"	6.8	2.38	0.00	2.38
	3	"	"	6.5	2.27	0.00	2.27
	4	"	"	6.0	2.10	0.00	2.10
September	1	"	0.60	5.6	2.35	0.00	2.35
	2	"	"	5.1	2.14	0.00	2.14
	3	"	"	4.6	1.93	0.00	1.93
	4	"	"	4.0	1.68	0.53	1.15
October	1	"	0.65	3.6	1.64	0.78	0.86
	2	"	"	2.7	1.23	1.76	–
	3	"	"	2.5	1.14	1.35	–
	4	"	"	2.2	1.00	0.85	–

Eddy Covariance

The technique of eddy covariance, also called eddy flux or eddy correlation, was first described by Swinbank (1951), though commercial instrumentation allowing the application of this method for calculating ET_c is only recently available. The technique is described in detail by Rosenberg, Blad and Verma (1983), among others. The method is based on continuous measurement of wind speed and air vapor pressure by complex instrumentation able to measure those variables in short periods of time. Evapotranspiration is calculated with a high temporal resolution, but the instrumentation is complex and requires well-trained people. In addition, the technique behaves well only in large, flat areas. The only work we know in which this technique has been used to estimate the olive ET_c is that of Villalobos et al. (1998). They used the technique in an olive orchard of LAI = 1.4 and 40% ground cover, measuring above the trees to estimate ET_c, and below the trees to estimate E_s. No details on the performance of the technique were given.

Sap Flow Measurements

The short-term water-use dynamics of the olive tree can be evaluated from sap flow measurements in the trunk, branches, or roots. This sort of information is essential for a better control of the high frequency irrigation systems normally used in olive orchards. Detailed descriptions of the existing methods for estimating sap flow and their applicability to different species and purposes were given by Swanson (1994), Smith and Allen (1996) and Braun (1997). All these methods use heat as a tracer for sap movement. They can be grouped into heat-balance and heat-pulse methods. In the heat-balance methods, the conductive organ is heated electrically by an attached gauge, and the mass flow is calculated as a function of the heat taken up by the moving sap stream. In the heat-pulse methods, heater and temperature probes are located in the sapwood of the conductive organ, and the mass flow is calculated from the travelling speed of heat pulses sent every certain time by the heater probe. All methods measure transpiration fluxes in real time, enabling the highest time resolution in the determination of E_t.

Since 1994, we have been using at *La Hampa* the compensation heat-pulse (CHP) technique described by Green and Clothier (1988). The papers by Moreno et al. (1996) and Fernández et al. (1996) are the first published work we are aware of in which sap flows were measured in olive. They measured sap flow in main branches and roots of 25-year-old 'Manzanilla' olive trees under different water regimes. A good agreement was found between the transpiration determined by the CHP technique and that predicted by the Penman-Monteith equation (Figure 12). Major features of the hydraulic behavior of the olive tree during recovery from water stress were also identified from the sap flow data. For instance, an immediate water uptake by the roots

FIGURE 12. The calculated and predicted transpiration from one of the two main branches of a 25-year-old 'Manzanilla' olive tree in *La Hampa*. Transpiration was calculated from sap flow measurements made by the compensation heat-pulse technique. The Penman-Monteith expression was used for estimating transpiration. The tree was pond-irrigated once a week. The last irrigation before the measurements was on day of the year 252. Shown here are the terms of the Penman-Monteith expression; namely the radiation term, the vapor pressure deficit driven term associated with the lit leaves, and this term for the shaded leaves (adapted from Moreno et al., 1996).

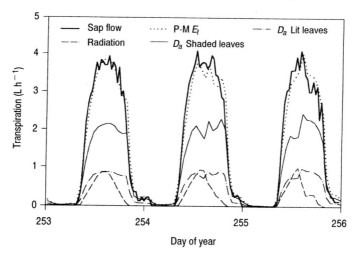

once water was finally available in the soil was observed, though recovery was not completed due to the presence of air emboli in xylem vessels of the sapwood's outer annuli. We have calibrated the CHP technique for olive to obtain accurate estimates of the volume of sap flow from the measured sap velocities. Preliminary results on calibration were published by Fernández et al. (1997), and a more detailed paper is in preparation. Daily values of E_t for two 28-year-old 'Manzanilla' olive trees, one irrigated and the other without irrigation, were determined in *La Hampa* by Díaz-Espejo et al. (1998). They found clear differences in water consumption between treatments. In addition, they monitored Ψ_l, g and A in the same trees in which sap flows were measured, to study the relations between those measurements at leaf level and the amount of water consumed by the tree. Fernández et al. (1998b) and Palomo et al. (1998a,b) evaluated the applicability of the CHP technique for estimating ET_c in an olive orchard. As mentioned before, they obtained an empirical equation for estimating E_s in the olive orchard of *La Hampa* (Figure 11). Fernández et al. (1998a) reviewed the performance of the CHP technique, outlining its advantages and limitations. They concluded that the CHP technique is reliable and of low maintenance, suitable for use in commercial olive orchards for long periods of time. It provides information on the dynamics of both water uptake by roots (Figure 3) and tree transpiration (Figure 13), and is useful for determining the effect of meteorological conditions and soil water status on both processes. The capability of the technique for estimating E_t is limited, however, by the considerable heterogeneity of the conductive area in mature olive trees, which may cause a high variability in the sap flow measurements, depending on probe location. We are currently using the CHP technique in further analysis of the hydraulic behavior of the olive tree to obtain information for the diagnosis of the onset or severity of tree water stress. The main hydraulic features we are using for that purpose are the delay in water uptake by roots observed early in the morning in stressed trees (root S versus root N in Figure 3), and the occurrence of cavitation in the xylem vessels of the sapwood's external annuli (Figure 14). Apart from our work with the CHP technique, the only published work we have found related to the measurement of sap flows in olive is that of Dichio et al. (1998). They used a heat-balance method to determine E_t in 2-year-old 'Coratina' olive plants grown in containers, and compared the results with the data obtained by weighting. They reported a good agreement between the two ways of estimating E_t, with maximum differences of ± 8%.

FIGURE 13. Daily transpiration fluxes determined from the sap flow measured with the compensation heat-pulse technique in the trunk of a 29-year-old 'Manzanilla' olive tree at *La Hampa*, and vapor pressure deficit of the air measured on the same days (adapted from Fernández et al. 1998a).

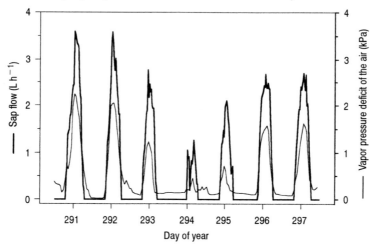

FIGURE 14. The diurnal rise and fall in the radial pattern of sap velocity by the compensation heat-pulse technique in the trunk of (a) a well-watered 29-year-old 'Manzanilla' olive tree at *La Hampa* on 6 August 1997. The FAO-Penman calculation of water use for a well-watered short crop for this day was 6.1 mm. The tree received daily irrigation, and the surface soil water content (0-1 m) around the tree was on average 0.27 $cm^3\,cm^{-3}$; and (b) a water-stressed olive tree of the same orchard, on 15 August 1997, in which the FAO-Penman evaporation was 7.7 mm. The surface soil water content (0-1 m) was below 0.14 $cm^3\,cm^{-3}$. This lower value of water use on a day of higher FAO-Penman evaporation indicates the severity of the water stress.

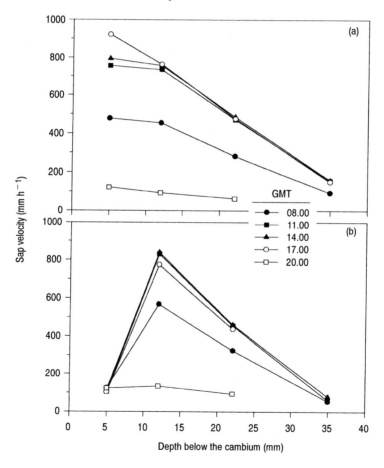

Modeling

Very few modeling exercises for assessing water consumption by the olive tree have been published. This may be due to the difficulty in accurately estimating LAI, one of the variables to which the models are more sensitive. Bongi and Palliotti (1994) mentioned the CRIS model for estimating the irrigation requirements of olive orchards. They reported an estimated annual consumption of 603 mm. They outlined the low ET_c estimated for olive, as compared with that estimated by the model for peach (1088 mm yr^{-1}) and *Citrus* (904 mm yr^{-1}). Moreno et al. (1996) calculated, from sap flow measurements, the amount of water transpired by the leaves of one of the two main branches of a 25-year-old 'Manzanilla' olive tree at *La Hampa*, and compared the results with the transpiration estimated by the Penman-Monteith equation. The values of leaf stomatal conductance and leaf-canopy boundary-layer conductance given to the model were determined from measurements in the orchard. The authors found a good agreement between the calculated and the simulated transpiration amounts (Figure 12). They used the model to estimate separately the three terms of the Penman-Monteith expression, namely the radiation term, the vapor pressure deficit driven term associated with the lit leaves, and this term for the shaded leaves. Recently, Villalobos et al. (1998) fitted an empirical model to estimate canopy conductance in two irrigated olive orchards in Córdoba, southern Spain. They were able to estimate ET_c in the orchards by combining the model with that of Ritchie (1972) for estimating E_s. They estimated average annual values of ET_c, E_s and E_t of 812, 291 and 521 mm, respectively, for an intensive orchard with LAI 1.4 and ground cover 40%. For a less intensive orchard with the same LAI and ground cover of 30%, the estimated values were 748 mm for ET_c, 293 mm for E_s and 454 mm for E_t.

IRRIGATION STRATEGIES WHEN WATER IS SCARCE

The amount of water supplied by deficit irrigation (DI) is lower than the crop requirements for an optimum development. Despite not allowing maximum crop performance to be achieved, deficit-irrigation strategies are interesting for olive, since water is scarce in many areas where the species is cultivated. Of the different DI modalities described in the literature, one of the most widely used is controlled deficit irrigation (CDI), or regulated deficit irrigation (RDI), in which the water supply is reduced or even interrupted except for the stages in which the crop is most sensitive to water stress. There are references in the literature to supplementary (Abdel-Rahman and El-Sharkawi, 1974) and complementary (Lavee et al., 1990) irrigation, in which water is supplied by irrigation a limited number of times throughout the crop season. An extreme case of this sort of DI is that reported by Pastor and Orgaz (1994) for an orchard with deep soils of high retention capacity, in which water was applied just once in winter, when water for irrigation was available. In a recent paper, Goldhamer (1997) studied the influence of RDI on the performance of mature 'Manzanilla' olive trees. He had a control treatment with 900 mm and three RDI treatments, with reductions of 12.9, 20.8 and 39.5% from the control. Fruit size was smaller in the most-severe RDI treatment than in the control. No significant differences were found in flesh oil content on a fresh or dry weight basis in the fruits harvested in January for oil. The author concludes that RDI saving about 20% (185 mm) of ET_c may be enough to maintain top yields and quality. Michelakis (1990) and d'Andria et al. (1998) observed no differences in crop production between trees irrigated with enough water for replacing control ET_c and those with doses 30 to 50% lower. The effect of DI on the leaf water status and gas exchange of mature 'Manzanilla' olive trees was studied in *La Hampa* by Fernández et al. (1997). They had a control treatment with no irrigation and two irrigation treatments, one with weekly irrigation to replace ET_c for maximum production (I treatment) and the other with 1/3 of this (I/3 treatment). Values of Ψ_l and g of the I/3 trees were quite close to those measured in the I trees, the differences with the control trees being much greater. The success of DI in the olive seems to be based, therefore, on the marked crop response to reduced water supplies. A correct DI strategy requires knowing the biological stages in which the olive is most sensitive to water stress. For 'Manzanilla', complementary irrigations must be applied just before flowering, at pit hardening, and a fortnight before harvesting. Alegre and Girona (1997), however, reported that fruits affected by water stress during pit hardening recovered with later water supplies. The recommended DI strategy may be different for table and oil cultivars. Alegre and Girona (1997) cited a paper by Romero et al. (1997) in which it was reported that oil quality was better in 'Arbequina' olive trees irrigated with DI than in those irrigated with enough water to meet the crop water requirements.

As mentioned before, we have used in some olive orchards at *La Hampa* a deficit irrigation designed for the trees to use part of the water stored in the soil at the end of the rainy season (Moreno et al., 1988; Fernández 1989). The use of this approach in an olive orchard has been reported by Pastor and Orgaz (1994), among others. It allows a certain depletion of soil moisture in the root zone, normally established as a percentage of the available water in the soil calculated as the difference between the water in the soil at field capacity and that at wilting point. The value of such percentage can be calculated by balancing the reduction of marketable product with the increased cost of a more generous irrigation. A difficulty in establishing the allowed depletion amount is the determination of what is the soil water content for wilting point in the olive tree. We have already mentioned that it can be significantly lower than for other fruit trees growing in Mediterranean areas. The available water in the olive orchards of *La Hampa* is about 3,000 m^3 ha^{-1}. An allowable water depletion of 25%, recommended by Pastor and Orgaz (1994), means 750 m^3 ha^{-1}. The average ET_c for maximum crop performance in the area is about 3,800 m^3 ha^{-1} (Table 4). Therefore, with this DI approach, about 20% of irrigation water can be saved.

IRRIGATING WITH LOW QUALITY WATER

The lack of water in the areas where the olive is grown and the increase in recent years of the area under cultivation enhance the use of low quality waters for irrigating olive orchards. This is favored by the moderate-to-high tolerance of the olive tree to the presence of salts in the soil (Troncoso et al., 1983; Tattini et al., 1995). Even municipal wastewater (Saavedra, Troncoso and Arambarri, 1984) and dilute sea-water (El-Gazzar, El-Azab and Sheata, 1979; Bongi and Loreto, 1989) have been considered for irrigating the olive tree. The effects of salinity on olive performance have been reported by many authors. It is not our aim to detail them here; they have been summarized by Fernández (1997) and Gucci and Tattini (1997). Regarding water for irrigation, reduction in crop performance has been reported when water with electrical conductivity (EC) greater than 5.5 dS m^{-1} was used (Freeman, Uriu and Hartmann, 1994). The limit of salt content in irrigation water for the olive tree has been established as 8 g L^{-1} of solid residue (Zarrouk and Cherif, 1981). The sodium adsorption ratio (SAR) should not exceed 9 for maximum production (Freeman, Uriu and Hartmann, 1994). Mature olive trees tolerate higher SAR than young trees. Al-Saket and Aesheh (1987) reported young olive trees as tolerating an SAR of 18 in the growing medium. Loreti and Natali (1981) reported mature trees as tolerating irrigation with SAR less than 26. Leaching requirements for avoiding salt accumulation in the soil depend on the salt concentration in the irrigation water, on the rainfall regime and on the irrigation system. For water containing 0.014, 0.028 and 0.069 mol L^{-1} salt, leaching requirements have been calculated as 15, 30 and 70% of volume applied, respectively (Ayers and Westcot, 1985; FAO, 1993). These percentages may be lower if high-frequency irrigation is used, and in areas where rainfall in the wet season is enough for washing out the salts accumulated during the irrigation season.

The use of municipal wastewater for irrigating the olive tree is an interesting option. The volume of municipal wastewater treated as a result of regulations continues to increase, making enough available for the irrigation of many hectares. The olive seems to behave well when irrigated with municipal wastewater. Saavedra, Troncoso and Arambarri (1984) grew 1-year-old

'Manzanilla' plants in pots, irrigating them for eight months with untreated wastewater. No toxicity symptoms or any other anomaly was observed. For health reasons, however, municipal wastewater must be treated before it can be used for irrigation. Different countries have different quality standards for wastewater used for irrigation, which have to be taken into account.

A significant amount of wastewater is produced every year by the olive industry, from both olive mills and factories processing olives for table consumption. Part of this water is used for irrigating olive orchards. Most of the existing literature refers to the use of olive mill wastewater (OMW). Generally, OMW contains 83-94% of water, 4-16% of organic matter, and 0.4-2.5% of mineral salts (Ramos-Cormenzana, 1986), though the chemical composition varies with the oil extraction procedure, the cultivar, and the degree of ripeness, among other factors. OMW has a high potential for causing pollution, but up to 800 m^3 ha^{-1} can be used for direct irrigation of soil (Fiestas Ros de Ursinos, 1986; García-Ortíz et al., 1993). The current production of OMW in Mediterranean countries is estimated to be around 10-12 $\times$ 10^6 m^3 yr^{-1}, though this amount has tended to decrease with new extraction methods (Cabrera et al., 1996). The high content in organic matter may be a problem when using certain irrigation systems. However, OMW can even be used for drip irrigation if it is taken from the ponds where it is usually stored, diluted with fresh water, and minimum precautions are taken to prevent dripper blockage.

We have not found any published work on irrigation of olive orchards with table olive industry wastewater (TOW). We are currently involved in a research project designed to evaluate the use of TOW for irrigating olive orchards. Some preliminary results on the effects on leaf water relations and photosynthesis of mature olive trees irrigated with TOW are shown in Figure 15. An immediate decrease in the values of Ψ_l, g and A was observed in the trees irrigated with TOW as compared with those irrigated with fresh water. A reduction in yield of about 30% was also recorded. EC values of the TOW used were between 4 and 6 dS m^{-1}, which is about the limit for a reduction in olive performance, and the sodium adsorption ratio was between 70 and 80, very much over the permitted limit. We observed a high temporal variability in the TOW composition, even when the water came from the same factory, which is not uncommon (Kopsidas, 1994). Though we analyzed TOW regularly, such variability could cause peak values of EC and SAR–or any other water characteristic affecting the results–to be undetected.

MANAGING IRRIGATED ORCHARDS

It is clear that irrigation is a required practice in intensive orchards where maximum productivity is the aim. The positive effects of irrigation are not

FIGURE 15. Leaf water potential, stomatal conductance, and photosynthesis rate measured in July and September in leaves of 20-year-old 'Manzanilla' olive trees near Morón de la Frontera (Seville, Spain). The trees were drip irrigated, some of them with good quality water and some with waste water from the table olive industry with electrical conductivity values from 4 to 6 dS m^{-1} and sodium adsorption ratio between 70 and 80. The irrigation season lasted from the beginning of July to mid-September. Two measurement sets per day were taken on July 30 and September 10, one in the morning and the other in the afternoon. Each point represents the mean of 10 values per treatment. Vertical bars indicate twice the standard error. The arrow in the more negative value of leaf water potential means that we reached the maximum pressure in our pressure pump (4 MPa) without reaching the equilibrium point.

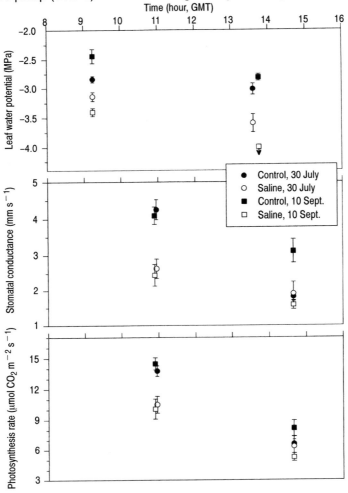

going to be fully achieved, however, if fertilization, pruning, or any other agricultural practice in the orchard is carried out incorrectly. Recommended practices for a correct design and management of intensive olive orchards are summarized below.

The cultivar must be well adapted to the environmental conditions in the orchard and must produce high quality products, either fruit for table consumption or oil. Plants of about 1 year old from self-rooted cuttings raised in mist houses are recommendable for propagation. More than one cultivar with concurrent anthesis must be planted in the orchard for increasing yield by cross-pollination. Plant density must range from 250 to 400 trees ha^{-1}, depending on vigor and plant formation. Current experiments with higher plant densities are giving good results in the first years of the orchard life, but they have not been long enough for definitive conclusions. Tree rows must follow the east-west direction. Trees must have a single trunk with main branches at a minimum height of 0.8-1 m above the ground surface, to allow for machine-harvesting. Canopy shape must favor leaf illumination. Reduced tillage is recommendable, especially for orchards in sloping areas. This not only reduces erosion, but also, combined with a correct weed management by the use of herbicides, reduces runoff and increases water stored in the soil. Pruning must be done every year, when growth has ceased. The height of the trees must be controlled by pruning, especially for the table cultivars, which cannot be machine-harvested. Mechanical pruning is an option. The orchard must be fertirrigated. A high-frequency localized irrigation system is recommendable, perhaps with one pipe per row of trees and 4-6 emitters per tree. Drippers are preferred to microjets in most cases. If enough water is available for irrigation, the volume of soil wetted by the irrigation system must be as great as possible, to avoid restrictions in crop performance due to root signaling. Fertilizers can be added to the irrigation water once or twice per week throughout the irrigation season. Programmable devices can be used for controlling fertirrigation. The nutrient status of the trees can be monitored by leaf analysis every 1-2 years in mid-summer. Soil analysis is advisable only at the beginning of planting, except in cases of difficult soils. Pest control must be carried out always under technical advice. The use of models for the evaluation of pest populations and disease conditions is helpful. The incidence of diseases associated with a high plant density (*Cycloconium oleaginum*) or with high soil water content (*Verticillium dahliae*) must be especially considered.

REFERENCES

Abd-El-Rahman, A.A., A.F. Shalaby and M.S. Balegh. (1966). Water economy of olive under desert conditions. *Flora, Abt. B, Bd.* 156: 202-219.

Abdel-Rahman, A.A. and H.M. El-Sharkawi. (1974). Response of olive and almonds orchards to partial irrigation under dry-farming practices in semi-arid regions: II.

Plant-soil water relations in olive during the growing season. *Plant and Soil* 41: 13-31.

Agabbio, M., S. Dettori and M. Azzena. (1983). Primi resultati sulle variazioni giornaliere e stagionali del potenziale idrico fogliare nella cultivar d'olivo "Ascolana tenera" sottoposta a differenti regimi idrici. *Rivista della Ortoflorofrutticoltura Italiana* 67: 317-328.

Alegre, S. and J. Girona. (1997). Riego deficitario controlado en olivo. *Fruticultura profesional* 88: 70-78.

Al-Saket, I.A. and I.A. Aesheh. (1987). Effect of saline water on growth of young olive trees. *Dirasat* 14: 7-17.

Angelopoulos K., B. Dichio and C. Xiloyannis. (1996). Inhibition of photosynthesis in olive trees (*Olea europaea* L.) during water stress and rewatering. *Journal of Experimental Botany* 47(301): 1093-1100.

Arambarri, P. and L. Madrid. (1974). Efecto de la localización del fertilizante fosfatado sobre su asimilación por olivos. *Anales de Edafología y Agrobiología* 33(5-6): 467-476.

Aspinall, D. (1980). Role of abcisic acid and other hormones in adaptation to water stress. In *Adaptation of Plants to Water and High Temperature Stress*. eds. N.C. Turner and P.J. Kramer, John Wiley & Sons, New York, pp. 155-172.

Ayers, R.S. and D.W. Westcot. (1985). Water quality for agriculture. In *FAO Irrigation and Drainage Paper 29*, Rev. 1, Rome.

Badr, S.A., and H.T. Hartmann. (1971). Effect of diurnally fluctuating vs. constant temperatures on flower induction and sex expression in the olive (*Olea europaea*). *Physiologia Plantarum* 24: 40-45.

Baldy Ch., J.C. Lhotel and J.F. Hanocq. (1985). Efectos de la radiación solar en la función fotosintética del olivo (*Olea europea* L.). *Olivae* 8: 18-23.

Barranco, D., D. Fernández-Escobar and L. Rallo. (1998). *El Cultivo del Olivo*. Junta de Andalucía y Grupo Mundi Prensa.

Bartolini G., G. Roselli and G. Di Milla. (1979). Relazione tra densità stomatica e vigoria dell'olivo. *Rivista della Ortoflorofrutticoltura Italiana* 63: 391-397.

Beede, R.H. and D.A. Goldhamer. (1994). Olive irrigation management. In *Olive Production Manual*. University of California, Publication 3353, pp. 61-68.

Beltrán, G., A. Jiménez and M. Uceda. (1995). Efecto del régimen hídrico del cultivo sobre la fracción fenólica del aceite de oliva de la variedad 'Arbequina'. Proceedings of the *I Simposi de l'Olivera Arbequina a Catalunya*. Borges Blanques (Lleida), pp. 153-155.

Böhm, W. (1979). *Methods of Studying Root Systems*. Springer-Verlag.

Bongi, G. and F. Loreto. (1989). Gas-exchange properties of salt-stressed olive (*Olea europea* L.) leaves. *Plant Physiology* 90: 1408-1416.

Bongi, G. and M. Lupatteli. (1986). Fluorescence properties and photosynthesis in *Olea europaea* (L.) leaves as modified by salt stress. Proceedings of the *26th Congress S.I.F.V.*, Rome, pp. 3.

Bongi G., M. Mencuccini and G. Fontanazza. (1987). Photosynthesis of olive leaves: effect of light flux density, leaf age, temperature, peltates, and H_2O vapor pressure deficit on gas exchange. *Journal of the American Society for Horticultural Science* 112(1): 143-148.

Bongi G. and A. Palliotti. (1994). Olive. *In Handbook of Environmental Physiology of Fruit Crops. Volume I: Temperate Crops,* eds. B. Schaffer and P.C. Andersen, CRC Press, Inc., pp. 165-187.

Bongi, G., P. Rocchi and A. Palliotti. (1994). Drought effects on chlorophyll fluorescence. *Médecine Biologie Environnement* 22: 75-82.

Braun, P. (1997). Sap flow measurements in fruit trees–Advantages and shortfalls of currently used systems. *Acta Horticulturae* 449: 267-272.

Burman, R. and L.O. Pochop. (1994). *Evaporation, Evapotranspiration and Climatic Data.* Developments in Atmospheric Science, 22, Elsevier Science B.V.

Cabrera, F., R. López, A. Martínez-Bordiú, E. Dupuy de Lome and J.M. Murillo. (1996). Land treatment of olive oil mill wastewater. *International Biodeterioration & Biodegradation* 38: 215-225.

Castel, J.R. and E. Fereres. (1982). Responses of young almond trees to two drought periods in the field. *Journal of Horticultural Science* 57: 175-187.

Celano G., B. Dichio, G. Montanaro, V. Nuzzo, A.M. Palese and C. Xiloyannis. (1999) Distribution of dry matter and amount of mineral elements in irrigated and non-irrigated olive trees. *Acta Horticulturae* 474: 381-384.

Celano G., V. Nuzzo, A.M. Palese, M. Romano and B. Dichio. (1998). Dinamica di crescita dell'apparato radicale di giovani piante di olivo allevate in differenti condizioni di disponibilità idrica del suolo. Proc. of the *Convegno Nazionale "Irrigazione e Ricerca: Progressi nell'uso della risorsa acqua."* Bari, 1-2 October, pp. 225-229.

Cimato, A., C. Cantini and G. Sani. (1990). Climate-phenology relationships on olive cv Frantoio. *Acta Horticulturae* 286: 171-174.

COI. (1996). *Enciclopedia Mundial del Olivo.* Plaza & Janés editores, S.A.

Cruz-Conde, J., and M. Fuentes-Cabanas. (1986). Riego por goteo del olivar: dosis de agua. *Olea* 17: 203-205.

d'Andria, R., G. Morelli, P. Giorio, M. Patumi, G. Vergari and G. Fontanazza. (1998). Yield and oil quality of young olive trees grown under different irrigation regimes. Proc. of the *Third International Symposium on Olive Growing.* Chania (Crete), 22-26 September, pp. 8.

Denney J.O. and G.R. McEachern. (1983). An analysis of several climatic temperature variables dealing with olive reproduction. *Journal of the American Society for Horticultural Science* 108(4): 578-581.

Denney J.O. and G.R. McEachern. (1985). Modeling the thermal adaptability of the olive (*Olea europaea* L.) in Texas. *Agricultural and Forest Meteorology* 35: 309-327.

Dettori, S. (1987). Estimación con los métodos de la FAO de las necesidades de riego de los cultivos de aceitunas de mesa de Cerdeña. *Olivae* 17: 30-35.

Díaz-Espejo, A., M.J. Palomo, J.E. Fernández, I.F. Girón and F. Moreno. (1998). Water relations and sap flow in mature olive trees. Proc. of the *Fifth Congress of the European Society for Agronomy,* Nitra, 28 June-2 July. Short Communications Volume II, pp. 332-333.

Dichio B., V. Nuzzo and C. Xiloyannis. (1997). Drought stress-induced variation of pressure-volume relationships in *Olea europaea* L. cv. "Coratina". *Acta Horticulturae* 449: 401-409.

Dichio B., C. Xiloyannis, G. Celano, M. Arcieri and A.M. Palese. (1998). Flussi xilematici e consumi idrici in olivo in condizioni di diversa disponibilità idrica. Proc. of the *Convegno Nazionale "Irrigazione e Ricerca: Progressi nell'uso della risorsa acqua."* Bari, 1-2 October, pp. 84-90.

Doorenbos, J. and W.O. Pruitt. (1977). Guidelines for predicting crop water requirements. *FAO Irrigation and Drainage Paper No. 24,* 2nd ed., FAO, Rome.

Drossopoulos, J.B. and C.A. Niavis. (1988). Seasonal changes of the metabolites in the leaves, bark and xylem tissues of olive tree (*Olea europaea,* L.) II. Carbohydrates. *Annals of Botany* 62: 321-327.

Durán-Grande, M. (1977). Estudio morfológico de los estomas y de las áreas estomáticas del fruto de *Olea europaea* L. *Boletín de la Real Sociedad Española de Historia Natural (Biología)* 75: 23-41.

El-Gazzar, A.M., E.M. El-Azab and M. Sheata. (1979). Effect of irrigation with fractions of sea water and drainage water on growth and mineral composition of young grapes, guavas, oranges and olives. *Alexandria Journal of Agricultural Research* 1: 207-219.

Evenari, M. and R. Richter. (1937). Physiological ecological investigation in the Wilderness of Judaca. *J. Linn. Soc.* 51: 333-381.

Fahmi, I. (1958). Changes in carbohydrate and nitrogen content in 'Souri' olive leaves in relation to alternate bearing, *Proceedings of the American Society of Horticultural Science* 78: 252-256.

FAO. (1993). The use of saline waters for crop production. In *Food and Agriculture Organization Technical Paper 48,* Rome.

Fereres, E. (1995). El riego del olivar. Proceedings of the *VII Simposio Científico-Técnico Expoliva '95,* pp. 18.

Fereres, E., A. Moriana and C. Ruz. (1998). Relaciones hídricas en olivo durante el invierno. Proc. of the *4th Simposium Hispano-Portugués "Relaciones Hídricas en las Plantas,"* Murcia, 2-3 November, pp. 49-52.

Fereres, E., W.O. Pruitt, J.A. Beutel, D.W. Henderson, E. Holzapfel, H. Shulbach and K. Uriu. (1981). ET and drip irrigation scheduling. In *Drip irrigation Management,* ed. E. Fereres, University of California, Div. of Agric. Sci. No. 21259, pp. 8-13.

Fereres, E., C. Ruz, J. Castro, J.A. Gómez and M. Pastor. (1996). Recuperación del olivo después de una sequía extrema. Proceedings of the *XIV Congreso Nacional de Riegos,* Aguadulce (Almería), 11-13 June, pp. 89-93.

Ferguson, L., G.S. Sibbett and G.C. Martin. (1994). *Olive Production Manual.* University of California, publication 3353.

Fernández, J.E. (1989). Comportamiento del olivo (*Olea europaea* L., var. manzanillo) sometido a distintos regímenes hídricos, con especial referencia a la dinámica del sistema radicular y de la transpiración. Ph. D., University of Córdoba, Department of Agronomy.

Fernández, J.E. (1997). Olive production under saline conditions. Proc. of the First Trans-National Meeting on *Salinity as a limiting factor for agricultural productivity in the Mediterranean basin.* Naples, 24-25 March, pp. 177-188.

Fernández, J.E., M.J. Palomo, A. Díaz-Espejo and I.F. Girón and F. Moreno. (1998a). Measuring sap flow in olive trees: potentialities and limitations of the compensation heat-pulse technique. Proc. of the *4th Workshop on Measuring Sap Flow in Intact Plants*. Zidlochovice, Check Republic, 3-4 November, pp. 16.

Fernández, J.E., A. Díaz-Espejo, M.J. Palomo, I.F. Girón and F. Moreno. (1998b). Riego y fertilización del olivar en la comarca de El Aljarafe (Sevilla). *Folleto divulgativo*, pp. 32.

Fernández, J.E., F. Moreno, F. Cabrera, P. Arambarri and J. Martín-Aranda. (1987). Experiencias con P^{32} para el seguimiento de la extracción de agua por el olivo regado por goteo. Proceedings of the *VII Congreso Nacional de Química*, Sevilla 12-17 October, pp. 6.

Fernández, J.E., F. Moreno, F. Cabrera, J.L. Arrúe and J. Martín-Aranda. (1991). Drip irrigation, soil characteristics and the root distribution and root activity of olive trees. *Plant and Soil* 133: 239-251.

Fernández, J.E., F. Moreno, B.E. Clothier and S.R. Green. (1996). Aplicación de la técnica de compensación de pulso de calor a la medida del flujo de savia en olivo. Proc. of the *XIV Congreso Nacional de Riegos*. Aguadulce (Almería), 11-13 June, pp. 1-7.

Fernández, J.F., F. Moreno, I.F. Girón, and O.M. Blázquez. (1997). Stomatal control of water use in olive tree leaves. *Plant and Soil* 190:179-192.

Fernández, J.E., F. Moreno and J. Martín-Aranda. (1990). Study of root dynamics of olive trees under drip irrigation and dry farming. *Acta Horticulturae* 286: 263-266.

Fernández, J.E., F. Moreno and J. Martín-Aranda. (1993). Water status of olive trees under dry-farming and drip-irrigation. *Acta Horticulturae* 335: 157-164.

Fernández, J.E., F. Moreno, J. Martín-Aranda and E. Fereres. (1992). Olive-tree root dynamics under different soil water regimes. *Agricoltura Mediterranea* 122: 225-235.

Fernández, J.E., F. Moreno, J. Martín-Aranda and H.F. Rapoport. (1994). Anatomical response of olive roots to dry and irrigated soils. *Advances on Horticultural Science* 8: 141-144.

Fernández, J.E., M.J. Palomo, A. Díaz-Espejo and I.F. Girón. (1997). Calibrating the compensation heat-pulse technique for measuring sap flow in olive. Proceedings of the *3rd International Symposium on Olive Growing*. Chania (Crete), 22-26 September (in press), pp. 8.

Fernández-Escobar, R., and M. Benlloch. (1992). The time of floral induction in olive. *Journal of American Society for Horticultural Science* 117(2): 304-307.

Fernández-Escobar, R., M. Benlloch, C. Navarro and G.C. Martin. (1992). The time of floral induction in the olive. *Journal of the American Society for Horticultural Science* 117(2): 304-307.

Fernández-Escobar, R. and G. Gómez-Valledor. (1985). Cross-pollination in 'Gordal Sevillana' olives. *HortScience* 20(2): 191-192.

Fiestas Ros de Ursinos, J.A. (1986). Vegetation water used as fertilizer. Proceedings of the *International Symposium on Olive By-Products Valorization*. FAO, UNDP, Sevilla, Spain, pp. 321-330.

Fischer, A. (1967). The after effect of water stress on stomatal function in tobacco (*Nicotiana tabacum*) and bean (*Vicia faba*). *Plant Physiology Supp.* 42: 5-18.

Freeman, M., K. Uriu and H.T. Hartmann. (1994). Diagnosing and correcting nutrient problems. In *Olive Production Manual*, University of California, Publication 3353, eds. L. Ferguson, G.S. Sibbett and G.C. Martin, pp. 77-86.

García-Ortiz, A., J.V. Giráldez-Cervera, P. González-Fernández and R. Ordóñez-Fernández. (1993). El riego con alpechín. Una alternativa al lagunaje. *Agricultura* 730: 426-431.

Gimenez, C., E. Fereres, C. Ruz and F. Orgaz. (1996). Water relations and gas exchange of olive trees: diurnal and seasonal patterns of leaf water potential photosynthesis and stomatal conductance. *Acta Horticulturae* 449: 411-415.

Goldhamer, D.A. (1997). Regulated deficit irrigation for olives. Proceedings of the *3rd International Symposium on Olive Growing*. Chania (Crete), 22-26 September, pp. 8.

Goldhamer, D.A., J. Dunai and F. Ferguson. (1993). Water use requirements of manzanillo olives and responses to sustained deficit irrigation. *Acta Horticulturae* 335: 365-371.

Goldhamer, D.A., J. Dunai and F. Ferguson. (1994). Irrigation requirements of olive trees and responses to sustained deficit irrigation. *Acta Horticulturae* 356: 172-175.

Gowing, D.J.G., W.J. Davies and H.G. Jones. (1990). A positive root-sourced signal as an indicator of soil drying in apple, *Malus* × *domestica* Borkh. *Journal of Experimental Botany* 233: 1535-1540.

Grammatikopoulos, G., G. Karabourniotis, and A. Kyparissis. (1994). Leaf hairs of olive (*Olea europaea*) prevent stomatal closure by ultraviolet-B radiation. *Australian Journal of Plant Physiology* 21:293-301.

Green, S.R., and B.E. Clothier. (1988). Water use of kiwifruit vines and apple trees by the heat-pulse technique. *Journal of Experimental Botany* 198: 115-123.

Griggs, W.H., H.T. Hartmann, M.V. Bradley, B.T. Iwakini and J. Whisler. (1975). Olive pollination in California. *Calif. Agric. Exp.*, Stn. Bull. 869.

Gucci, R. and M. Tattini. (1997). Salinity tolerance in olive. *Horticultural Reviews* 21: 177-214.

Guerrero, A. (1997). *Nueva olivicultura*. Grupo Mundi-Prensa.

Hackett, W.P. and H.T. Hartmann. (1964). Inflorescence formation in olive as influenced by low temperature, photoperiod, and leaf area. *Botanical Gazette* 125(1): 65-72.

Hammouda, M.A. (1954). Studies in the water relations of Egyptian desert plants. *Ph. D.* thesis, Faculty of Sciences, University of El Cairo.

Hartmann, H.T. (1953). Effect of winter chilling on fruitfulness and vegetative growth in the olive. *Proceedings of the American Society for Horticultural Science* 62: 184-190.

Hartmann H.T. and K.W. Opitz. (1980). *Olive Production in California*. University of California leaflet 2474.

Hinckley, T.M., M.O. Schroeder, J.E. Roberts and D.N. Bruckerhoff. (1975). Effect of several environmental variables and xylem pressure potential on leaf surface resistance in white oak. *For. Sci.* 21: 201-211.

Jarvis, P.G. (1976). The interpretation of the variation in leaf water potential and stomatal conductance found in canopies in the field. *Philosophical Transactions of the Royal Society, London Ser.* B 273: 593-610.

Jensen, M.E., R.D. Burman and R.G. Allen. (1990). *Evapotranspiration and Irriga-*

tion Water Requirements. ASCE Manuals and Reports on Engineering Practice No. 70.

Jorba, J., L. Tapia, and D. Sant. (1985). Photosynthesis, leaf water potential, and stomatal conductance in olea europea under wet and drought conditions. *Acta Horticulturae* 171: 237-246.

Karabourniotis, G., A. Kyparissis and Y. Manetas. (1993). Leaf hairs of *Olea europaea* protect underlying tissues against ultraviolet-B radiation damage. *Environmental and Experimental Botany* 3: 341-345.

Kopsidas, G.C. (1994). Wastewater from the table olive industry. *Water Research* 28(1): 201-205.

Larcher, W. J.A.P.V. Moraes and H. Bauer. (1981). Adaptative responses of leaf water potential, CO_2-gas exchange and water used efficiency of *Olea europea* during drying and rewatering. In *Components of Productivity of Mediterranean-Climate Regions. Basic and Applied Aspects*, eds. N.S. Margaris and H.A. Mooney, The Hague, Boston, London, pp. 77-.

Larsen, F.E., S.S. Higgins, and A. Al Wir. (1989). Diurnal water relations of apple, apriot, grape, olive and peach in an arid environment (Jordan). *Scientia Horticulturae* 39: 211-222.

Lavee, S. (1985). *Olea europea.* In *CRC Handbook of Flowering, Volume III*, ed. A.H. Halevy, CRC Press, Inc., pp. 423-434.

Lavee, S. (1986). Olive. In *CRC Handbook of Fruit Set and Development*, ed. S.P. Monselise, CRC Press, Inc., pp. 261-276.

Lavee, S. (1996). Biology and physiology of the olive. In *World Olive Encyclopaedia*, ed. COI, pp. 59-110.

Lavee, S., M. Nashef, M. Wodner and H. Harshemesh. (1990). The effect of complementary irrigation added to old olive trees (*Olea europaea* L.) cv. Souri on fruit characteristics, yield and oil production. *Advances in Horticultural Science* 4: 135-138.

Lavee, S. and M. Wodner. (1991). Factors affecting the nature of oil accumulation in fruit of olive (*Olea europaea* L.) cultivars. *Journal of Horticultural Science* 66(5): 583-591.

Leon, J.M. and M.J. Bukovac. (1978). Cuticle development and surface morphology of olive leaves with reference to penetration of foliar-applied chemicals. *Journal of the American Society for Horticultural Science* 103(4): 465-472.

Leva, A.C. (1977). Influenza dello stato nutritivo sul numero degli stomi nelle foglie di olivo. *Rivista della Ortoflorofrutticoltura Italiana* 61: 405-408.

Lo Gullo, M.A. and S. Salleo. (1988). Different strategies of drought resistance in three Mediterranean sclerophyllous trees growing in the same environmental conditions. *The New Phytologist* 108: 267-276.

Loreti, F. and S. Natali. (1981). Irrigazione. In *L'Olivo*, eds. E. Baldini and F. Scaramuzzi, Ramo Editoriale Agricoltori, Rome, pp. 93-118.

Loreto, F., and T. D. Sharkey. (1990). Low humidity can cause uneven photosynthesis in olive (*Olea europaea* L.) leaves. *Tree Physiology.* 6: 409-415.

Loukas, M. and C.B. Krimbas. (1983). History of olive cultivars based on their genetic distances, *Journal of Horticultural Science* 58: 121-127.

Loussert, R. and G. Brousse. (1978). *L'Olivier.* G.P. Maisonneuve & Larose, Paris.

Magliulo, V., R. d'Andria, G. Moreli and F. Fragnito. (1998). Proceedings of the *Third International Symposium on Olive Growing*. Chania, Creta, Greece, 4 pp.

Manrique, T. (1996). Desarrollo de la aseituna bajo diferentes regímenes de riego. *Trabajo Profesional Fin de Carrera*, University of Córdoba, Spain, pp. 113.

Mantovani, C.E. (1994). Desarrollo y evaluación de modelos para el manejo del riego: estimación de la evapotranspiración y efectos de la uniformidad de aplicación del riego sobre la producción de los cultivos. *Ph.D. Thesis*, Department of Agronomy, University of Córdoba.

Mariscal, M.J., F. Orgaz and F.J. Villalobos. (1998). Radiation use efficiency by a young olive orchad (*Olea europaea* L.).Proceedings of the *Fifth Congress of European Society for Agronomy*. Short Communication, Volume II, 28 June-2 July, pp. 324-323.

Martín-Aranda, J., J.L. Arrúe-Ugarte, J.L. Muriel-Fernández. (1975). Evapotranspiration regime and water economy physical data in olive grove soils in SW Spain. *Agrochimica*. 1: 82-87.

Martín-Aranda, J., D. Nuñez, F. Moreno, J.L. Arrúe and M. Roca. (1982). Localized fertigation of no-tillage intensive young olive fields. Proceedings of the *9th Conference of the International Soil Tillage Research Organization*, pp. 357-362. Osijek, Yugoslavia.

Messing, I. and N. Jarvis. (1993). Temporal variation in the hydraulic conductivity of a tilled clay soil as measured by tension infiltrometers. *Journal of Soil Science* 44: 11-24.

Michelakis, N. (1986). Olive bahaviour under various irrigation conditions. *Olea*. 207-209.

Michelakis, N. (1990). Yield response of table and oil olive varieties to different water use levels under drip irrigation. *Acta Horticulturae*. 286: 271-275.

Michelakis, N. and E. Vougioucalou. (1988). Water used, root and top growth of the olive trees for different methods of irrigation and levels of soil water potential. *Olea* 19: 17-31.

Michelakis, N., E. Vouyoucalou, and G. Clapaki. (1996). Water use and the soil moisture depletion by olive trees under different irrigation conditions. *Agricultural Water Management* 29: 315-325.

Migahid, A.M. and A.A. Abd-El-Rahman. (1953). Studies in the water economy of Egyptian desert plants. III. Observations on the drought resistance of desert plants. *Bull. Inst. Des. d'Egypte* 3: 58-83.

Milella, A. and S. Dettori. (1987). Regimi idrici ottimali e parziali per giovani olivi da mensa. *Rivista di Frutticoltura* 8:65-69.

Moreno, F. (1986). Variabilidad de la infiltración de agua en el suelo durante lluvias intensas. Proceedings of the *II Simposio sobre el Agua en Andalucía*, Vol. I, pp. 173-184. Granada, Spain.

Moreno, F., J.E. Fernández, B.E. Clothier and S.R. Green. (1996). Transpiration and root water uptake by olive trees. *Plant and Soil*. 184: 85-96.

Moreno, F., J.E. Fernández, M.J. Palomo, I.F. Girón, J.M. Villau, and A. Diaz-Espejo. (1998). Measurements of soil hydraulic properties in an olive orchard under drip irrigation. In *Water and the Environment: Innovation Issues in Irrigation and Drainage*, eds. L.S. Pereira and J. Gowing, E & FN Spon Publisher, pp. 390-395.

Moreno, F., F. Pelegrín, J.E. Fernández and J.M. Murillo. (1997). Soil physical properties, water depletion and crop development under traditional and conservation tillage in southern Spain. *Soil and Tillage Research* 41: 25-42.

Moreno, F., G. Vachaud and J. Martín-Aranda. (1983). Caracterización hidrodinámica de un suelo de olivar. Fundamentos teóricos y métodos experimentales. *Anales de Edafología y Agrobiología* 42: 695-721.

Moreno, F., G. Vachaud, J. Martín-Aranda, M. Vauclin and E. Fernández. (1988). Balance hídrico de un olivar con riego gota a gota. Resultados de cuatro años de experiencias. *Agronomie* 8(6): 521-537.

Natali, S., C. Bignami, and A. Fusari. (1991). Water consumption, photosynthesis, transpiration and leaf water potential in olea europea L., cv. "Frantoio", at different levels of available water. *Agricotura Mediterranea* 121: 205-212.

Natali, S., C. Xiloyannis and P. Angelini. (1985). Water consumptive use of olive trees (*Olea europaea*) and effect of water stress on leaf water potential and diffusive resistance. *Acta Horticulturae* 171:341-351.

Nuñez-Aguilar, I., J.L. Arrúe-Ugarte, F. Moreno, and J. Martín-Aranda. (1980). Sustracción de humedad en la zona radicular del olivo (variedad manzanillo). Técnicas de seguimiento y primeros resultados obtenidos. Proceedings of the *VII Simposio de Bioclimatología*. Avances sobre la Investigación en Bioclimatología, pp. 515-524.

Nuzzo, V., B. Dichio and C. Xiloyannis. (1995). Effetto cumulato negli anni della carenza idrica sulla crescita della parte aerea e radicale della cv. Coratina. Proceedings of the congress on *L'olivicoltura Mediterranea: Stato e Prospettive della Coltura e della Ricerca*. Rende, 26-28 January, pp. 323-331.

Nuzzo, V., C. Xiloyannis, B. Dichio, G. Montarano and G. Celano. (1997). Growth and yield in irrigated and non-irrigated olive trees cultivar Corotina over four years after planting. *Acta Horticulturae*. 449: 75-82.

Orgaz, F. and E. Fereres. (1997). Riego. In *El Cultivo del Olivo,* 1st ed.. Junta de Andalucía y Grupo Mundi Prensa, pp. 251-272.

Palliotti, A., G. Bongi, and P. Rocchi. (1994). Peltate trichomes effects on photosynthetic gas exchange of Olea europea L. leaves. *Plant Physiology* 13: 35-44.

Palomo, M.J., A. Díaz-Espejo, J.E. Fernández, I.F. Girón and F. Moreno. (1998). Using sap flow measurements to quantify water consumption in the olive tree. In *Water and the Environment: Innovative Issues in Irrigation and Drainage*, eds. L. S. Pereira and J.W. Gowing. E & FN Spon, pp. 205-212.

Palomo, M.J., A. Díaz-Espejo, J.E. Fernández, I.F. Girón and F. Moreno. (1998). Sap flow in different conductive organs of the olive and the influence of the environmental variables. Proceedings of the *IV Simposium Hispano-Portugués sobre Relaciones Hídricas en las Plantas*, pp. 154-157.

Pastor, M., J. Castro and V. Vega. (1997). Programación del riego de olivar en Andalucía. Proceedings of the *VIII Simposium Científico-Técnico. Expoliva '97*. 2-3 October 19 pp.

Pastor, M., J. Castro, V. Vega and J. Salas. (1995). Influencia del riego sobre la composición y calidad del aceite de oliva. Proceeding of the *VII Simposium Científico-Técnico Expoloiva '95*. 5-6 May. 7 pp.

Pastor, M. and F. Orgaz. (1994). Riego deficitario del olivar. *Agricultura* 746: 768-776.

Patumi, M., R. d'Andria, G. Fontanazza, G. Morelli and G. Vergari. (1998). Effetto dell'irrigazione sulla produzione e sullo sviluppo vegetativo di un giovane oliveto. *Olivo and Oio* 1: 36-37.

Pisanu, G., G. Corrias. (1971). Osservazioni sui sistemi radicali dell'Olivo in coltura intensiva. *Studi Sassar.* 3(19): 247-263.

Poli, M. (1986a). La vecería en la producción del olivo (estudio bibliográfico). *Olivae* 10: 11-33.

Poli, M. (1986b). La vecería en la producción del olivo. *Olivae* 12: 7-27.

Punthakey, J.F., M.J. McFarland and J.W. Worthington. (1984). Stomatal responses to leaf water potentials of drip irrigated peach (*Prunus persica*). *Transactions of the American Society for Agriculture Engineering* 27: 1442-1450.

Rallo, L. (1995). Fructificación y producción en olivo. *Agricultura*, Suplemento Mayo, pp. 13-16.

Rallo, L. (1998). Fructificación y producción. In *El cultivo del olivo*. Junta de Andalucía y Grupo Mundi-Prensa, pp. 107-136.

Rallo, L. and R. Fernández-Escobar. (1985). Influence of cultivar and flower thinning within the inflorescence on competition among olive fruit. *Journal of American Society for Horticultural Science* 110(2): 303-308.

Ramos-Cormenzana, A. (1986). Physical, chemical, microbiological and biochemical characteristics of vegetation water. Proceedings of the *International Symposium on Olive By-Products Valorization*, FAO, UNDP, Sevilla, Spain, pp. 19-40.

Rhizopoulos, S., M.S. Meletiou-Christou and S. Diamantagiou. (1991). Water relations for sun and shade leaves of four Mediterranean evergreen schlerophylls. *Journal of Experimental Botany* 42: 627-635.

Ritchie, J.T. (1972). Model for predicting evaporation from a row crop with incomplete cover. *Water Resources Research* 8: 1204-1213.

Romero, M.P., P. Plaza, A. Arnalot and M.J. Motilva. (1997). Influencia de las aplicaciones estratégicas de riego en olivo (cv *Arbequina*) sobre el rendimiento en aceite y su estabilidad. Primeros resultados. Proceedings of the *XV Congreso Nacional de Riegos*, Lleida, Spain (in press).

Roselli, G., G. Benelli, and D. Morelli. (1989). Relationship between stomatal density and winter hardiness in olive (*Olea europaea* L.). *Journal of Horticultural Science* 64: 199-203.

Rosenberg, N.J., B.L. Blad and S.B. Verma. (1983). *Microclimate–The Biological Environment*. John Wiley and Sons, New York,

Rouschal, E. (1938). Zur Ökologie der Macchien. *Jb. f. wiss. Bot.* 87: 436-532.

Saavedra, M., A. Troncoso and P. de Arambarri. (1984). Utilización de aguas fuertemente contaminadas en el riego del olivo. *Anales de Edafología y Agrobiologolía* 43(9-10): 1449-1466.

Salleo, S., M.A. Lo Gullo and F. Oliveri. (1985). Hydraulic parameters measured in 1-year-old twig of some mediterranean species with diffuse-porous wood: changes in hydraulic conductivity and their possible functional significance. *Journal of Experimental Botany*. 36: 1-11.

Sánchez-Raya, A.J. (1988). Algunos efectos de la sequía en la fisiología del olivo *(Olea europaea, L.). Agrochimica* 32(4): 301-309.

Scaramella, P. and A. Ricci. (1988). La longevità di *Olea europaea* L. e la presenza del floema interxilare. *Bolletino quadrimestrale dell'Accademia Nazionale dell'Olivo* 2: 45-49.

Scharpenseel, H.W., A. Essafi and A. Bouguerra. (1966). Etude de la puissance d'absorption des ions nutritifs par les racines de l'olivier par application d'acide phosphorique ou de superphosphate marque avec ^{32}P. Proceedings of the *Colloque sur l'emploi des radioisotopes et des rayonnements dans les études phyto pédologiques.* Ankara (Turquie), 28 June-2 July, pp. 3-27.

Schulte, P.J. (1992). The units of currency for plant water status, *Plant Cell and Environment* 15: 7-10.

Schwabe, W.W., and S.M. Lionakis. (1996). Leaf attitude in olive relation to drought resistance. *Journal of Horticultural Science* 71(1): 157-166.

Schweingruber, F.H. (1990). *Anatomy of European Woods.* Paul Haupt Berne and Stuttgart Publishers.

Smit, A.L., G. Bengough, C. Engels, M. van Noordwijk, S. Pellerin and S. van de Geijn. (1999). *Root Growth and Function: a Handbook of Methods.* Springer Verlag, Heidelberg, Germany (in preparation).

Smith, D.M. and S.J. Allen. (1996). Measurement of sap flow in plant stems. *Journal of Experimental Botany* 47: 1833-1844.

Sorrentino, G., P. Giorio and R. d'Andria. (1998). Leaf water status of field-grown olive trees *(Olea europaea* L., cv Kalamata) under three water regimes. Proceedings of the *Third International Symposium on Olive Growing*, Crete, 22-26 September, pp. 4.

Spiegel, P. (1955). The water requirements of the olive tree, critical periods of moisture stress and the effect of irigation upon the oil content of its fruits. Proceedings of the *14th International Congress on Horticultural Science,* Wageningen, The Netherlands, pp. 1363-.

Suárez, M.P., R. Fernández-Escobar and L. Rallo. (1984). Competition among fruits in olive II. Influence of inflorescence or fruit thinning and cross-pollination on fruit set components and crop efficiency. *Acta Horticulturae* 149: 131-143.

Swanson, R.H. (1994). Significant historical developments in thermal methods for measuring sap flow in trees. *Agricultural and Forest Meteorology* 72: 113-132.

Swinbank, W.C. (1951). The measurement of vertical transfer of heat and water vapour by eddies in the lower atmosphere. *Journal of Meteorology* 8: 135-145.

Tardieu, F. and W.J. Davies. (1993). Integration of hydraulic and chemical signalling in the control of stomatal conductance and water status of droughted plants. *Plant, Cell and Environment* 16: 341-349.

Tattini, M., R. Gucci, M.A. Coradeschi, C. Ponzio and J.D. Everard. (1995). Growth, gas exchange and ion content in *Olea europaea* plants during salinity stress and subsequent relief. *Physiology Plantarum* 95: 203-210.

Thompson, R.G., M.T. Tyree, M.A. Lo Gullo and S. Salleo. (1983). The water relation of young olive trees in a Mediterranean winter: measurements of evaporation from leaves and water conduction in wood. *Annals of Botany* 52: 399-406.

Tombesi, A., P. Proietti and G. Nottiani. (1986). Effect of water stress on photosynthesis, transpiration, stomata resistance and carbohydrate level in olive trees. *Olea* 17: 35-40.

Trigui, A. (1984). Propiedades biofísicas y ópticas de las hojas del olivo (*Olea europaea*). Contribución al estudio bioclimático de la especie. *Olivae* 2: 45-47.

Troncoso, A., J.M. Murillo, R. Gallardo, M. Barroso and J.M. Hernández. (1983). Influencia de la salinidad del medio en el desarrollo y estado nutritivo de plantas jóvenes de olivo. Proceedings of the *V Reunión de la Sociedad Española de Fisiología Vegetal*. Murcia, 27-30 Septiembre, p. 251.

Troncoso, A., J. Prieto and J. Liñán. (1978). Observaciones sobre las pérdidas de flores y frutos jóvenes en el olivo "Manzanillo" en Sevilla. *Anales de Edafología y Agrobiología* 37(11-12): 1121-1129.

Tyree, M.T. and J.S. Sperry. (1989). Vulnerability of xylem to cavitation and embolism. *Annual Review of Plant Physiology and Plant Molecular Biology* 40: 19-38.

Vanderlinden, K., D. Gabriels and J.V. Giráldez. (1998). Evaluation of infiltration measurements under olive trees in Córdoba. *Soil and Tillage Research* 48: 303-315.

Villagra, M.M., O.O.S. Bacchi, R.L. Tuon and K. Reichardt (1995). Difficulties of estimating evapotranspiration from the water balance equation. *Agricultural and Forest Meteorology* 72: 317-325.

Villalobos, F.J., F. Orgaz and L. Mateos. (1995). Non-destructive measurement of leaf area in olive (*Olea europaea* L.) trees using a gap inversion method. *Agricultural and Forest Meteorology* 73: 29-42.

Villalobos, F.J., F. Orgaz, L. Testi and E. Fereres. (1998). Measurement and modelling of evaporation of olive (*Olea europaea* L.) orchards. Proceedings of the *Fifth ESA Congress*, Nitra (The Slovak Republic), 28 June-2 July, pp. 2.

Vitagliano, C., A. Minnocci, L. Sebastiani, A. Panicucci and G. Lorenzini. (1997). Physiological response of two olive genotypes to gaseous pollutants. Proceedings of the *Third International Symposium on Olive Growing*, Crete, 22-26 September, pp. 4.

Xiloyannis, C., B. Dichio, V. Nuzzo and G. Celano. (1996). L'olivo: pianta esempio per la sua capacità di resistenza in condizioni di estrema siccità. *Seminari di Olivicoltura*. Spoleto, 7 and 28 June. pp. 79-111.

Yáñez, J. and M. LaChica. (1971). El olivar. Revisión de las condiciones de cultivo. *Anales de Edafología y Agrobiología* 30(7-8): 789-816.

Zarrouk, M. and A. Cherif. (1981). Effect of sodium chloride on lipid content of olive trees (*Olea europaea* L.). *Zeitschrift fur Pflanzenphysiol.* 105: 85-92.

Zimmermann, M.H. (1978). Hydraulic architecture of some diffuse-porous trees. *Canadian Journal of Botany* 56: 2286-2295.

Canopy Position and Leaf Age Affect Stomatal Response and Water Use of Citrus

T. M. Mills
K. T. Morgan
L. R. Parsons

SUMMARY. We investigated the responses of citrus stomata to leaf age, leaf position, canopy size and local microclimate. Large differences existed in the stomatal response due to leaf age within a 15-year-old canopy. Much smaller differences due to leaf age were evident in 7-year-old and 2-year-old trees. Likewise, position in the canopy played a significant role in determining stomatal response within a 15-year-old tree, but had little influence in the smaller trees. Stomatal conductance data were in good agreement with values calculated using a previously published stomatal-response function based on incident photosynthetically active radiation and the vapor pressure deficit of air. The model of stomatal conductance was used to calculate total daily plant water use of 2-year-old trees based on the Penman-Monteith model. Calculated values of plant water use were close to actual values for these 2-year-old trees as measured using weighing lysimeters. These results confirm the robustness of the Penman-Monteith model for the

T. M. Mills is affiliated with the Horticulture & Food Research Institute of New Zealand, Ltd., Palmerston North, New Zealand. K. T. Morgan and L. R. Parsons are affiliated with the Citrus Research and Education Center, University of Florida, 700 Experiment Station Road, Lake Alfred, FL 33850 USA.

Address correspondence to: T. M. Mills, The Environment and Risk Management Group, HortResearch, Private Bag 11030, Palmerston North, New Zealand.

The authors would like to thank Marjie Cody for her technical support during the experiment. The authors are also grateful to Drs. S. R. Green, A. S. Walcroft, J. P. Syvertsen and M. H. Behboudian for critical review of the manuscript.

[Haworth co-indexing entry note]: "Canopy Position and Leaf Age Affect Stomatal Response and Water Use of Citrus." Mills, T. M., K. T. Morgan, and L. R. Parsons. Co-published simultaneously in *Journal of Crop Production* (Food Products Press, an imprint of The Haworth Press, Inc.) Vol. 2, No. 2 (#4), 1999, pp. 163-179; and: *Water Use in Crop Production* (ed: M. B. Kirkham) Food Products Press, an imprint of The Haworth Press, Inc., 1999, pp. 163-179. Single or multiple copies of this article are available for a fee from The Haworth Document Delivery Service [1-800-342-9678, 9:00 a.m. - 5:00 p.m. (EST). E-mail address: getinfo@haworthpressinc.com].

evaluation of citrus water use. We then scaled our calculations up to the 15-year-old field grown trees. Our study highlights the differences in stomatal response between various populations of leaves within a citrus canopy and indicate the important role stomatal conductance has in determining water use of citrus trees. *[Article copies available for a fee from The Haworth Document Delivery Service: 1-800-342-9678. E-mail address: getinfo@haworthpressinc.com <Website: http://www.haworthpressinc.com>]*

KEYWORDS. Diurnal stomatal conductance, plant water use

INTRODUCTION

The response of stomata to environmental and physiological influences has been of considerable interest for many decades (Kaufmann and Levy, 1976). Primarily this is because stomatal conductance (g_s) regulates transpiration and photosynthesis and therefore impacts directly on the water use and productivity of plants (Jones, Lakso and Syvertsen, 1985). Given that 85% of the total global consumption of fresh water is due to irrigation (van Schilfgaarde, 1994), and the fact that good quality water is becoming an increasingly scarce resource worldwide (Postal, 1993), more effective irrigation management strategies need to be developed in order to utilise most efficiently the water resources available. Understanding how the various environmental and physiological factors regulate g_s is a prerequisite to determining how much water plants are using.

Stomata are sensitive to many interacting environmental variables such as light, CO_2, plant water status, and ambient temperature and vapor pressure deficit (VPD) (Jarvis and McNaughton, 1986; Jones, 1998). Often it is difficult to separate the stomatal response to any one variable because the observed responses are an integration across all stimuli. A model can be used to integrate data so that we may study the effect of one variable while holding all others constant. The model that we use to describe stomatal response in this study is primarily driven by light and the vapor pressure deficit of the air calculated from temperature and humidity data. Since ambient CO_2 concentrations are generally stable, and all experimental trees were fully watered, we may ignore these two factors. Additionally, stomata may acclimate to environmental conditions, for example, low light levels or change because of physiological factors such as leaf age. By increasing our knowledge of how g_s in citrus respond to both environmental and physiological stimuli we seek to further our understanding of water use in citrus.

Citrus species have a number of unique attributes which conspire to make it a crop with a relatively low water requirement (Brakke and Allen, 1995). Firstly, citrus leaves are thick and waxy (Levy and Syvertsen, 1981). Therefore the cuticular resistance to water loss is high. Citrus also tend to have a low maximal stomatal conductance (3-5 mm s^{-1}), even in the height of summer (Kriedemann, 1971; Meyer and Green, 1981; Syvertsen, 1982; Lloyd and Howie, 1989; Syvertsen and Lloyd, 1994). However, citrus are also evergreen, so they require water throughout the year. In addition, citrus leaves are long-lived, and like most evergreen tree species, the stomatal response is influenced by leaf age (Syvertsen, Smith and Allen, 1981; Syvertsen, 1982). It is therefore important to understand the physiological implications of leaf age when investigating stomatal conductance as it relates to water use.

Citrus develop a dense outer canopy as they mature. Thus we may expect that the light environment of the leaves will also differ as the trees mature. A large fraction of citrus leaves are commonly found within the inner portions of the canopy where light levels are typically 10% or less than values recorded on the outermost leaves (Cohen et al., 1987). Citrus is thought to have evolved as an understory plant in subtropical rainforests and so has a high tolerance to shade (Syvertsen and Lloyd, 1994). But, because a significant number of citrus leaves are shaded, it is important to evaluate whether the sun and shade leaves differ in their stomatal response (Leuning et al., 1995). Mature citrus in commercial groves are pruned as a hedgerow canopy and rows may be orientated in an east-west direction. During the winter months leaves on the north facing side of trees growing in the northern hemisphere are heavily shaded throughout the day. The importance of row orientation on stomatal function is therefore addressed here.

We use our observations of stomatal conductance to develop a leaf model which, in conjunction with the Penman-Monteith model, is used to calculate plant water use. Information from this study will improve our understanding of stomatal function of citrus leaves and how it varies in response to canopy position, leaf age and canopy size.

MATERIALS AND METHODS

Site Description

This study was carried out from February 1998 at the Citrus Research and Education Center, University of Florida, Lake Alfred, Florida (lattitude 28.6 N, longitude 81.43 W, elevation 42 m). We investigated three mature (15-year-old) field grown 'Hamlin' orange trees grafted on *Carrizo citrange* rootstock, plus three small (7-year-old) field grown 'Hamlin' orange trees on *Swingle citrumelo* rootstock, and three juvenile (2-year-old) pot grown 'Hamlin' orange trees on *Carrizo citrange* rootstock. These will hereby be referred to as 15-year-old, 7-year-old and 2-year-old trees. The predominant soil type within the research block was Apopka sand (U.S. Department of Agriculture, Soil Conservation Department), while the soil used in the pots was classified as a Candler fine sand (U.S. Dept. of Agriculture, Soil Conservation Department). Common commercial practices were used to manage the field-grown trees. The 15-year-old trees had an east-west row orientation, the 7-year-old trees, a north-south orientation. Potted trees were periodically given liquid nutrient solution in the irrigation water. The 1500 L plastic pots containing the 2-year-old trees were placed upon industrial scales (maximum capacity 2270 ± 0.05 kg) and their average weight was recorded once every 15 minutes using a Campbell data logger (Campbell Scientific Ltd., Logan, Utah).

Data Collection

Stomatal conductance (g_s) was measured using a LI-COR LI-1600 steady state porometer (LI-COR, Lincoln, Nebraska, USA) over several days during late winter and early spring 1998. For the 15-year-old trees, the g_s of shade leaves growing within the canopy and sun leaves growing outside the canopy were recorded. We also separated our measurements into leaves facing north and south due to heavy leaf shading on the north facing side.

Citrus leaves live up to three years or more. The stomatal responses of both old (fully expanded, hardened) and new (fully expanded, non-hardened) leaves were studied. Inner (shade) and outer (sun) leaves as well as old and new leaves were differentiated for 7-year-old trees. With the 2-year-old trees, which had a sparse canopy (leaf area < 2 m^2), all leaves were initially assumed sunlit. Age was the only factor to differentiate leaves on the 2-year-old trees.

All trees were fully irrigated throughout the experimental period. To check that no plant water-deficit developed in any of the experimental trees, midday stem xylem water potential (Ψ_x) of fully expanded, hardened mature leaves, growing on the outside of the canopy and 1 to 2 m above the ground, was periodically measured using a Scholander-type pressure chamber. Leaves were bagged first in black plastic, and then covered in aluminum foil and left to equilibrate overnight (Begg and Turner, 1970). Stem xylem water potential rather than leaf water potential was measured because the former is known to be less prone to fluctuating environmental conditions (McCutchan and Shackel, 1992).

Meteorological data were collected using an automated weather station (Adcon Telemetry, Boca Raton, Florida, USA) installed within the 7-year-old citrus grove. The weather station was about 10 m away from the 2-year-old trees and 200 m away from the 15-year-old trees. Fifteen minute averages of air temperature, relative humidity, wind speed, rainfall and global short-wave radiation were recorded. A net radiometer (Q*7, Radiation and Energy Balance Systems, Inc., Seattle, WA, USA) was installed on a separate tower 2 m above the ground overlooking the citrus foliage.

Soil moisture data were collected from the top 1 m of the rootzone of the 15-year-old trees using capacitance probes (EnviroSCAN, Sentek Pty LTD, Australia) located at depths of 0.1, 0.2, 0.4 and 0.8 m.

Model Calculations

Leaf stomatal conductance (g_s) was modelled as a function of vapor pressure deficit of the air and photosynthetic photon flux density using the parametric model of Jarvis (1976), *viz.*

$$g_s = g_m \left(\frac{R_p}{R_p + \beta} \right) (1 - \alpha D_a) \qquad [1]$$

Where g_m is maximum stomatal conductance; α expresses the influence of vapor pressure deficit; D_a, represents vapor pressure deficit (Pa); R_p the photosynthetic photon flux density (PPFD, μmol m^{-2} s^{-1}) incident upon the leaf and β the PPFD for $0.5g_m$. The PPFD is assumed as half of short-wave global radiation. Although this is an estimate of PPFD it is satisfactory for this calculation as the g_s of citrus reaches a maximum at low light levels. The g_m is evaluated from a scatter of stomatal conductance plotted against photosynthetically active radiation (PAR, μmol m^{-2} s^{-1}) at the time of measurement. We have used this stomatal response model to predict the diurnal pattern of stomatal conductance on selected days. We use such calculations to evaluate the robustness of the stomatal response function when compared to actual stomatal conductance data.

Weight loss of the 2-year-old potted trees was compared against calculations of tree water use, (L h^{-1}) obtained via a modified Penman-Monteith equation, *viz.*

$$\lambda E = \sum_i a_i \left[\frac{sR_{n,i} + \rho c_p D_a g_{b,i}}{s + \gamma(2 + g_{b,i}/g_{s,i})} \right] A_t \qquad [2]$$

In this equation the summation is made over a set of i uniform leaves each being a fraction, a_i, of the total leaf area. λ is the latent heat of vaporisation (2.454 kJg^{-1}); E is the transpiration rate (L h^{-1}); s is the slope of the saturation vapor pressure curve (Pa$^\circ$C^{-1}) at the ambient air temperature T_a ($^\circ$C); $R_{n,i}$ is the net radiation flux density (W m^{-2}); ρ is the air density (kg m^{-3}); c_p is the specific heat capacity of air at a constant pressure (J kg^{-1} K^{-1}); D_a is the vapor pressure deficit of air (Pa); $g_{b,i}$ is the leaf boundary-layer conductance calculated from mean leaf dimensions, d (m), and windspeed, u (m s^{-1}) in the canopy, $(g_{b,i} = 6.62 \, (u/d)^{0.5})$, (Landsberg and Powell, 1973); γ is the psychometric constant (66.1 Pa); $g_{s,i}$ is the leaf stomatal conductance (mm s^{-1}) and A_t is the total leaf area of the tree (Green and McNaughton, 1997). Total leaf area (A_t) of each 2-year-old tree was calculated periodically throughout the experimental period by determining total leaf number. Average area per leaf was measured using a LI-COR LI 3000 portable area meter (LI-COR, Lincoln, Nebraska, USA) from a sample of leaves picked from unused 2-year-old trees. The A_t of the 15-year-old trees was estimated using canopy volume calculations (Wheaton et al., 1995) and leaf numbers within a quadrant (0.5 × 0.5 × 0.5 m^3).

RESULTS AND DISCUSSION

Leaf Age, Canopy Size and Canopy Position

It has been previously reported that during January the younger (3-4 month old, summer flush) leaves of mature citrus trees in Florida have a higher g_s than older leaves (Syvertsen, 1982). Our results indicate that from March to early May (DOY 60-126) new leaves indeed have a higher mean value of g_s than old leaves irrespective of time of day, or time of year. Higher g_s values for young leaves is typical for many species (Jones, 1983).

Early in the season, before leaf emergence in early March (DOY 60), the old leaves of the 15-year-old trees had a relatively high level of g_s at midday. As the season progressed, however, g_s of these leaves declined (Figure 1c). In contrast, the old leaves on both the 7-year-old and 2-year-old trees had higher conductances during the summer months compared to the values recorded in spring (Figure 1a, b). The midday g_s of new leaves of mature citrus was greater than old leaves during the summer months (Figure 1f). The midday g_s values of old and new leaves of 7-year-old and 2-year-old trees were similar throughout the experimental period (Figure 1d, e). This may be due to the overriding influence of the environment on g_s within these smaller trees. Characteristically small trees are tightly coupled to the environment due to their rough, well ventilated canopy and exposed position. Large trees within a mature grove are more sheltered, and therefore less closely coupled to their environment (Jarvis, 1985; Jarvis and McNaughton, 1986; Monteith and Unsworth, 1990).

During the winter months the old leaves of mature citrus showed significantly higher g_s values than those of 7-year-old and 2-year-old trees. However by late February, values of g_s of old leaves were similar among the three tree types, and remained so throughout the summer months (Figure 1a, b, c). New leaves emerged later, and showed a lower g_s value in the 7-year-old and

FIGURE 1. Midday stomatal conductance (g_s) values for old (a, b, c) and new (d, e, f) leaves on 2-year-old (□), 7-year-old (Δ) and 15-year-old (○) citrus trees. Bars indicate standard error of the mean.

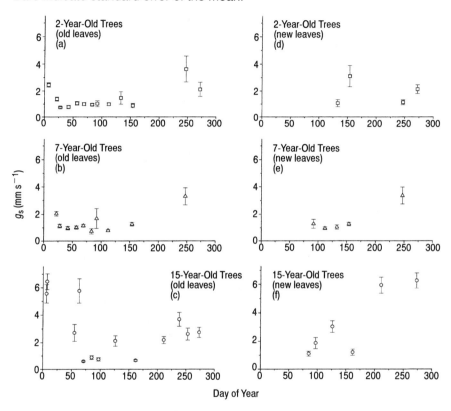

2-year-old trees. The g_s of new leaves on the mature citrus was significantly higher than values recorded for new leaves on the 7-year-old and 2-year-old trees (Figure 1d, e, f). While the 15-year-old trees had growth flushes earlier than the 7-year-old trees, their new leaves also exhibited an increase in g_s as the season progressed. Similar trends are suggested for 7-year-old trees, but not for 2-year-old trees.

These data illustrate important differences in stomatal response between the old and new leaves of different-sized canopies. These differences highlight the complex nature of the stomatal systems involved. Canopy dynamics of large and small trees clearly has an influence on the stomatal response of all leaves. The more exposed canopy of the smaller trees resulted in little or no difference in g_s between new and old leaves. Conversely the relatively sheltered leaf environment of the large canopy highlights differences in g_s between leaves of different ages. Light levels incident upon individual leaves may also contribute to the difference or lack of between leaves of different ages. Typically older leaves within a mature canopy are more shaded than new ones. The old leaves that were once sunlit and which have acclimated to high light levels require a greater light level for saturation than leaves that have always been shaded. Likewise the relatively sparse canopy of the 2-year-old trees ensures all leaves are exposed to relatively high light levels at some time of the day.

Stem xylem water potential measurements indicated that 2-year-old trees had slightly lower plant water status than either the 7-year-old, or the 15-year-old trees. Mean values of Ψ_x from measurements made periodically throughout the March to May period were -1.05 ± 0.088, -0.97 ± 0.048 and -0.81 ± 0.022 MPa for the 2-year-old, 7-year-old and 15-year-old trees. These values of Ψ_x indicate plants were not water stressed throughout the experimental period (Lloyd and Howie, 1989; Syvertsen, 1982).

Leaves of the inner canopy are more shaded in the 15-year-old trees and they have lower g_s than outer leaves of the same tree (Figure 2a, b, c). This is interesting because we found that citrus stomata have a low light saturation level at around 200 μmol m^{-2} s^{-1}. This is somewhat lower than the value of 500 μmol m^{-2} s^{-1} reported by Syvertsen and Lloyd (1994). Measurements of incident PAR showed that light levels of the inner leaves were, on average, only 10% of the PAR reaching the outer leaves. However, it is likely that shade leaves reach light saturation at lower light levels than sun leaves do. Because a large proportion of citrus leaves on mature trees are located within the dense, shaded parts of the canopy their different stomatal response must be considered when calculating plant water use. Figure 2 also highlights large differences in g_s values between DOY 63 and DOY 85 for the 15-year-old trees. Leaf temperature was, on average, 8°C cooler on DOY 63 than on DOY 85. This may help explain the large differences in g_s between these two days, considering that both PAR and VPD for these two days were similar.

The 7-year-old trees showed a trend towards a lower g_s in the inner leaves when compared to outer leaves but this difference was not significant (Figure 2d, e, f). The light levels for the inner leaves of the 7-year-old trees were approximately 20% of that incident upon the outer leaves, a slightly higher light level than that recorded for the inner leaves of the 15-year-old canopy. We may assume that leaves in the inner canopy of these 7-year-old trees received sufficient light to reach saturation. It is important to note that shade acclimated leaves require lower light levels to open fully than do sun acclimated leaves. It appears that we may treat all leaves on a 7-year-old canopy as being outer leaves. The mean overall g_s of outer leaves of the 15-year-old

FIGURE 2. Measured diurnal stomatal conductance (g_s, mm s^{-1}) values for inner and outer leaves of 15-year-old (a, b, c) and 7-year-old trees (d, e, f). Note differences in scale between 15-year-old and 7-year-old trees. Bars indicate standard errors of the mean.

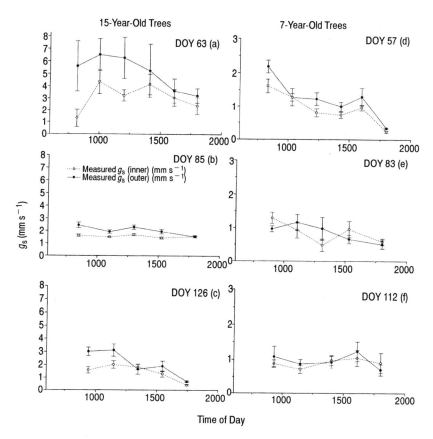

trees was higher than the g_s on the outer leaves of the 7-year-old canopy. The well ventilated canopy of the smaller trees may mean an increase in VPD of the air surrounding these trees when compared to VPD levels in the sheltered mature grove. Increased VPD leads to reduced g_s. This difference may also be related to leaf temperature. A reduction in leaf temperature reduces leaf-to-air vapor pressure difference and therefore increased g_s. The g_s of the inner leaves was comparable between canopies despite more light reaching the inner leaves of the 7-year-old canopy. The differences in stomatal conduc-

FIGURE 3. Diurnal measurement of stomatal conductance (g_s, mm s^{-1}) on the north and south sides of 15-year-old citrus trees. Bars indicate standard errors of the mean.

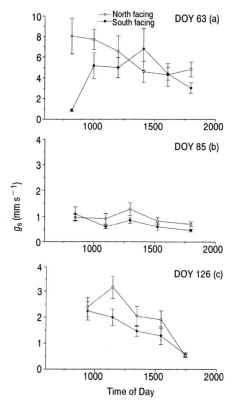

tance between canopies of different sizes indicates that microclimates must be taken into account when investigating stomatal response.

Figure 3 illustrates the difference in g_s levels between the north and south side of the 15-year-old trees. The heavily shaded north side gives frequently higher values of diurnal g_s over the three days. The ability of citrus stomata to open under low-light levels has already been mentioned. The leaves on the north facing side received enough light to open fully. It may be that leaves on the shaded outer portion of the canopy have a greater level of conductance because they do not experience the short term water-deficit that fully sunlit leaves do. Mean midday stem xylem-water potentials on the north facing side of the tree were found to be consistently higher (less negative) than values recorded on the south side. Values were -0.87 ± 0.03 and -0.76 ± 0.03.

Additionally, leaf temperatures were significantly lower for the north facing section of the canopy. Values of mean daily leaf temperature from February to April 1998 were 24.16 ± 0.42 and 25.39 ± 0.40°C for the north and south sides, respectively.

Model Predictions

Diurnal measurements of g_s during late winter through to early spring (DOY 60-126) consistently showed that the maximum g_s of all trees occurred early in the day when VPD was low (Figures 4 and 5). These data also confirm that citrus leaves have very low values of g_s during the late winter-early spring months. Values of g_s presented here are similar to late winter-early spring values presented by Syvertsen (1982). Values of calculated g_s from Equation 1 were in good agreement with the measured g_s on both the 2-year-old and 7-year-old trees (Figure 4). However, the model output was not so convincing for the 15-year-old trees if we used the g_s/VPD response developed using our porometer data (Figure 5). The scatter in g_s values from the 15-year-old trees was greater than the leaf-to-leaf variation in g_s on the smaller trees. For the purpose of modelling the g_s/VPD response, a much better prediction of g_s was obtained using the function of Syvertsen and Salyani (1991) which was derived for mature grapefruit leaves in Florida. Their response was derived using a LI-COR 6200 (LI-COR, Lincoln, Nebraska, USA) portable photosynthesis system used in controlled environmental conditions, and they developed leaf-to-air vapor pressure differences.

The general trend was higher g_s for 15-year-old trees than for 7-year-old or 2-year-old trees. It is unlikely that this greater g_s is due to the slightly elevated water status of the leaves of the 15-year-old trees. Values such as those recorded for the 7-year-old and 2-year-old trees indicate no water stress over the experimental period (Kaufmann and Levy, 1976). Rather it may be due to the lower leaf temperatures recorded on the 15-year-old trees when compared to leaf temperatures of the 7-year-older trees due to the shading of leaves within the mature canopy. A reduction in leaf temperature will reduce the leaf to air vapor pressure difference and therefore increase g_s of leaves from the large trees. Figures 4 and 5 show the model of the stomatal response function does a reasonable job of predicting g_s of old citrus leaves. Further model parameterisation is required to establish the stomatal response of new leaves.

Calculated and Measured Water Use

Having established the stomatal response function for these citrus trees we may now calculate plant water use via the modified Penman-Monteith equation. A comparison between the calculated and measured plant water use of

FIGURE 4. Measured and calculated diurnal stomatal conductance (g_s, mm s^{-1}) values for 2-year-old and 7-year-old citrus trees and the corresponding VPD (kPa) throughout the day. Bars indicate standard errors of the mean.

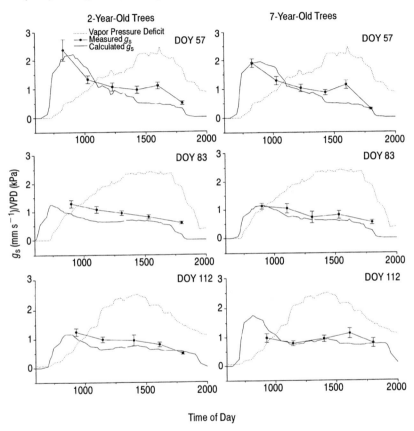

the 2-year-old pot grown trees is presented in Figure 6b. Calculated daily values of water use (E_p) show good agreement with measured changes in pot weight (E_w). We assume the fraction of sunlit leaves as 0.65 of the total canopy at any particular time. If we assume all leaves within the 2-year-old canopy are continuously sunlit as we initially suspected, the model gives an overestimation of water use. As expected, the daily transpiration rates in these 2-year-old trees was low ($E < 1.5$ L d^{-1}, Figure 6a, 6b) because the trees had a small leaf area ($A_t < 2$ m^2), and the g_s were low ($g_s < 2$mm s^{-1}, Figure 4). For the purpose of calculation we assume all leaves are the same age.

FIGURE 5. Measured, calculated (using the leaf-to-air pressure difference response function developed by Syversten and Salyani, 1991) and calculated (using the VPD response developed for our 15-year-old trees) diurnal stomatal conductance (g_s, mm s^{-1}) values and the corresponding VPD throughout the day. Bars indicate standard errors of the mean.

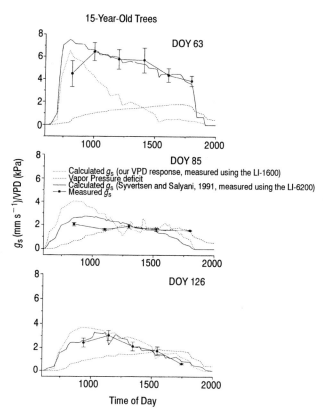

Having an indication of the accuracy of the model we may now scale up so as to calculate daily water use from a mature tree. Figure 6c illustrates calculated daily water use of the 15-year-old trees over the same days. These calculations indicate daily water use in these 15-year-old trees of about 70 L d^{-1}. We have assumed a ground area per tree of 27 m^2 (spacing 4.5 × 6.0 m^2), a leaf area density of 4 (m^2 m^{-3}, from canopy volume estimates, Wheaton et al., 1995) and a sunlit leaf area of 25% of the total canopy leaf area measured to be between 80-90 m^2. Water use per tree of about 70 L d^{-1} equates to 2.6 mm d^{-1} transpiration from the tree. If we add soil evaporation

FIGURE 6. (a) Calculated, via modified Penman-Monteith, values of water use (E, L h^{-1}) of 2-year-old trees, (b) comparison of actual daily weight change (E_w actual) and calculated daily water use via modified Penman-Monteith (E_p calculated) of 2-year-old trees, and (c) calculated daily water use of 15-year-old trees. Note differences in scale between the three graphs.

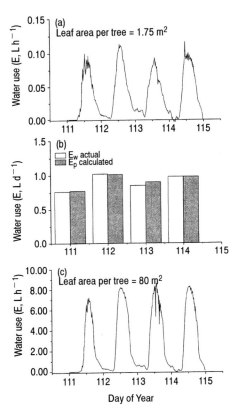

as a maximum of 1.5 mm d^{-1}, we then have a total of 4.1 mm d^{-1} at this time of the year. Despite having assumed a high level of evaporation from the soil surface (Jones, 1983, p. 95) this values of water use is still lower than previously published values for mature Florida grown citrus (Smajstrla et al., 1984). Additionally, soil moisture measurements made within the top 1 m of the root zone of the 15-year-old trees showed that between 2-3 mm of water was missing from the rootzone each day. These soil-based values of evapotranspiration are similar to our evapotranspiration levels predicted from the modified Penman-Monteith equation. These results thus indicate that current guidelines for irrigation of citrus in Florida may be too high. This could be an important consideration as much of Florida citrus is grown on very free draining sands overlying aquifers which supply upwards of 85% of Florida's drinking water. If excess water is passing through the rootzone so are nutrients and contaminants.

CONCLUSIONS

We have determined that the new leaves on mature citrus have a greater g_s than old leaves; this was not true for smaller trees. This difference in g_s related to canopy size and tree age. This will, in turn, impact upon water use of trees. Citrus stomata have low g_s values which tend to peak early in the day when VPD is low. The model we used for g_s did a reasonable job of predicting g_s for old leaves on both small and large trees. However, further work is needed in order to parameterise the model better for different populations of leaves. To do this we will need a greater understanding of the microclimate within the different sized canopies. This would help to unravel the stomatal response differences between the inner and outer leaves of both the 7-year-old and 15-year-old canopies. The consistently lower g_s values on the south (sun-facing) side of the canopy when compared to the north facing canopy suggests that citrus leaves require only low light levels to open fully. Further it suggests that short-term reductions in stem xylem water potential and/or increased temperature of the south facing leaves appears to have an influence on g_s. Finally, the agreement between measured and calculated plant water use of the 2-year-old trees illustrates the suitability of the Penman-Monteith model for the prediction of water use in citrus. From this model we may scale up from small to large and therefore estimate water use of 15-year-old field grown trees.

This study highlights the differences between leaves of differing ages and positions within the citrus canopy and thus the influence this is likely to have on the plant water use of citrus. With better knowledge of plant water use, irrigation may be more effectively managed.

REFERENCES

Begg, J. E. and N. C. Turner. (1970). Water potential gradients in field tobacco. *Plant Physiology* 46:343-346.

Brakke, M. and L. H. Allen, Jr. (1995). Gas exchange of citrus seedlings at different temperatures, vapor-pressure deficits, and soil water contents. *Journal of the American Society for Horticultural Science* 120(3):497-504.

Cohen, S., M. Fuchs, M. S. Moreshet and Y. Cohen. (1987). The distribution of leaf area, radiation, photosynthesis and transpiration in a shamouti orange hedgerow orchard. Part II. Photosynthesis, transpiration, and the effect of row shape and direction. *Agricultural and Forest Meteorology* 40:145-162.

Green, S. R. and K. G. McNaughton. (1997). Modelling effective stomatal resistance for calculating transpiration from an apple tree. *Agricultural and Forest Meteorology* 83:1-26.

Jarvis, P. G. (1976). The interception of the variations in leaf water potential and stomatal conductance found in canopies in the field. *Philosophical Transactions of the Royal Society of London B Biological Sciences* 273:593-610.

Jarvis, P. G. (1985). Coupling of transpiration to the atmosphere in horticultural crops: the omega factor. *Acta Horticulturae* 171:187-205.

Jarvis, P. G. and K. G. McNaughton. (1986). Stomatal control of transpiration: Scaling up from leaf to region. *Advances in Ecological Research* 15:1-49.

Jones, H. G. (1983). *Plants and Microclimate: A Quantitative Approach to Environmental Plant Physiology.* Cambridge University Press, Cambridge, U.K.

Jones, H. J. (1998). Stomatal control of photosynthesis and transpiration. *Journal of Experimental Botany* 49(special issue):387-398.

Jones, H. G., A. N. Lakso, and J. P. Syvertsen. (1985). Physiological control of water status in temperate and subtropical fruit trees. *Horticultural Reviews* 7:301-344.

Kaufmann, M. R. and Y. Levy. (1976). Stomatal response of *Citrus jambhiri* to water stress and humidity. *Physiologia Plantarum* 38:105-108.

Kriedemann, P. E. (1971). Photosynthesis and transpiration as a function of gaseous diffusive resistance in orange leaves. *Physiologia Plantarum* 24:218-225.

Landsberg, J. J. and D. B. B. Powell. (1973). Surface exchange characteristics of leaves subject to mutual interference. *Agricultural Meteorology* 12:169-184.

Levy, Y. and J. P. Syvertsen. (1981). Water relations of citrus in climates with different evaporative demands. *Proceedings of the International Society of Citriculture* II:501-503.

Leuning, R., F. M. Kelliher, D. G. G. De Pury, and E. D. Schulze. (1995). Leaf nitrogen, photosynthesis, conductance and transpiration: Scaling from leaves to canopies. *Plant, Cell and Environment* 18:1183-1200.

Lloyd, J. and H. Howie, (1989). Salinity, stomatal responses and whole-tree hydraulic conductivity of orchard 'Washington Navel' orange, *Citrus sinesis* (L.) Osbeck. *Australian Journal of Plant Physiology* 16:169-179.

McCutchan, H. and K. A. Shackel. (1992). Stem-water potential as a sensitive indicator of water stress in prune trees (*Prunus domestica* L. cv. French). *Journal of the American Society for Horticultural Science* 17:607-611.

Meyer, W. S. and G. C. Green. (1981). Comparison of stomatal action of orange, soybean and wheat under field conditions. *Australian Journal of Plant Physiology* 8:65-76.

Monteith, J. L. and M. Unsworth. (1990). *Principles of Environmental Physics, 2nd ed.,* Edward Arnold, U.K.

Postal, S. (1993). Running dry. *The Unesco Courier,* May 1993.

Smajstrla, A. G., G. A. Clark, S. F. Shih, F. S. Zazueta, and D. S. Harrison. (1984). Characteristics of potential evapotranspiration in Florida. *Soil and Crop Science Society of Florida* 43:40-46.

Syvertsen, J. P. (1982). Minimum leaf water potential and stomatal closure in citrus leaves of different ages. *Annals of Botany* 49:827-834.

Syvertsen, J. P. and M. Salyani. (1991). Petroleum spray oil effects on net gas exchange of grapefruit leaves at various pressures. *HortScience* 26:168-170.

Syvertsen, J. P., M. L. Smith Jr., and J. C. Allen. (1981). Growth rates and water relations of citrus leaf flushes. *Annals of Botany* 47:97-105.

Syvertsen, J. P. and J. J. Lloyd. (1994). Citrus. In: *Handbook of Environmental Physiology of Fruit Crops, Volume II: Sub-Tropical and Tropical Crops*. Schaffer, B. and P. C. Andersen (eds.) Boca Raton, FL: CRC press Inc., USA.

van Schilfgaarde, J. (1994). Irrigation–a blessing or a curse? *Agricultural Water Management* 25:203-219.

Wheaton, T. A., J. D. Whitney, W. S. Castle, R. P. Muraro, H. W. Browning, D. P. H. Tucker (1995). Citrus scion and rootstock, topping height and tree spacing affect tree size, yield, fruit quality, and economic return. *Journal of the American Society for Horticultural Science* 120:861-870.

Long-Term Reuse of Drainage Waters of Varying Salinities for Crop Irrigation in a Cotton-Safflower Rotation System in the San Joaquin Valley of California– A Nine Year Study: I. Cotton (*Gossypium hirsutum* L.)

Sham S. Goyal
Surinder K. Sharma
Donald W. Rains
André Läuchli

Sham S. Goyal, Surinder K. Sharma, Donald W. Rains, and André Läuchli are affiliated with the Department of Agronomy and Range Science, Department of Land, Air, and Water Resources, University of California at Davis, Davis, CA 95616.

Address correspondence to: Sham S. Goyal at the above address (E-mail: ssgoyal@ ucdavis.edu).

The current address for Surinder K. Sharma: Central Soil Salinity Research Institute, Karnal 132001, Haryana, India.

The authors wish to express their appreciation to their major cooperator, J. G. Boswell Company in Corcoran, California, and their staff at The El Rico Ranch for providing field-work related physical facilities, including land and water, and for their help with the field operations. The authors would also like to thank their colleagues namely, James Biggar, Donald Nielsen, James Oster, Dennis Rolston, and Kenneth Tanji for their valuable contributions to discussions during the tenure of the study. The funds for the study were provided by the Office of the Dean, College of Agricultural and Environmental Sciences, University of California at Davis; the Tulare Lake Drainage District, Corcoran; J. G. Boswell Company, Corcoran; University of California Salinity/Drainage Taskforce; and The Water Resource Center of the University of California.

[Haworth co-indexing entry note]: "Long-Term Reuse of Drainage Waters of Varying Salinities for Crop Irrigation in a Cotton-Safflower Rotation System in the San Joaquin Valley of California–A Nine Year Study: I. Cotton (*Gossypium hirsutum* L.)." Goyal, Sham S. et al. Co-published simultaneously in *Journal of Crop Production* (Food Products Press, an imprint of The Haworth Press, Inc.) Vol. 2, No. 2 (#4), 1999, pp. 181-213; and: *Water Use in Crop Production* (ed: M. B. Kirkham) Food Products Press, an imprint of The Haworth Press, Inc., 1999, pp. 181-213. Single or multiple copies of this article are available for a fee from The Haworth Document Delivery Service [1-800-342-9678, 9:00 a.m. - 5:00 p.m. (EST). E-mail address: getinfo@haworthpressinc.com].

SUMMARY. Use of saline drainage water for crop irrigation was evaluated as a means of decreasing its volume. Results of a nine-year crop rotation (cotton-cotton-safflower, × 3) in which only the cotton was irrigated with drainage water of 400, 1,500, 3,000, 4,500, 6,000, and 9,000 ppm total dissolved salts are presented. The different salinity levels of irrigation waters were achieved by mixing nonsaline canal water (400 ppm) and saline drainage water. Cotton lint yields were not affected by increased salinity level of the irrigation water for the first two years. Detrimental effects became evident in the third cotton crop with increasing severity in later years. In the fifth year of cotton (seventh year of the study), lint yields were adversely affected by waters of salinity greater than 3,000 ppm. However, fiber quality remained unaffected at all levels of irrigation water salinity. The reductions in lint yield appeared to be a function of time and the salinity level of applied water. Shoot height and biomass were reduced by the irrigation water salinity before lint yields. Stand establishment appeared to be the most sensitive to salinity and was perhaps the main reason for yield reduction. Increase in irrigation water salinity increased Na^+ content of leaf blades and petioles and decreased K^+/Na^+ ratio of leaf blades and petioles. The study showed that irrigation waters of up to 3,000 ppm salinity may be used for four years without any yield reductions, as long as some leaching occurs through preplant irrigations with low salinity water. Data on crop growth and development and ionic content collected over the nine year period are presented. *[Article copies available for a fee from The Haworth Document Delivery Service: 1-800-342-9678. E-mail address: getinfo@haworthpressinc.com <Website: http://www.haworthpressinc.com>]*

KEYWORDS. Cotton, fibre quality, water salinity, shoot height, crop growth, nutrient content

Disposal of sub-surface drainage water in the Central Valley of California has been a controversial issue. Its disposal into the rivers or into the San Francisco Bay via the Delta has been dismissed because of concern for down stream urban usage of water and environmental reasons (Tanji and Enos, 1994). Therefore, the growers are forced to store the drainage water in on-farm or district-wide evaporation ponds. However, some evaporation ponds may represent an environmental hazard due to their high concentrations of salts and of certain hazardous elements, such as selenium (Tanji et al., 1986 and Page et al., 1990). Moreover, according to estimates, the evaporation ponds may occupy 10-15% of the drained land area (K.K. Tanji, personal communication).

Use of drainage water for crop irrigation has been considered in the San Joaquin Valley of California as a means to decrease the volume of drainage water (Imhoff, 1990) and to decrease the need for nonsaline irrigation water, which many times has to be brought in from longer distances. Studies conducted in the Imperial Valley of California showed that saline water at levels up to 3,000 ppm total dissolved solids (TDS) can be used for irrigation of salt tolerant crops as long as some leaching occurred through preplant irrigations with nonsaline water or irrigation of salt-sensitive crops in the rotation with nonsaline water (Rhoades et al., 1988a,b). Ayars et al. (1986a,b) were also able to get reasonable yields of cotton, wheat, and sugarbeets in the San Joaquin Valley for three years using water of about 6,000 ppm level accompanied with preplant irrigations using nonsaline water. In a study on the cyclic use of drainage water (one or two years out of three), saline drainage water with an average EC of 7.4 dS m^{-1} (approximately 5,300 ppm TDS) was used to irrigate a tomato-cotton rotation system after the plants were established using nonsaline water. The tomato and cotton yields were little affected until the last year of the study (Shennan et al., 1995). However, the potential of long-term use of these waters for irrigation is not clear because most experiments have not proceeded long enough to test fully the impacts on crop productivity and soil degradation. Hence, the currently available information is not sufficient to recommend such a practice. Also, much of the salinity research in the past has been concerned with the issues of salt-affected soils and much less work has been done with the idea of using saline water for irrigation. In-depth research is needed to establish the levels of potential long-term reuse of drainage water and its possible detrimental effects on soil properties and crop productivity. Results of this research will provide the growers, irrigation and drainage managers, and water resource planners with information to help them in dealing with the present problem of drainage water disposal. This study hopes to make a valuable contribution in addressing these long-term issues.

The objectives of this research were (a) to evaluate the effects of long-term use of drainage waters of varying salinity/sodicity levels applied as irrigation on the germination and stand establishment and the subsequent growth, development and yield response of crops grown in a typical cropping rotation in the San Joaquin valley (cotton-cotton-safflower) and (b) to evaluate the effects of drainage water of varying salinity levels applied as irrigation on plant ion and nutrient characteristics. Results on cotton are reported in this paper.

MATERIALS AND METHODS

Experimental Site

The experimental site consisted of an approximately 8 ha site (195 m × 405 m) in the Tulare Lake Basin in Kings County of the San Joaquin Valley in Central California. The soil type was partially-drained Tulare Clay and is classified as fine, montmorillonitic (calcareous), thermic, Vertic Endoaquolls. The site was divided into 26 plots of 15.5 m × 195 m out of which 24 plots were used for 6 treatments replicated 4 times (Figure 1). One plot was left on each side as borders. The plots were large enough so that all operations including harvesting could be done using commercial equipment. Prior to the beginning of the experiment in 1984, three parallel tile drains running the entire length of the experimental site were installed at a depth of about 3 m (Figure 1).

Experimental Treatments

The experimental treatments consisted of irrigation with waters of six salinity levels. The salinity levels of the waters used for irrigation were approximately 400 (nonsaline control), 1,500, 3,000, 4,500, 6,000 and 9,000 ppm total dissolved salts (TDS). Various salinity levels of irrigation water were achieved by mixing nonsaline water (400 ppm TDS) from the California Aqueduct with saline drainage water from evaporation ponds. Appropriate metering valves to achieve the desired blends of the two waters in large volumes were installed. The salinity levels of the blended waters were closely monitored by using an EC meter (Jenco Electronics/Chemtrix, Inc., Hillsboro, OR 97124). To minimize the cross-effects of irrigation waters, the treatments within a block were arranged such that a plot of a particular salinity level was neighbored only by the next salinity level up or down. For example, the control plots were flanked by the plots irrigated with waters of 1,500 and 3,000 ppm. The sequence of irrigation was such that the lowest salinity plots were irrigated first followed by the next higher salinity level up and so on. Sometimes the three higher salinity level (4,500, 6,000, and 9,000 ppm) plots had to be irrigated at different times than the lower three. In such instances also the lower salinity level plots, within the group, were irrigated first followed by the next higher level. This irrigation scheme helped to keep the cross effects to a minimum. The saline drainage water was notably high in Na^+, Cl^-, SO_4^{-2}, NO_3^-, and B (Table 1). Consequently, the concentration of these ions increased in the irrigation waters with increasing salinity. In contrast, K^+ content changed only slightly with increasing salinity of the irrigation waters. All plots received a pre-plant irrigation with nonsaline water (400 ppm TDS).

Cultural Practices

In general, typical cultural practices of commercial agriculture in the area were followed. The entire site was preirrigated every year in January with

FIGURE 1. An infra-red aerial photograph of the field-site showing various treatments and location of tile-drains. Photo was taken on August 18, 1985.

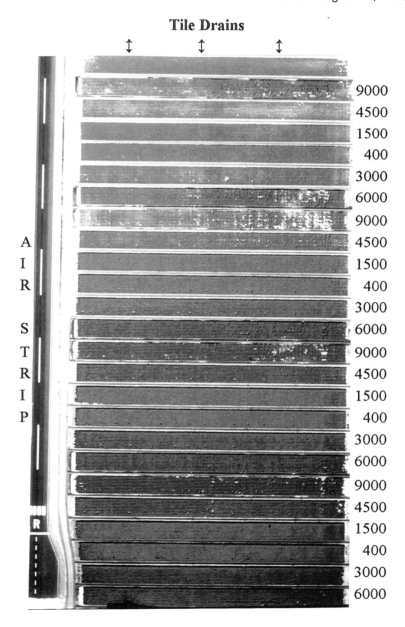

TABLE 1. Chemical analysis of typical irrigation waters; samples taken at outlet before application to plots.

Item	Analysis at treatment (total dissolved solids, ppm)					
	400	1,500	3,000	4,500	6,000	9,000
Electrical conductivity (EC), dS/m	0.95	2.46	4.95	6.95	9.34	11.60
pH	8.70	8.80	8.80	9.00	8.50	8.30
Sodium (Na^+), meq/L	4.79	17.50	42.0	63.50	87.40	109.00
Calcium (Ca^{+2}), meq/L	1.60	2.01	3.45	5.90	6.11	8.82
Magnesium (Mg^{+2}), meq/L	2.30	4.02	10.90	14.80	14.10	16.30
Potassium (K^+), meq/L	0.35	0.34	0.30	0.26	0.38	0.39
Bicarbonate (HCO_3^-), meq/L	3.10	3.20	4.40	4.30	4.70	5.20
Carbonate (CO_3^{-2}), meq/L	0.80	1.60	1.80	2.30	1.60	1.10
Chloride (Cl^-), meq/L	2.05	5.19	11.30	15.60	51.10	70.60
Sulphate (SO_4^{-2}), meq/L	3.66	15.00	37.60	59.00	59.40	74.80
Nitrate (NO_3^-), ppm	1.40	3.30	7.60	10.00	25.00	22.00
Boron (B), ppm	0.64	1.10	1.90	2.90	4.30	4.50

Reproduced with permission from Rains et al., 1987. *Calif. Agric.* 41 (Sept.-Oct.):24-26.

about 8 inches of nonsaline water from the California Aqueduct (wetting about top 24 inches of soil profile). In early March, the land was disked several times to obtain a good seed-bed and about two feet high berms separating the plots were raised. Twelve rows of cotton of cultivar SJ-2 at 1 m row spacing were planted in each plot along the length of the plot in late March or early April at a seeding rate of 20 kg per hectare. Nitrogen was applied pre-plant in the fall at the rate of 125 kg per hectare as anhydrous ammonia. In 1988, 30 kg per hectare of additional N was flown-on as ammonium nitrate. Treflan herbicide was applied pre-plant in the preceding fall for weed control. However, some cultivation was also performed during the growing season, as necessary.

Crop Growth and Development Parameters

The crop rotation consisted of 2 years of cotton followed by 1 year of safflower and three complete rotations (9 years total) were completed during the course of this study. For cotton all irrigations after the pre-plant irrigation were made with about 6 inches (15 cm) waters of various salinity discussed previously (Table 1). The safflower crop was grown only with a pre-plant irrigation with nonsaline water and no additional irrigation was necessary. The first cotton crop was planted in 1984. Representative samples of cotton plants were harvested throughout the growing season over different years from the treated plots. Plants from a 1-m row length were sampled at different times from the third row on the south side of the plots. Each treatment was sampled at three sites chosen randomly across the entire plot. Results are thus based on 12 samples (three sites times four replications) from 1-m of crop row. Data for shoot and root biomass, shoot and root length, number of bolls, seed cotton, and lint yields were recorded throughout the study.

Plant Analysis

Samples of leaf petioles and blades were collected from the plants harvested for the measurements of growth parameters. The samples were oven dried and ground to pass 40 mesh using a Wiley mill and analyzed for Na^+, K^+, Ca^{+2}, Mg^{+2}, Cl^-, NO_3^-, PO_4^{-3} and SO_4^{-2}. The cations in plant material were analyzed using atomic absorption spectrophotometry (Perkin Elmer, Inc., Norwalk, Conn., Model 5000) following digestion with nitric and perchloric acids (wet ashing) according to the AOAC method (Issac, 1990). The inorganic anions of Cl^-, NO_3^-, PO_4^{-3} and SO_4^{-2} were assayed using suppressed ion-chromatography (Dionex Corporation, Sunnyvale, CA) following extraction with distilled de-ionized water.

Soil Analysis

Electrical conductivity (EC), exchangeable sodium percentage (ESP), Cl^-, and B were determined on the saturation extracts. Nitrate, P, K^+, and SO_4^{-2} in soils were extracted with a saturated solution of calcium sulfate with 3.5 g/L silver sulfate, 0.5 M sodium bicarbonate (pH 8.5), 1.0 M ammonium acetate (pH 7.0), and 0.1 M lithium chloride, respectively. Nitrate and P in soil extracts were determined colorimetrically by the phenoldisulphonic acid

(Bremner, 1965) and the ammonium phosphomolybdate (Olsen and Sommers, 1982) procedures, respectively. Sulfate was determined turbimetrically (Beaton et al., 1968) and K^+ was measured using flame emission spectrophotometry. Boron in extracts was measured colorimetrically by the azomethine-H procedure (Wolf, 1974).

Cotton Harvest

The crop was chemically defoliated prior to harvest, using either Paraquat or Triphos. The center 6 rows out of the total 12 rows per plot were harvested using commercial three or six-row spindle (cotton) pickers.

Fiber Quality

Micronaire and fiber strength were determined by using Zellwerger Uster High Volume Instruments (HVI) and methods D-4604 as described in the Annual Book of ASTM Standards (Baldini et al., 1995).

The data reported are means of 12 or more observations. Standard errors are reported as " $\pm$ " values in tables and as vertical bars in figures.

RESULTS

The results have been summarized in two sections; the first section presents the results obtained over six cotton crops and includes effects on shoot and root growth, lint yield, and changes in Na^+ and K^+ content in leaf blades and petioles. The second section includes detailed changes in these parameters and changes in Ca^{+2}, Mg^{+2}, total-N, NO_3^-, Cl^-, SO_4^{-2} and PO_4^{-3} content within a single growing season.

Water and Soil Salinity

Analyses of the irrigation waters applied to plots are presented in Table 1. The concentrations of NO_3^-, Cl^-, SO_4^{-2} and Na^+ increased substantially as the proportion of drainage water increased. Ca^{+2}, Mg^{+2}, and B concentrations also increased, but the increase was relatively smaller than that of NO_3^-, Cl^-, SO_4^{-2} or Na^+. Potassium concentrations did not change significantly in response to proportion of drainage water.

An earlier paper reported changes in soil salinity in various plots in this study (Rolston et al., 1988). The soil profile was very saline at the deeper depths with EC as high as 11 dS/m in the 90-120 cm depth zone at the

TABLE 2. Changes in properties of the upper 30 cm of soil in plots where cotton was irrigated with saline drainage water of different salinity levels.

Treatment ppm	$EC^†$ dS/m	$ESP^‡$	NO_3^-	P ppm	SO_4^{-2}	Cl^- meq/L	B ppm	K^+ ppm
400	1.6 ± 0.2	8.0 ± 2.3	3.2 ± 1.0	4.5 ± 2.0	84	4.6 ± 0.9	0.7 ± 0.2	517
1500	2.6 ± 1.2	12.4 ± 3.1	2.0 ± 1.1	3.6 ± 0.7	1980	4.7 ± 0.7	1.1 ± 0.1	524
3000	4.6 ± 2.2	16.8 ± 2.5	3.5 ± 0.9	4.4 ± 1.2	3650	4.2 ± 0.9	1.4 ± 0.2	502
4500	7.3 ± 2.6	20.4 ± 4.1	2.7 ± 0.7	4.2 ± 1.4	5142	5.5 ± 1.3	1.5 ± 0.2	485
6000	9.8 ± 2.8	26.1 ± 4.2	26.1 ± 4.2	4.4 ± 1.9	6811	18.4 ± 1.6	1.6 ± 0.5	491
9000	10.4 ± 3.5	27.0 ± 6.1	19.3 ± 4.4	5.0 ± 1.7	5654	16.0 ± 0.2	1.9 ± 0.6	499

[†] EC = Electrical conductivity
[‡] ESP = Exchangeable sodium percentage

beginning of trial. Salinity was still high in most of the treatment plots at the 90-120 cm depth in 1985. Many areas still had ECs greater than 12 dS/m. This was largely due to a relatively high saline-water table before the start of the experiment.

The analyses of soil samples taken in 1993 suggest that considerable changes in the soil profile were caused by the use of saline drainage water for irrigation during the course of the experiment (Table 2). Although the soil samples taken in 1993 (one year after conclusion of the main study) were after a soil treatment with 2,000 kg/ha sulphuric acid followed by ponding with nonsaline water to help leach sodium, the salinity/sodicity of plots irrigated with saline waters was still significantly higher than control plots (Table 2). For example, EC values in top 30 cm ranged from 1.6, 7.3, 9.8, to 10.4 dS/m in plots irrigated with 400, 4,500, 6,000, and 9,000 ppm waters, respectively. Corresponding ESP values in these plots were 8, 20, 26, and 27, respectively (Table 2). The increase in ESP coupled with the increase in EC of the soil profile increases the availability of sodium in the ionic form, which was reflected in sharp increases in the sodium content of cotton plants (to be discussed later). Significant accumulation of Cl^-, NO_3^-, and SO_4^{-2} in soil also took place during this time (Table 2).

Crop Growth and Development

Plant density was not affected by the increase in salinity of irrigation waters in the first year, but was affected the second year onwards (Figure 2a). In 1985 (the second year of cotton) the plant density was reduced only by the irrigation water of the highest salinity (9,000 ppm). Plant density was significantly reduced in 1987 (the third cotton planting) by saline irrigation waters

FIGURE 2. Effect of irrigation water salinity on cotton (a) plant density, (b) shoot height, and (c) shoot dry weight represented as percent of 400 ppm (control) over six cropping years. Each data point is a mean of 12 independent observations and standard errors are shown as bars.

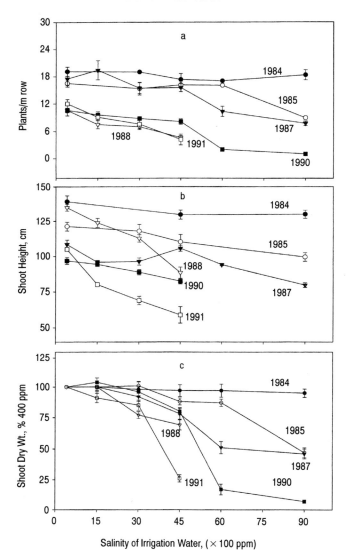

of 4,500 ppm and above. A decrease of 50% in plant density was observed in plots irrigated with 6,000 and 9,000 ppm waters as compared to the control. Plant densities in the years 1988 and 1991 (fourth and sixth cotton crops) were significantly reduced in plots irrigated with waters of 4,500 ppm salinity and were affected so drastically in plots irrigated with waters of 6,000 and 9,000 ppm salinity that there was essentially no stand establishment. In 1990 (fifth cotton year), a minimal stand establishment (only 1-2 plants per meter row) in plots irrigated with the two highest salinity waters was observed whereas the 4,500 ppm waters did not appear to be as detrimental as in 1988 and 1991 (Figure 2a).

Shoot height was not affected due to the salinity of the irrigation waters in 1984 (first cotton year) and started to be influenced from 1985 onwards (second cotton year, Figure 2b). In general, the shoot heights decreased as the salinity of the irrigation waters increased, and the magnitude of effect became progressively larger with the number of years of irrigation with saline water. For example, in 1985 and 1991 (second and sixth cotton crop) the shoot heights in plots irrigated with 4,500 ppm waters decreased by 10 and 44%, respectively, as compared to the control.

The shoot biomass was also affected from 1985 onwards (second cotton year) by the salinity of irrigation waters (Figure 2c). In 1985, the shoot biomass was decreased drastically only in plots irrigated with 9,000 ppm salinity (53% of the control). In 1987 and 1990 (third and fifth cotton crops), plots irrigated with waters of 6,000 and 9,000 ppm salinity showed drastic reductions in shoot biomass and in 1988 and 1991 (fourth and sixth cotton years). Significant reductions in shoot biomass were also observed in plots irrigated with 4,500 ppm waters (Figure 2c). In 1991, the effect was so drastic that the shoot biomass in plots irrigated with 4,500 ppm waters was almost the same as in the plots irrigated with waters of 6,000 or 9,000 ppm salinity in the previous year (Figure 2c).

Root biomass was not affected significantly in 1984 and 1985 (first two cotton crops), but was reduced from 1987 onwards (third cotton crop, Figure 3a). In 1987 and the subsequent years, root weight in all plots irrigated with waters of different salinity levels, including 400 ppm, was lower than in 1984 or 1985. However, within a given year the root biomass decreased as the salinity of the irrigation water increased, and the effect became progressively worse with the years of irrigation with saline waters. For example, in plots irrigated with waters of 4,500 ppm salinity, the root weight was 57 and 15 g/m row in the years 1987 and 1991, respectively. The pattern of root length, as affected by the salinity of the irrigation waters, was similar to that of root biomass (Figure 3b). Reductions in root length were observed a year earlier than root biomass. Overall, the root growth was affected more than the shoot growth.

FIGURE 3. Effect of irrigation water salinity on cotton (a) root dry weight and (b) root length over six cropping years.

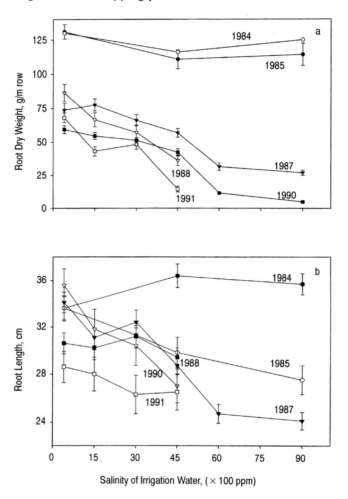

Salinity of Irrigation Water, (× 100 ppm)

Number of bolls per plant remained unaffected in 1984, the first cotton year. In 1985 and 1987 (second and third cotton crops), the number of bolls per plant increased with increasing salinity of irrigation waters (Figure 4a). A marginal increase in the number of bolls in plots irrigated with 1,500, 3,000 and 4,500 ppm waters over the 400 ppm plots was also observed in the years 1988 and 1990. However, in 1991 (sixth cotton crop), the increase in the

FIGURE 4. Effect of irrigation water salinity on cotton (a) number of bolls and (b) lint yields over six cropping years shown as percent of 400 ppm (control).

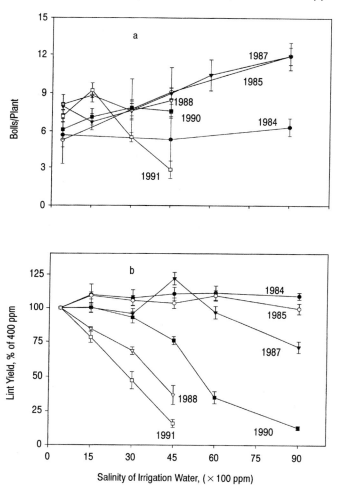

number of bolls was recorded only at 1,500 ppm level and a decline was observed at 3,000 and 4,500 ppm levels. The plots irrigated with waters of 6,000 and 9,000 ppm waters did not have a stand establishment in 1988 and 1991 (Figure 2a); hence the boll count from these treatments could not be recorded.

Lint yields were not reduced by the increase in salinity of irrigation water in 1984 or 1985 (first two cotton crops). Instead, irrigation with waters of 1,500 ppm and higher salinity promoted lint yields as compared to irrigation with water of 400 ppm (Figure 4b). It was only in 1987 (third cotton crop) that the detrimental effects of irrigation with saline waters became evident. The detrimental effects on lint yield became more severe in the later years, i.e., 1988 and 1991, when no harvests were made at 6,000 and 9,000 ppm levels due to lack of stand establishment (Figure 2a). In 1990 (fifth cotton crop), lint yields were adversely affected by salinity of irrigation waters greater than 3,000 ppm with a 24, 65, and 83% reduction in yield at 4,500, 6,000, and 9,000 ppm levels, respectively. Overall, irrigation with saline waters caused relatively more detrimental effects on vegetative growth than reproductive growth of plants.

Sodium and Potassium Accumulation in Cotton

In general, irrigation with saline drainage water increased the Na^+ content of leaf blades and petioles, and the increases were in direct proportion to the salinity of the irrigation water. In 1984 (first cotton crop), increase in Na^+ content due to the increase in salinity levels of the irrigation water was marginal. However, from 1985 onwards (second cotton crop) higher Na^+ content, both in leaf blades and petioles, was observed in plots irrigated with saline waters (Figures 5 and 6). The highest Na^+ levels in a given year, especially in leaf blades, were observed in the first sampling, which declined with plant age (Figures 5 and 6, a-f). Highest Na^+ contents were observed in 1987 and 1990, which probably corresponded with the salinity that might have accumulated in the soil due to the continuous use of saline waters for irrigation. Higher concentrations of Na^+ were observed in the leaf blades than the leaf petioles at all stages and in all years. Leaf petioles appeared to retain only a small part of the Na^+ taken up by the plant.

During the entire course of the study, K^+ content of leaf blades and petioles showed varying responses to irrigation with saline waters. In 1984, irrigation with saline drainage water caused only a marginal decrease in the K^+ content of leaf blades. However, as the years progressed the magnitude of decrease became larger. For example, in 1985, 1987, and 1990 (second, third, and fifth cotton crops) decreases of 15-30% in the K^+ content of leaf blades from plots irrigated with 9,000 ppm relative to those irrigated with 400 ppm waters were observed. In 1988 and 1991 (fourth and sixth cotton crop), however, a reverse pattern was noticed; the K^+ contents of leaf blades appeared to increase with the increasing salinity of the irrigation water. Leaf petioles consistently had 4-6 times higher K^+ content as compared to leaf blades. In general, the K^+ content of leaf petioles increased by 20-30% upon irrigation with waters of 9,000 ppm as compared to those with 400 ppm salinity (Figures 7 and 8). The changes in Na^+ and K^+ contents of leaf blades

FIGURE 5. Effect of irrigation water salinity on tissue Na+ concentration of cotton leaf blades and petioles in the years 1984 (a, b), 1985 (c, d), and 1987 (e, f).

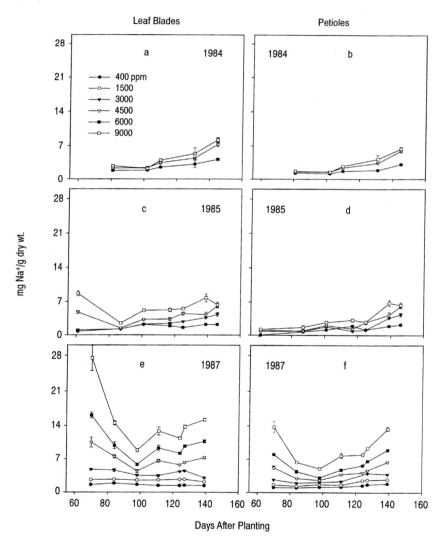

FIGURE 6. Effect of irrigation water salinity on tissue Na+ concentration of cotton leaf blades and petioles in the years 1988 (a, b), 1990 (c, d) and 1991 (e, f).

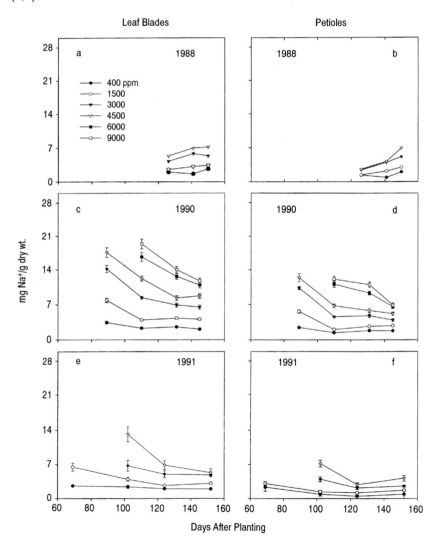

FIGURE 7. Effect of irrigation water salinity on tissue K⁺ concentration of cotton leaf blades and petioles in the years 1984 (a, b), 1985 (c, d), and 1987 (e, f).

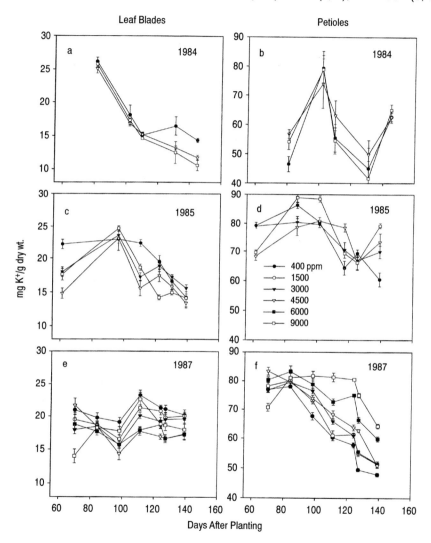

FIGURE 8. Effect of irrigation water salinity on tissue K⁺ concentration of cotton leaf blades and petioles in the years 1988 (a, b), 1990 (c, d) and 1991 (e, f).

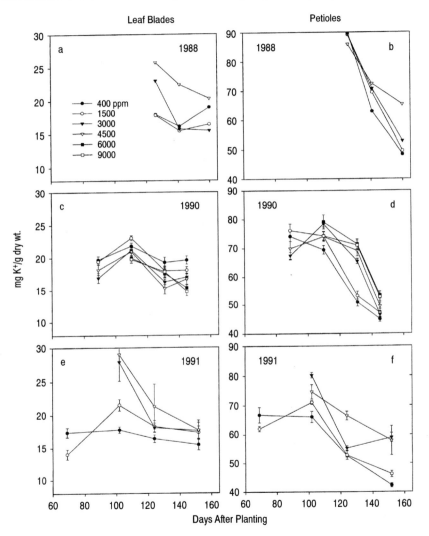

and petioles led to reductions in K^+/Na^+ ratios due to the use of saline water for irrigation (Figures 5, 6, 7, 8, and 10).

DETAILS OF A TYPICAL YEAR: 1987

Because the significant effects of irrigation water salinity on crop growth and development were first observed in 1987 (third cotton crop), data detailing the changes within that growing season are presented below:

Crop Growth and Development

Plant density and growth were affected severely by the salinity of irrigation waters. For example, plant density in the 6,000 and 9,000 ppm treatments was reduced by more than 50% of that in the control (400 ppm). However, no significant changes in stand establishment occurred as the season progressed (Figure 9a). Reductions in the shoot height and weight were directly proportional to the increase in salinity of applied waters. The shoot height was reduced by 10-25% due to the use of saline irrigation waters, as compared to the control (Figure 9b). During the earlier growth phase, shoot weight of plants in the plots irrigated with waters of the three highest levels of salinity was lower as compared to those irrigated with lower levels of salinity (Figure 9c). However, as the season progressed the shoot weight of affected plants started increasing rapidly and appeared to achieve similar growth rates as of those irrigated with lower levels of salinity.

Root weight and length were also suppressed due to the application of saline irrigation waters and the magnitude of reduction was proportional to the salinity of irrigation waters. Reductions of 63 and 30% in root weight and length were observed in plots irrigated with waters of 9,000 ppm as compared to those irrigated with 400 ppm salinity (Figure 9d, e). The root weight appeared to be more responsive to the adverse effect of saline irrigation water than root length.

The number of bolls per plant was increased by irrigation water salinity of 4,500 ppm and above; an increase of about 60% in the number of bolls was observed in plots irrigated with waters of 9,000 ppm as compared to those irrigated with 400 ppm salinity (Figure 9f). Lint yields in 1987 were 1,020, 1,020, 974, 1,247, 986 and 725 Kg/hectare at 400, 1,500, 3,000, 4,500, 6,000 and 9,000 ppm levels of irrigation waters, respectively. Apparently, the lint yields were reduced only in the plots irrigated with the waters of 9,000 ppm salinity, where a decrease of 29% was recorded as compared to that from plots irrigated with 400 ppm water. Interestingly, an increase of 22% in lint yield was observed in the plots irrigated with 4,500 ppm waters as compared to those irrigated with 400 ppm water salinity.

FIGURE 9. Effect of irrigation water salinity on cotton (a) plant density, (b) shoot height, (c) shoot dry weight, (d) root dry weight, (e) root length, and (f) number of bolls.

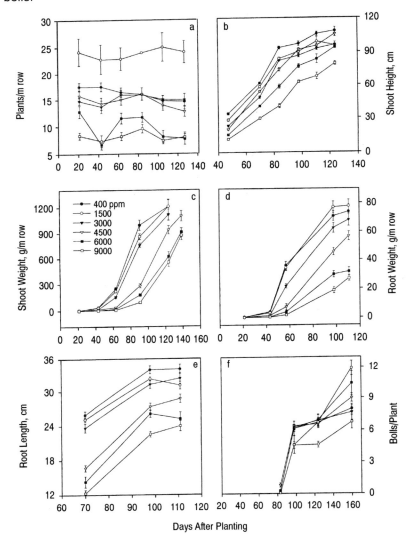

Sodium and Potassium Accumulation

Irrigation with saline drainage water during the third cropping season (1987) led to a sharp increase in Na^+ contents of leaf blades and petioles; blades showed a greater increase than petioles (Figure 5e, f). Increases in Na^+ content of both leaf blades and petioles were directly proportional to the increase in salinity of the applied waters. Leaf blades and petioles from plots irrigated with waters of 9,000 ppm salinity exhibited 7-10 and 4-5 fold higher Na^+ content, respectively, as compared to those from irrigated with 400 ppm. The Na^+ content of leaf blades and petioles increased significantly even when waters of lower salinity (e.g., 1,500) were used for irrigation (Figure 5e and f); this shows that Na^+ content is a sensitive indicator of salinity in cotton. Use of saline water for irrigation decreased the K^+ content of leaf blades but caused an increase in petioles (Figure 7e, f). This is in contrast to the observations made for Na^+ where the content increased in both plant parts in response to the use of saline waters for irrigation. This probably reflects the composition of the irrigation waters, which showed little change in K^+ content with increasing salinity while Na^+ increased. The changes in Na^+ and K^+ contents resulted in drastic alterations in K^+/Na^+ ratios of both leaf blades and petioles, which decreased progressively with the increase in irrigation water salinity. For instance, it decreased from about 14 to 0.5 and from 75 to 5 in leaf blades and petioles, respectively, with the increase in irrigation water salinity from 400 to 9,000 ppm (Figure 10a, b).

Other Cations and Anions

Application of saline irrigation water reduced the Ca^{+2} content of both leaf blades and petioles (Figure 10c, d). The magnitude of decrease appeared larger during the initial growth stages and narrowed with the plant age. For example, 21 days after planting, the Ca^{+2} content of blades decreased by about 100% in plots irrigated with waters of 9,000 ppm as compared to those irrigated with 400 ppm salinity. However, 139 days after planting the decrease narrowed down to only 25% (Figure 10c). In general, Ca^{+2} content of leaf blades was higher than those of petioles and increased with plant age. Irrigation with saline waters seemed to have no major effects on Mg^{+2} content of leaf blades or petioles (Figure 10e, f).

In general, Cl^- concentrations of leaf blades and petioles increased in response to salinity of irrigation waters and were higher in petioles than blades (Figure 11a, b). The overall Cl^- concentrations in the leaves declined, and the differences due to various treatments narrowed during the later stages of growth. Irrigation with saline drainage waters also tended to increase SO_4^{-2}-S content of leaves and petioles (Figure 11c, d).

Total nitrogen content was not influenced by the salinity levels of the

FIGURE 10. Effect of irrigation water salinity on tissue K^+/Na^+ ratio (a, b) and Ca^{+2} (c, d) and Mg^{+2} (e, f) concentrations in cotton leaf blades and petioles in the year 1987.

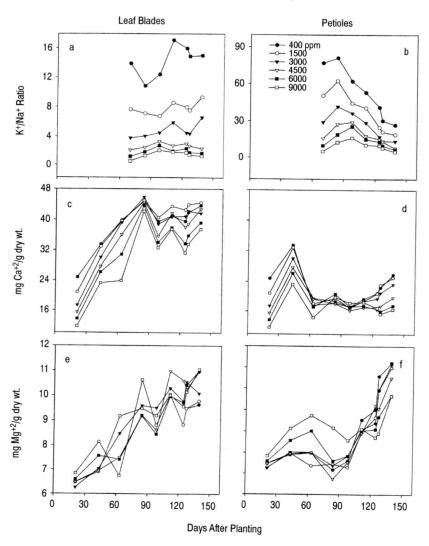

FIGURE 11. Effect of irrigation water salinity on tissue Cl$^-$ (a, b), SO$_4$$^{-2}$ (c, d), and NO$_3$$^-$ (e, f) concentrations in cotton leaf blades and petioles in the year 1987.

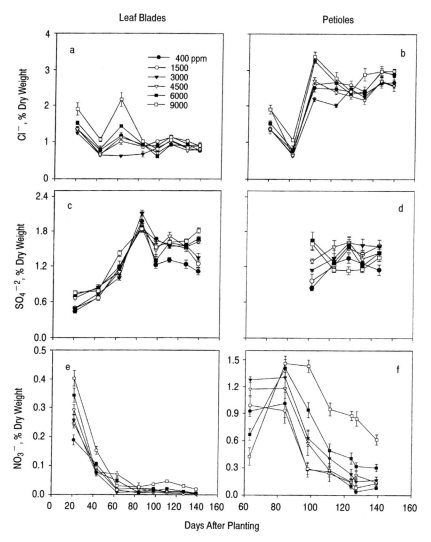

applied waters and also remained unaffected with plant age (results not shown). However, accumulation of NO_3^- in leaf blades and petioles increased with increasing salinity of the irrigation waters; the effects were more pronounced in the petioles than blades. This corresponded with the increasing NO_3^- concentration of irrigation waters with increasing salinity levels (Table 1). Generally, the NO_3^- content of leaf blades and petioles was higher in the beginning of the season and declined as the season progressed (Figure 11e, f). The NO_3^- content of petioles stayed consistently higher than those of leaf blades.

Fiber Quality

Effects of irrigation with saline waters on the fiber quality were measured in terms of strength and micronaire. Micronaire values increased with increase in the salinity levels of the irrigation waters, and an increase of about 20% was observed in fibers from plots irrigated with the waters of the highest salinity (9,000 ppm) as compared to those from control plots (Figure 12). Fiber strength was not affected by the salinity of irrigation waters (Figure 12).

DISCUSSION

Application of saline water for irrigation did not affect germination, plant density, and root and shoot growth of cotton plants in the first year. This was

FIGURE 12. Effect of irrigation water salinity on cotton fiber micronaire and fiber strength.

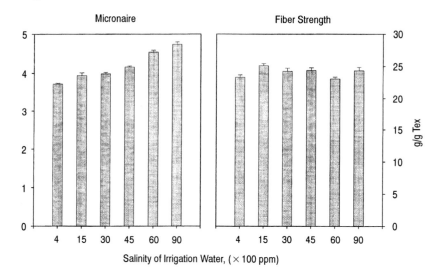

largely because the first irrigation with saline water was 82 days after planting by which time the seedlings had been fully established. However, the effects became evident from the second year. The reduction in plant density in plots irrigated with waters of 9,000 ppm salinity starting in the second year, and the others irrigated with waters of lower salinity in subsequent years, demonstrates the deleterious effects of saline water on stand establishment over longer periods. In 1988 (fourth cotton crop) heavy rainfall fell three days after planting. Due to the clay nature (Tulare clay) of the soil, significant crusting occurred and severely affected the seedling emergence in all plots. The stand establishment in plots previously irrigated with 6,000 and 9,000 ppm waters was affected so severely that the plots were not even worth harvesting (Figure 2a). Irrigation with saline water may have a lesser effect on soil permeability because increased soil salinity can compensate for increased sodicity (Thellier et al., 1990). However, problems are more likely to occur during or after rainfall or irrigation with nonsaline water on a field that was previously irrigated with a saline-sodic water as in this study. Obviously, the crust formation due to rainfall confounded the effects of soil salinity in the year 1988. Major effects on soil structure also were observed in plots irrigated with the two highest salinity waters which may have contributed to the poor stand. In 1990, the plots irrigated with 6,000 and 9,000 ppm waters had only about 2 and 1 plants/meter row, respectively, as compared to 10-11 plants/meter row in those irrigated with nonsaline canal water (Figure 2a). Despite an additional year of irrigation with saline waters, the plant density observed in 1990 was relatively higher than those observed in 1988. There are two possible explanations: (1) the crust formation due to rain in 1988 lowered the seedling emergence below the expected levels, and (2) 1989 was the safflower year in the crop rotation and no saline water was applied during that year. This coupled with nonsaline water preirrigation and winter rainfall may have provided some leaching of the top zone for the 1990 cotton crop. This is further supported by the observations made on next year's (1991) cotton crop in which the lowest plant densities were recorded for respective treatments; essentially no plants established in the plots irrigated with waters of the two highest salinity levels (Figure 2a). These observations show that for cotton, the primary effect of salinity may be on seed germination and stand establishment. Despite being categorized as a salt tolerant crop (Carter, 1975; Maas, 1986), cotton is known to be sensitive to salinity at germination (Shannon and Francois, 1977; Kent and Lauchli, 1985). Reductions in plant stand of cotton due to salinity have been reported earlier (Rains et al., 1987; Rhoades et al., 1988a,b; Shennan et al., 1995).

The adverse effects of saline water irrigation on the growth of cotton plants (as evidenced by decreases in shoot and root length and weight) appeared to be a function of the salinity level of the irrigation water and the number of years of application. The actual effects may be governed by the integrative amounts of salts added to the soil over the duration. The root growth, however, was affected earlier and more than the shoot growth. The shoot biomass in plots irrigated with 9,000 ppm waters was affected during the second year, whereas it remained almost unaffected in those irrigated throughout with waters up to 3,000 ppm salinity. In contrast, the root length of cotton plants in plots irrigated with waters of as low as 1,500 ppm salinity was affected during the third year (Figure 3b). Plant roots are directly exposed to the saline environment and are, therefore, affected the most, probably due to alterations in soil structure, water relations, and specific ion effects. The responses in root growth may have also been due to the effects of the saline water table and higher salinity at deeper layers of soil (see Rolston et al., 1988 for soil data) or their combined effects.

In the early years of experiment, when increases in soil salinity due to irrigation with saline waters were marginal, an increase in the number of bolls per plant was observed. However, continuous irrigation with saline waters over time resulted in the accumulation of salts in soil (Table 2 and Rolston, et al., 1988) which eventually led to a reduction in the number of bolls per plant even at 3,000 and 4,500 ppm levels as observed in 1991. It appears that boll numbers increased with increasing salinity up to a threshold beyond, which they declined. Pasternak et al. (1979) and Twerski et al. (1973) also observed similar increases in number of bolls of cotton irrigated with saline water of 4.2-8.0 dS/m EC. Saline water also increased the number of bolls at the lower nodes without reducing boll weight. Bernstein and Hayward (1958) observed a marked decline in vegetative growth and vigor of cotton plants in response to increase in soil salinity with little effect on the seed cotton yield. Lower to moderate levels of irrigation water salinity probably restricted the vegetative growth and caused the plants to shift into reproductive phase which promoted the number of bolls. Also, plant biomass was reduced much earlier in time than the lint yield, showing that vegetative plant growth was more sensitive than the reproductive growth. This suggests that the plants growing under salinity were able to redistribute their dry matter to favor reproductive growth (Pasternak and Malach, 1994). Changes in dry matter partitioning due to salinity have been reported in cotton (Pasternak et al., 1979) and barley (Grattan et al., 1994). Water deficit, a component of salt stress, is known to have primary effects on source activity due to reduction in leaf area and photosynthetic rate and not on sink activity (Ackerson et al., 1977). Ella and Shalaby (1993), on the contrary, reported no effect on the number of bolls per plant by waters of 3,200 and 6,400 ppm salinity levels, applied to cotton plants grown in pots. Continuation of flowering and boll formation in cotton is known to be a function of vegetative growth, which produces sites for additional fruiting branches and additional nodes on existing branches (Mauney, 1986). Severe reduction of vegetative growth of plants (Figure 2c) due to continuous irrigation with saline waters, and the resulting high soil salinities in these plots in later years, would have eventually led to the observed reduction in the number of bolls.

Higher concentrations of Na^+ in the irrigation waters, accompanied with increase in electrical conductivity (EC) and sodicity (ESP) of the soil, led to higher Na^+ uptake by the cotton plants, which was correlated with responses in crop growth and development (Table 2 and Figures 4 to 6). Relatively higher concentrations of Na^+ were observed during early growth stages, which tended to decline to lower levels during later stages (Figures 5 and 6). The decline in Na^+ content with increase in plant age may have either been due to higher rates of shoot growth (Figure 9c), which provided dilution effects and/or decrease in Na^+ uptake rates.

The Na^+ content patterns exhibited by cotton plants in this study are characteristic of Na^+-includer species as discussed by Leidi et al. (1992). Rains (1969) observed that cotton translocates both Na^+ and K^+ to the tops. He proposed the lack of a Na^+ regulation mechanism because the cotton plants translocated both Na^+ and K^+, whereas bean plants translocated considerable amounts of K^+ but little Na^+. The higher Na^+ concentrations of leaf blades than those of petioles show that the cotton leaf petioles did not have much regulatory control over Na^+ unloading into the leaf blades. Retention of Na^+ in the leaf petioles of *Lycopersicon* (tomato) was reported by Besford (1978). In the latter species, xylem parenchyma transfer (XPTs) cells of the leaf petioles removed Na^+ from the xylem stream before it entered the leaf lamina.

The observed increase in K^+ content of leaf petioles and decrease in that of leaf blades, in response to irrigation with saline water, may have been caused by some disruption (probably due to higher Na^+) of K^+ unloading from petioles to blades. The increases in Na^+ and decreases in K^+ contents of leaf blades led to sharp reductions in K^+/Na^+ ratios (Figures 5-8 and 10), which were mainly due to increases in Na^+ content. Decreases in K^+/Na^+ ratios in leaf blades almost paralleled the reductions in K^+/Na^+ ratios in the saline waters applied to the respective plots. Leidi et al. (1992) also reported proportional increases in Na^+ and Cl^- and decreases in K^+ and K^+/Na^+ ratios in cotton in response to increase in salinity levels of 50 to 300 mM in greenhouse experiments. Rather (1982) found that in solution culture experiments a Na^+/K^+ of 9 or greater reduced the concentration of K^+ in the leaves of a salt-sensitive cotton cultivar (Dandara) more than that in a salt-tolerant cultivar (Giza). In saline drainage waters used for irrigation in this study, increase in sodium at constant potassium concentrations increased Na^+/K^+ ratios from 14 to 279. Increased Na^+ coupled with decreased K^+ contents of leaf blades can be harmful to biological systems, as higher levels of sodium in the cytoplasm are reported to interfere with K^+ metabolism and many enzymes are sensitive to high Na^+ concentrations (Cheeseman, 1988).

Reduction in Ca^{+2} content of both leaf blades and petioles at higher salinity levels may primarily be due to the increase in Na^+/Ca^{+2} ratio of the applied saline waters from 3.0 in 400 ppm water to 12.3 in 9,000 ppm waters. Many studies have shown that Na^+-dominated soils or solutions reduce K^+ or Ca^{+2} uptake by plants and/or affect the internal distribution of these elements (Grattan and Grieve, 1994). Hence, the decrease in Ca^{+2} contents perhaps represents the competitive inhibition of its uptake by sodium. Higher Na^+/Ca^{+2} ratios may restrict root growth and function in cotton (Cramer et al., 1986). This may also explain the higher reductions in root growth observed in this study. Cramer et al. (1985, 1987) concluded that the primary response to NaCl stress in cotton roots was the displacement of membrane-associated calcium by sodium, leading to increased membrane permeability and the loss of K^+/Na^+ selectivity.

Increased nitrate contents of leaf blades and petioles (Figure 11e, f) were most probably due to the additional nitrate added via irrigation waters, as the nitrate concentration of irrigation waters increased with increasing salinity levels (Table 1). This may also have contributed to the observed delayed development and maturity of plants in plots irrigated with saline drainage water.

Phosphate content was reduced by increase in salinity. Martinez and Läuchli (1994) found earlier that high Na^+ concentrations inhibited phosphate uptake by cotton seedlings. Cotton is known to have a low P requirement (Tucker, 1974). It has also been long known that salt-stressed plants resemble P-deficient plants (Hewitt, 1963). Salinity-induced reductions in PO_4^{-3} have been reported in many crops under field conditions, but there is no evidence of P deficiency in the crops. Hence, we do not believe that reduced PO_4^{-3} content of cotton plants affected plant growth and development. However, reduced P levels may make the plants more sensitive to salinity (Grattan and Grieve, 1994).

Increased Cl^- and SO_4^{-2} contents of leaf blades and petioles in response to increase in the salinity of irrigation waters (Figure 11a-d) may have been contributing factors in affecting crop growth and development, but the effects are not clearly understood. It is interesting to note that in 1987 plants from plots irrigated with 400 ppm waters had as high as 1.2% Cl^- in their leaf blades.

The B levels of irrigation waters increased with the increasing salinity of the waters mainly due to the high levels of the element in the evaporation pond waters (Table 1). During the course of this study, B concentrations in soil were 0.7 and 1.9 ppm in the plots irrigated with 400 and 9,000 ppm waters, respectively. The waters applied to these plots had B concentrations of 0.64 and 4.50 ppm, respectively. Thus, the B concentrations of these plots did not show much rise despite the use of these saline waters. Ayars et al. (1993) in a six year study of the use of saline drainage water on the west side of San Joaquin Valley reported gradual increases in B concentrations in the top 1 meter of soil from 2.2 to 6.8 mg/Kg soil. The increased concentration was due to the application of saline irrigation water with a B concentration ranging from 4 to 7 ppm in their study.

Changes in growth patterns and ionic composition of plants may influence fiber quality. Based on the observations made in this study, there was no evidence of any negative effects of irrigation water salinity on fiber quality. The observed increase in fiber micronaire due to the increase in irrigation water salinity shows that the development and maturity of cotton fiber was not affected (Figure 12).

The salt-tolerant nature of cotton was evident from the fact that it took three years of irrigating with saline waters of 9,000 ppm to cause a reduction of about 28% in lint yield (Figure 4b). Actually, there was evidence of some beneficial effect of irrigation with saline waters on the lint yields, especially in the second and third years. This agrees well with the general belief that cotton yields are enhanced under mild stress. Promotional or no effects of irrigation water salinity on lint yields were observed despite the fact that several growth and developmental parameters had been affected a year earlier even in plots irrigated with waters of lower salinities. This suggests that, within limits, many of these parameters (e.g., shoot height and weight and root length and weight) may not have a direct bearing on the lint yield (Figures 2b, c, and 3a, b). The first detrimental effect of saline water irrigation on the lint yield was observed in the year 1987 in the plots irrigated with waters of 9,000 ppm salinity. This also happened to be the first time when the plant density fell to 7.7 plants/m row in any treatment. Plant densities below 9 plants per meter row at 1 meter row spacing (90,000 plants/ha) have been reported to result in reduced cotton yields under normal conditions (Kerby and Hake, 1996). When areas with 9 or more plants per meter row, from the plots irrigated with waters of 9,000 ppm salinity, were selectively harvested in 1987, yields similar to those irrigated with 400 ppm waters were obtained (data not shown).

The effects of irrigation with saline water on the fourth cotton crop (1988) and the resulting lint yields were confounded by the rainfall shortly after planting. The plant densities this year were severely affected and fell below the threshold value of 9 plants/m row in all plots except those irrigated with waters of 400 ppm salinity (Figure 2a). The effect of reduced plant densities was readily reflected as lint yields lesser than those from the control were observed, even from the plots irrigated with waters of 1,500 ppm salinity. This is further supported by the observations made in the fifth cotton crop in 1990;

significant reduction in lint yields occurred only in plots irrigated with waters of 4,500 ppm salinity or higher. This is in contrast with the year 1988 where lint yields were significantly reduced in all treatments. The plant densities fell below 9 plants/m row in plots irrigated with waters of 1,500 ppm or higher in 1988, whereas the same level of reductions occurred only in plots irrigated with 4,500 ppm or higher salinity in 1990. The reductions in plant density, in response to the irrigation water salinity, was the greatest in 1991 and so were the reductions in lint yield. Consistent with other years, a direct and strong relationship between lint yield and plant density also was observed in 1991.

Evidently, the detrimental effects of irrigation with saline drainage water on lint yield were largely reflected through reduced plant density. Rhoades et al. (1988a) in their study on reuse of drainage water in the Imperial Valley reported reduction in cotton lint yields in the second year when irrigated throughout with Alamo River water having 3,000 ppm salts. The loss in lint yield resulted primarily from a decrease in the stand caused by higher salinity in the seedbed during the seedling establishment period. However, no significant loss in lint yield was caused by using this saline water for irrigation following seedling establishment with nonsaline water.

Historically, reductions in crop yields due to soil salinity mainly caused by continuous irrigation with "nonsaline" canal waters have been well known. In this study, the use of saline drainage water for irrigation reduced cotton lint yields. The yield reduction responses appeared to be a function of time and the salinity level of the applied water; the lower the salinity level the longer it took to cause yield reductions and vice versa. Our results showed that saline drainage waters having salts up to 3,000 ppm (4.7 dS/m) can be used for irrigation of cotton crops for four years as long as some leaching occurs through preplant irrigations with nonsaline water and grown in rotation with an unirrigated crop (e.g., safflower) or irrigated with nonsaline water. Waters of higher salinities may be used for shorter durations with similar results. An alternate strategy would be to use the water of a particular salinity level but only for one irrigation per year. This would probably allow the irrigation with saline water for much longer periods without affecting the yields. When considering the use of saline drainage water for irrigation, it is necessary to develop special management procedures to assure maximum yields. Stand establishment, which is the critical factor, should be carefully monitored and managed. Preparation of a finer seedbed also may be helpful in enhancing the seed germination and stand establishment under such situations. Proper understanding of plant mechanisms governing crop(s) responses at different growth stages under salinity will further help in utilizing saline waters for crop irrigation and thus reduce the volume of drainage water that needs to be disposed.

REFERENCES

Ackerson, R.C., D.R. Krieg. and N. Chang. 1977. Effects of plant water status on stomatal activity, photosynthesis and nitrate reductase activity of field grown cotton. *Crop Sci.* 17:81-84.

Ayars, J.E., R.B. Hutmacher, R.A. Schoneman, S.S. Vail, and D. Felleke. 1986a. Drip irrigation of cotton with saline drainage water. *Transactions of the ASAE* (Amer. Soc. Agric. Eng.) 29(6):1668-1673.

Ayars, J.E., R.B. Hutmacher, R.A. Schoneman, and S.S. Vail. 1986b. Trickle irrigation of sugarbeets with saline drainage water. pp. 5-6. In: Annual report of the Water Management Research Laboratory, USDA-ARS (U.S. Dep. Agr.–Agr. Res. Service), Fresno, CA.

Ayars, J.E., R.B. Hutchmaker, R.A. Sconeman, S.S. Vail, and T. Pflaum. 1993. Long term use of saline water for irrigation. *Irrig. Sci.* 14:27-34.

Baldini, N.C., L. Bernhardt, E.L. Gutman, S.L. Kaufmann, J.G. Kramer, C.M. Leinweber, and V.A. Mayer. (eds.). 1995. Standard test methods for measurement of cotton fibers by High Volume Instruments (HVI) (Motion Control Fiber Information System). Section 7–Textiles. Annual Books of ASTM Standards. Volume 07.02:451-461.

Beaton, J.D., G.R. Burns, and J. Platou. 1968. Determination of sulphur in soils and plant material. Tech. Bull. 14, The Sulphur Inst., Washington, DC.

Besford, R.T. 1978. Effect of replacing nutrient potassium by sodium on uptake and distribution of sodium in tomato plants. *Plant Soil.* 58: 399-409.

Bernstein, L., and H.E. Hayward. 1958. Physiology of salt tolerance. *Ann. Rev. Plant Physiol.* 9:25-46.

Bremner, J.M. 1965. Inorganic forms of nitrogen. In: C.A. Black et al. (eds.) Methods of Soil Analysis, Part 2. Agronomy 9:1179-1237. Am. Soc. of Agron., Inc., Madison, WI.

Carter, D.L. 1975. Problems of salinity in agriculture. pp. 25-30. In: M. Poljakoff and J. Gale (eds.) Plants in Saline Environments. Ecological Studies, Vol. 15. Springer, Berlin.

Cheeseman. J.M. 1988. Mechanisms of salinity tolerance in plants. Plant Physiol. 87:547-550.

Cramer, G.R., A. Lauchli, and V.S. Polito. 1985. Displacement of Ca^{+2} by Na^+ from the plasmalemma of root cells: A primary response to salt stress. *Plant Physiol.* 79: 207-211.

Cramer, G.R., A. Lauchli, and E. Epstein. 1986. Effects of NaCl and $CaCl_2$ on ion activities in complex nutrient solutions and root growth of cotton. *Plant Physiol.* 81: 792-797.

Cramer, G.R., J. Lynch, J.A. Lauchli, and E. Epstein. 1987. Influx of Na^+, K^+ and Ca^{+2} in to roots of salt-stressed cotton seedlings. Effects of supplemented Ca^{+2}. *Plant Physiol.* 83: 510-516.

Ella, M.K.A. and E.A. Shalaby. 1993. Cotton response to salinity and different potassium-sodium ratio in irrigation water. *J. Agron. Crop Sci.* 170: 25-31.

Grattan, S.R., A. Royo, and R. Aragues. 1994. Chloride accumulation and partitioning in barley as affected by differential root and foliar salt absorption under sprinkler irrigation. *Irrig. Sci.* 14:147-155.

Grattan, S.R. and Grieve, C.M. 1994. Mineral nutrient acquisition and response by plants grown in saline environments. pp. 203-226. In: M. Pessarakali (ed.) Handbook of Plant and Crop Stress. Marcel Dekker, Inc. New York.

Hewitt, E.J. 1963. The essential nutrient elements: Requirements and interactions in plants. pp. 137-360. In: F.C. Steward (ed.) Plant Physiology, a Treatise, Vol. III. Academic Press,

Imhoff, E.A. 1990. A Management Plan for Agricultural Subsurface Drainage and Related Problems on the Westside San Joaquin Valley. Final Report of the San Joaquin Valley Drainage Programme. pp. 1-183.

Issac, R.A. 1990. Metals in Plants. pp. 40-42. In K. Elrich. Official Methods of Plant Analysis. Vol. I. AOAC (Assoc. Offic. Anal. Chem), Inc. Arlington, Virginia.

Kent, L.M. and Lauchli, A. 1985. Germination and seedling growth of cotton: salinity-calcium interactions. *Plant Cell and Environment*. 8:155-159.

Kerby, T. and Hake, K. 1996. Monitoring cotton's growth. pp. 335-355. In: T. Kerby, K. Hake, and S. Hake (eds.) Cotton Production. Division of Agriculture and Natural Resources Publication, University of California, Oakland, California.

Leidi, E.O., M. Silberbush, M.I.M. Soares, and S.H. Lips. 1992. Salinity and nitrogen nutrition studies on peanut and cotton plants. *J. Plant Nutrition*. 15: 591-604.

Maas, E.V. 1986. Salt tolerance of plants. *Applied Agric. Res*. 1:12-26.

Martinez, V. and A. Läuchli. 1994. Salt-induced inhibition of phosphate uptake in plants of cotton (*Gossypium hirsutum* L.). New Phytol. 125:609-614.

Mauney, J.R. 1986. Vegetative growth and development of fruiting sites. In: J.R. Mauney and J.M. Stewart (eds.) Cotton Physiology. The Cotton Foundation, Memphis.

Olsen, S.R. and L.E. Sommers. 1982. Phosphorus. pp: 4043-430. In: A.L. Page et al., (eds) Methods of soil analysis. Part 2, Agronomy 9: 403-430. Amer. Soc. of Agron., Madison, WI.

Page, A.L., A.C. Chang, and D.C. Adriano. 1990. Deficiencies and toxicities of trace elements. pp. 138-160. In: K.K. Tanji (ed.) Agricultural Salinity Assessment and Management. ASCE (Amer. Soc. Chem. Eng.) Manual.

Pasternak, D., and Y. De. Malach. 1994. Crop Irrigation with Saline Water. pp. 600-621. In: M. Pessarakali (ed.) Handbook of Plant and Crop Stress. Marcel Dekker, Inc. New York.

Pasternak, D., M. Twersky, and Malach, Y. De. 1979. Salt resistance in agricultural crops. pp. 127-142. In: H. Mussel and R.C. Staples (eds.) Stress Physiology in Crop Plants. Wiley, New York.

Rains, D.W. 1969. Cation absorption by slices of stem tissue of bean and cotton. *Experientia*. 25:215-216.

Rains, D.W., S. Goyal, R. Weyrauch, and A. Lauchli. 1987. Saline drainage water reuse in a cotton rotation system. *Calif. Agric*. 41:24-26.

Rather, G. 1982. Influence of extreme K:Na ratios and high substrate salinity on plant metabolism of crops differing salt tolerance. *J. Plant Nutrition* 5: 183-193.

Rhoades, J.D., F.T. Bingham, J. Letey, A.R. Dedrick, M. Bean, G.J. Hoffman, W.J. Alves, R.V. Swain, P.G. Pacheco, and Lemert, R.D. 1988a. Reuse of drainage water for irrigation: results of Imperial Valley study. I. Hypothesis, experimental procedures, and cropping results. *Hilgardia* 56(5): 1-16.

Rhoades, J.D., F.T. Bingham, J. Letey, P.J. Pinter, Jr., R.D. LeMert, W.J. Alves, G.J. Hoffman, J.A. Replogle, R.V. Swain, and P.G. Pacheco. 1988b. Reuse of drainage water for irrigation: Results of Imperial Valley study. II. Soil salinity and water balance. *Hilgardia* 56(5): 17-44.

Rolston, D.E., D.W. Rains, J.W. Biggar. and A. Lauchli. 1988. Reuse of saline drain water for irrigation. pp. H1-H6. In: UCD/INIFAP Conference, Guadalajara, Mexico, March 1988.

Shennan, C., S.R. Grattan, D.M. May, C.J. Hillhouse, D.P. Schachtman, M. Wander, B. Roberts, S. Tafoya, R.G. Burau, C. McNeish, and L. Zelinski. 1995. Feasibility of cyclic reuse of saline drainage in a tomato-cotton rotation. *J. Environmental Qual.* 24:476-486.

Shannon, M.C. and Francois, L.E. 1977. Influence of seed treatments on salt tolerance of cotton during germination. *Agron. J.* 69:619-622.

Tanji, K.K., and Enos, C.A. 1994. Global water resources and agricultural use. pp. 3-22. In: K.K. Tanji and B. Yaron (eds.). Management of Water Use in Agriculture. Springer Verlag. Berlin.

Tanji, K.K., A. Lauchli, and A. Meyer. 1986. Selenium in the San Joaquin Valley. *Environment.* 28(6): 6-11.

Thellier, C., K.M. Holtzclaw, J.D. Rhoades, and J. Sposito. 1990. Chemical effects of saline irrigation water on a San Joaquin valley soil. II. Field soil samples. *J. Environment Qual.* 19:56-60.

Tucker, B.B. 1974. Efficient use of fertilizer for maximizing profits in cotton production. Proc. Beltwide Cotton Production Research Conference, pp. 158-162.

Twerski, M., D. Pasternak, and I. Borovic. 1973. Second year of cotton with brackish water (Hebrew), Hassadeh. 1016.

Wolf, B. 1974. Improvements in the azomethine-H method for the determination of boron. *Commun. Soil Sci. Plant Anal.* 5:39-44.

Long-Term Reuse of Drainage Waters of Varying Salinities for Crop Irrigation in a Cotton-Safflower Rotation System in the San Joaquin Valley of California– A Nine Year Study: II. Safflower (*Carthamus tinctorius* L.)

Sham S. Goyal
Surinder K. Sharma
Donald W. Rains
André Läuchli

Sham S. Goyal, Surinder K. Sharma, Donald W. Rains, and André Läuchli are affiliated with the Department of Agronomy and Range Science, Department of Land, Air, and Water Resources, University of California at Davis, Davis, CA 95616.

Address correspondence to: Sham S. Goyal at the above address (E-mail: ssgoyal@ ucdavis.edu).

Current address for Surinder K. Sharma: Central Soil Salinity Research Institute, Karnal 132001, Haryana, India.

The authors wish to express their appreciation to their major cooperator, J. G. Boswell Company in Corcoran, California, and their staff at The El Rico Ranch for providing field-work related physical facilities, including land and water and for their help with the field operations. The authors would also like to thank their colleagues namely, James Biggar, Donald Nielsen, James Oster, Dennis Rolston, and Kenneth Tanji, for their valuable contributions to discussions during the tenure of the study. The funds for the study were provided by the Office of the Dean, College of Agricultural and Environmental Sciences, University of California at Davis; the Tulare Lake Drainage District, Corcoran; J. G. Boswell Company, Corcoran; University of California Salinity/Drainage Taskforce; and The Water Resource Center of the University of California.

[Haworth co-indexing entry note]: "Long-Term Reuse of Drainage Waters of Varying Salinities for Crop Irrigation in a Cotton-Safflower Rotation System in the San Joaquin Valley of California–A Nine Year Study: II. Safflower (*Carthamus tinctorius* L.)." Goyal Sham S. et al. Co-published simultaneously in *Journal of Crop Production* (Food Products Press, an imprint of The Haworth Press, Inc.) Vol. 2, No. 2 (#4), 1999, pp. 215-227; and: *Water Use in Crop Production* (ed: M. B. Kirkham) Food Products Press, an imprint of The Haworth Press, Inc., 1999, pp. 215-227. Single or multiple copies of this article are available for a fee from The Haworth Document Delivery Service [1-800-342-9678, 9:00 a.m. - 5:00 p.m. (EST). E-mail address: getinfo@haworthpressinc.com].

SUMMARY. This paper reports the results on safflower crops grown in a nine-year study, conducted on a 8 ha site, to determine the feasibility of using drainage water for irrigation in a 2-year cotton/1-year safflower rotation system. The cotton crops were irrigated with waters of 400, 1,500, 3,000, 4,500, 6,000, and 9,000 ppm total dissolved salts, and safflower was grown only with a preplant irrigation with nonsaline water. The use of drainage water for crop irrigation may be a means of decreasing its volume. Even though safflower was never irrigated with saline drainage water directly, the residual effect of using saline water for cotton irrigation adversely impacted safflower growth and development. Safflower seed yields were reduced in plots previously irrigated with waters of 4,500 ppm or higher salinity and even more severe effects on crop growth were seen as the numbers of years of irrigation with the saline water increased. After irrigating six cotton crops, the safflower seed yield in plots irrigated with 9,000 ppm waters was reduced to only 14% of the control. The safflower oil content and quality were not affected. Impacts on plant density, shoot height, shoot biomass, and leaf ionic content also are discussed. *[Article copies available for a fee from The Haworth Document Delivery Service: 1-800-342-9678. E-mail address: getinfo@haworthpressinc.com <Website: http://www. haworthpressinc. com>]*

KEYWORDS. *Carthamus tinctorium*, safflower, drainage water, crop irrigation, plant density

Many soils in the San Joaquin Valley of California are characterized by high clay content and a high sub-surface water table. Cotton, the most important cash crop of the area, is commonly grown in rotation with safflower. Cotton is irrigated valley-wide with as much as 35 inches (89 cm) of water annually whereas safflower (*Carthamus tinctorius* L.) is generally grown without irrigation after a preirrigation prior to seeding. Rotation of cotton with safflower is believed to be agronomically helpful in a number of ways, including the lowering of sub-surface water table. Safflower is grown for its oil, which is considered superior due to the high percentage of unsaturated fatty acids.

As mentioned in the previous paper, the farmers in the San Joaquin Valley of California are having to consider the use of saline drainage water for crop irrigation. This is envisioned as a means of decreasing the volume of drainage water that otherwise needs to be disposed, while reducing the need for irrigation water brought in from long distances via canal net works. Although the drainage water may not be used directly for irrigation of safflower, it is important that the residual effects of using saline water for cotton irrigation be studied on safflower. Previous research has shown that drainage waters of up to 3000 ppm salinity may be used to irrigate cotton for two years without affecting the safflower yields (Rains et al., 1987). However, the potential of long-term use of these waters is not known. Therefore, additional in-depth research is needed before the use of saline drainage waters can be recommended as a practice for the cotton-safflower system.

The objective of this investigation was to study the long-term residual effects of using saline drainage water of various salinity levels for cotton irrigation on the growth, development, seed yield, and oil characteristics of safflower grown in rotation with cotton.

MATERIALS AND METHODS

The experimental site, plan, and treatments were detailed in the previous paper (Goyal et al., 1999). The aspects unique to the safflower crop are given below.

Cultural Practices

Safflower was grown in rotation with cotton and the study reported here comprised of three crop rotation cycles each of two years cotton followed by one year of safflower. The first safflower crop was grown during the winter-spring of 1986. One pre-plant irrigation with 20-30 cm of nonsaline water from California Aqueduct (400 ppm TDS) was applied before the seeding date and no more water was applied for the rest of the season. Therefore, the treatment effects on the growth and development of safflower were the result of using saline irrigation water on the preceding two cotton crops. Subsequent crops of safflower grown in 1989 and 1992 show the responses to the cumulative applications of saline waters to all the preceding cotton crops.

The safflower crops were seeded following the pre-irrigations with of 20-30 cm of nonsaline water. Water was ponded on the plots for 3-4 days in order to fully saturate the soil to a minimum depth of 75 cm. After most of the applied water had percolated down and only a thin layer of water was left on the soil surface, safflower (variety S-317) was sown by broadcasting the seeds using an airplane. This resulted in imbedding the seeds into the fully saturated soil (mud) at shallow depths. The seed rate was 67 Kg/ha. To obtain a better plant stand, the seed rate was doubled in the year 1992. Nitrogen was applied pre-plant at a rate of 200 Kg N/ha as NH_4NO_3, broadcast and disked in the fall. Additionally, 16.5 Kg N/ha and 78 Kg P_2O_5/ha (150 Kg/ha of 11-52-0) also was applied and disced in the fall.

In general, typical cultural practices of commercial agriculture in the area were followed. For seed yields, an area of 7.6 × 177 m^2 (leaving a border of 3.9 m on both sides) was harvested from each plot using a Case International harvester with a 7.6 m (25 foot) header.

Crop Growth and Development Parameters

The growth parameters were recorded at the early bloom stage. A frame with an inside area of 1 sq m was randomly placed in the plots. The plants that fell inside the frame were counted and the height on 10 plants was recorded. Such observations were recorded at a minimum of three different sites in each plot. For plant biomass, a ring of 0.1 sq m area was used, and all plants within the ring were harvested and dried for biomass. The youngest fully expanded leaves were collected and dried for chemical analysis.

Plant Analysis

Na^+, K^+, total-Ca, total-N, NO_3-N, and Cl^- in leaf samples were determined by the methods detailed in the previous paper (Goyal et al., 1999).

Oil Content and Composition

For total oil content, the seeds were refrigerated the evening of harvest until analyzed using the following procedure: 60 seeds were weighed and crushed using a quartz mortar and pestle and transferred quantitatively with hexane into an extraction thimble. The "crush" was refluxed with 300 ml hexane for 2.5 hours in a Soxhlet apparatus. The hexane-oil mixture was filtered, evaporated to a few milliliters using a rotary evaporator, quantitatively transferred to a tared beaker and held in a ventilated fume-hood overnight or until the hexane evaporated. The oil left behind in the beaker was weighed using an analytical balance.

For oil composition, six seeds were crushed using a quartz mortar and pestle with 10 ml of petroleum ether stored over sodium hydroxide pellets. The mixture was filtered using laboratory tissue paper into an Erlenmeyer flask. Petroleum ether was then removed using a rotary evaporator heated to about 80°C. A 10 ml portion of 1.25% thionyl chloride in dry methanol was added to the Erlenmeyer flask for esterification and was gently refluxed for 30 minutes. The solution was evaporated to dryness again and 300 mg of $NaHCO_3$ and 0.5 ml of hexane were added to bring the fatty acid esters into solution. The fatty acid esters were quantified using a gas chromatograph (Varian, model 3700) fitted with a flame ionization detector and an 8 foot (244 cm) glass column filled with 20% DEGS on Chromosorb W (80-100 mesh). The temperature of the column was maintained at 150°C and that of injector and detector at 245°C. The procedure was developed in the Analytical Laboratories of the Department of Agronomy and Range Science at UC Davis (Ruckman, unpublished).

The data are means of at least 12 observations, and ± values and bars in figures indicate standard errors.

RESULTS AND DISCUSSION

Growth and Development

Growth of plants in the first safflower crop (1986) was significantly affected by the saline water applied in two previous years to irrigate cotton crops, and the effects became more detrimental at the two highest levels of salinity. In this year, plant density was not affected by irrigation water salinity up to 4,500 ppm. However, a reduction of about 30% in plant density was observed at salinity levels of 6,000 and 9,000 ppm (Figure 1a). Soil salinity, measured up to a depth of 45 cm, also increased due to irrigation of two previous cotton crops with saline waters (discussed at the end of the paper). Considerably higher soil salinity in the plots previously irrigated with waters of 6,000 and 9,000 ppm was evident, which may explain the poor stand establishment in these plots in year 1986.

In the second safflower crop (1989, second crop rotation cycle), plant density was reduced by about 20% in plots previously irrigated with waters of 9,000 ppm as compared to those irrigated with 400 ppm (Figure 1b). In the third safflower crop (1992), surprisingly, an increase in plant density with increasing salinity of irrigation waters was observed (Figure 1c). The stand in plots previously irrigated with higher salinity waters would be expected to be lower than those irrigated with waters of lower salinity due to the accumulative soil salinity/sodicity. Increased sodium in irrigation water or soil can decrease soil hydraulic conductivity and thus water movement (Yaron and Frenkel, 1994). As mentioned earlier, the safflower crop was planted by flying the seeds over the plots that had been irrigated a day or so earlier and still fully muddy. Therefore, at the time of planting in 1992, the soil surface of plots previously irrigated with waters of lower salinity was probably drier and firmer than those irrigated with waters of higher salinities which may have been wetter. Consequently, the safflower seed, when flown over, was perhaps not properly imbedded in plots previously irrigated with waters of lower salinity and hence may have resulted in poorer germination.

The main study ended in 1992 and the recovery during the soil reclamation phase was studied in 1993. In the fall of 1992, the soil reclamation was started by applying a sulfuric acid treatment. Also, the use of saline water for irrigation ended after the 1991 cotton crop. This resulted in a major decrease in the soil salinity and sodicity during the 1993 safflower crop (Table 2 in Goyal et al., 1999). The effects of reduced soil salinity/sodicity were readily reflected in narrower differences in plant densities from plots previously irrigated with waters of 9,000 ppm salinity relative to those irrigated with 400 ppm (Figure 1d).

Reductions in the height of safflower plants with increase in the irrigation water salinity were recorded in all years, but the magnitude of reduction

FIGURE 1. Growth and development of safflower crops in different years as affected by the salinity of irrigation waters applied to previously grown cotton crops: Safflower plant density and biomass in 1986 (a), plant density and shoot height in 1989 (b), 1992 (c), and 1993 (d).

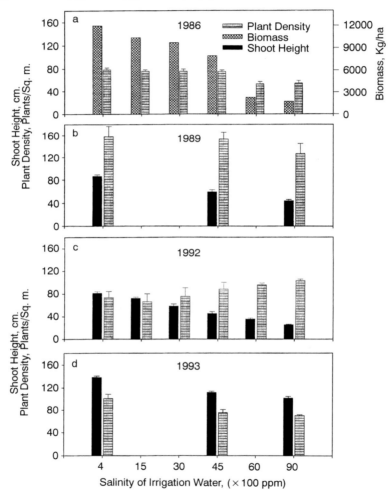

varied in different years. Plant height measurements could not be made in 1986, though a general reduction in plant height was evident (Rains et al., 1987). This is also supported by the reductions recorded in shoot biomass. The safflower shoot biomass recorded in 1986 showed significant effects of previously applied saline waters on the growth and development of the crop, which became more severe with the increase in salinity levels (Figure 1a). For example, the biomass decreased by 81 and 86% due to salinity levels of 6000 and 9000 ppm, respectively, as compared to the control.

Reductions in biomass were observed both on per plant and on per unit area basis (per plant basis data not shown). The biomass per unit area integrates the plant density and the individual plant weight and height. Salinity of irrigation waters had relatively much less effect on plant density than on plant biomass in 1986 (Figure 1a). The plant density in the 6,000 and 9,000 ppm treatments plots was approximately 70 percent of that of the control, whereas biomass per hectare was only about 17 percent of control. This suggests that, in safflower, soil salinity had greater effects on shoot biomass accumulation than on germination, emergence, and plant establishment. This is in contrast to cotton, where the plant density was affected more than the biomass accumulation (see Figures 1 and 8 in Goyal et al., 1999). Although safflower is rated "moderately tolerant" as compared to cotton, which is rated as "tolerant" to salinity (Maas and Hoffman, 1977), under the management regime safflower germinated and emerged more readily than cotton. This highlights the choice of crop and suitable management practices in the use of saline waters in agriculture.

In 1989, safflower shoot height was reduced by about 31 and 50%, respectively, in plots previously irrigated with waters of 4,500 and 9,000 ppm as compared to the control (Figure 1b). Reductions in shoot height due to the increase in salinity levels of the irrigation waters were even greater in the 1992 crop; shoot heights were reduced by about 44 and 68% in 4,500 and 9,000 ppm plots, respectively, as compared to the control (Figure 1c). Progressive decrease in shoot height with years of irrigation is consistent with the cumulative effects of the use of saline water. The shoot heights in 1993 were reduced only by 25 and 30% in 4,500 and 9,000 ppm plots, respectively, as compared to the control, which represents a substantial improvement over the previous safflower crops (Figure 1d). This was obviously due to the attempted soil reclamation with sulfuric acid preceding the 1993 crop.

Leaf Ion Content

Generally, the Na^+ content of safflower leaves increased in direct proportion to the salinity of waters applied to the previous cotton crops and progressively with years of irrigation (Figure 2a). In 1986, after the saline water had been used for irrigating for two cotton crops, the Na^+ content of leaves was about 0.37 and 5.81 mg/g from 400 and 9,000 ppm plots, respectively. Progressive increases in Na^+ content of leaves were observed in 1989 and 1992. The increase in Na^+ content of safflower leaves over the years was not nearly as great as seen with cotton (see Figure 4 in Goyal et al., 1999). Nevertheless, the inhibitory effects on safflower plant growth were much greater than those

FIGURE 2. Sodium (a) and potassium (b) content and potassium/sodium ratio (c) of safflower leaves in different years, as affected by the salinity of irrigation waters applied to previously grown cotton crops.

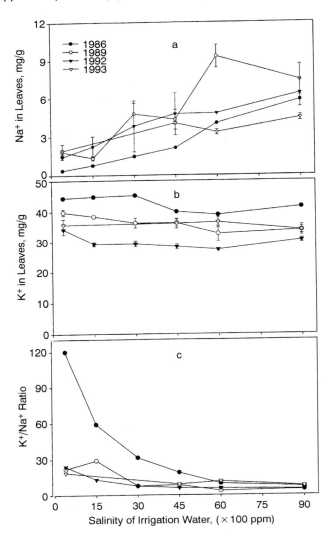

observed in cotton (Figure 1a). This suggests that safflower may be relatively more sensitive to soil salinity than cotton. The soil reclamation attempt of the fall of 1992 reduced the Na^+ content of safflower leaves in 1993 and increased plant growth (Figure 1d). Apparently, there was a strong negative relationship between the Na^+ content of leaves and growth of safflower plants.

In general, the K^+ content of leaves declined slightly in response to the increase in salinity levels of the irrigation waters. The K^+ contents of leaves in all plots decreased progressively with the number of years of irrigation with saline waters. In other words, K^+ content was the highest in 1986, lower in 1989, and the lowest in 1992, across all treatments. Again, the soil reclamation attempt in the fall of 1992 increased the K^+ contents of leaves in all treatments mostly to the levels recorded in 1989. It appears that the K^+ content of safflower leaves was very responsive to the salinity levels in the rhizosphere, which may be related to the inhibitory effects of Na^+ on K^+ uptake (Figure 2b).

Changes in the contents of Na^+ and K^+ led to reductions in K^+/Na^+ ratios in the leaves. Highest reductions in the K^+/Na^+ ratios were observed in the year 1986, when the ratios declined from 120 in 400 ppm (control) to 7.1 in 9,000 ppm plots. Reduced K^+/Na^+ ratios have been known to disrupt normal biochemical functioning of the cell (Läuchli and Epstein, 1990).

Total-N, total-Ca, NO_3^-, and Cl^- content of leaves were recorded in 1989. Increase in salinity levels of the irrigation waters reduced the Ca content of the leaves by 10 to 20% (Figure 3a), whereas little changes in the Cl^- (Figure 3b) and total-N content were observed (Figure 3c). However, the NO_3^- content increased with increasing salinity of the irrigation waters (Figure 3d), which may have been due to the NO_3^- present in the drainage water. The reduction in Ca content of leaves coupled with the increase in Na^+ and decrease in K^+ may have contributed to the reduced growth of plants.

Seed Yield

Safflower seed yield in plots irrigated with saline waters of 4,500 ppm and higher were reduced, and the reductions were proportional to the increase in salinity of the plots (Table 1). However, a small increase in yield in 1,500 ppm over control plots was observed in the years 1986 and 1989. The detrimental effects of salinity on safflower seed yields increased in the years 1989 and 1992, so much so that, in 1992, yields in the 9,000 ppm plots were reduced to only 14% of that obtained in the control plots. As mentioned before, the main study ended in 1992, and the soil reclamation phase using sulfuric acid was undertaken in the fall of 1992. A significant improvement in safflower seed yield in response to soil reclamation was observed in all treatments during the year 1993 (Table 1). In addition to the overall improvement in seed yields,

FIGURE 3. Calcium (a), chloride (b), total-N (c), and nitrate (d) content of safflower leaves in 1989, as affected by the salinity of irrigation waters applied to previously grown cotton crops.

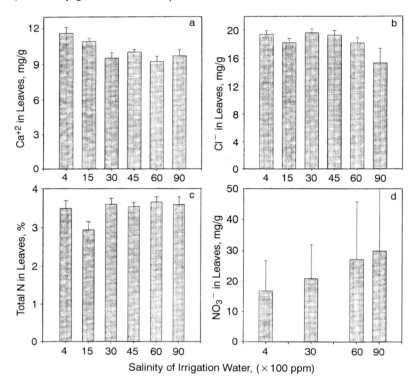

reductions in response to increase in salinity levels were much lower this year. For example, in 1993, plots irrigated with waters of 9,000 ppm yielded about 70% of that of control plots as compared to 14% in 1992. Significant improvements in plant growth (as evidenced by shoot height, Figure 1d) were also observed in 1993 following the soil reclamation attempt. The improvement in growth and increase in seed yield were probably due to reduction in salinity/sodicity of the plots.

Oil Content and Quality

The oil content of safflower seeds showed no influence of increasing salinity of the irrigation water in either 1986 or 1989 (Table 2). The safflower oil mainly contained three fatty acids: palmitic, oleic, and linoleic acids.

TABLE 1. Safflower seed yields grown in plots previously irrigated with saline drain water of various salinity levels. Each value represents an average of 4 replications and the ± values indicate standard errors.

Irrigation water salinity, ppm	Seed yield, Kg/ha			
	1986[a]	1989	1992	1993*
400	2,946 ± 38.5	3,468 ± 120	2,208 ± 371	4,241 ± 57
1,500	3,134 ± 74.4	4,037 ± 117	2,256 ± 275	4,111 ± 133
3,000	2,872 ± 93.9	3,265 ± 39	1,931 ± 332	3,805 ± 148
4,500	2,552 ± 93.9	3,017 ± 99	1,506 ± 199	3,482 ± 100
6,000	1,800 ± 77	2,200 ± 95	862 ± 112	3,177 ± 165
9,000	1,604 ± 67	1,239 ± 45	313 ± 27	2,978 ± 164

[a] 1986 yield data previously reported in Rains et al., 1987.
* Recovery year after soil reclamation with sulfuric acid treatment in Fall 1992.

TABLE 2. Total oil content and fatty acid composition of safflower oil as affected by the salinity levels of water used for irrigating the preceding cotton crops. Each value represents an average of 12 observations.

Irrigation Water Salinity, ppm	Fatty Acids %, 1986			Total Oil %	
	Oleic	Linoleic	Palmitic	1986	1989
400	77.6	16.1	6.3	39.1	38.7
1,500	77.9	15.9	6.4	39.8	40.4
3,000	78.0	15.6	6.4	39.0	41.0
4,500	78.2	15.4	6.4	38.6	39.0
6,000	77.8	15.8	6.4	38.3	41.4
9,000	78.8	14.9	6.3	38.5	39.6

Stearic acid was also detected in some random samples and ranged in concentration from 0.75 to 2.2 percent. Relatively higher content of linoleic acid in safflower oil is a major factor in its popularity and acceptance by the industry. The oil composition showed no effect of salinity treatments on the relative proportions of saturated and unsaturated fatty acids, and hence no change in oil quality (Table 2). Apparently, the salinity of irrigation waters had no influence on either oil content or composition. This agrees well with the earlier report of Yermanos et al. (1964), who also reported that soil salinity did not affect the chemical composition of the safflower oil.

In summary, waters of 4,500 ppm salinity or higher, used for irrigating the preceding cotton crops, increased soil electrical conductivity (Figure 4) and adversely affected the safflower seed yields. Biomass was impacted the most, while seed germination and stand establishment appeared relatively less sensitive. The reductions in biomass seemed related to the increased levels of Na^+ and decreased levels of K^+ in the leaves. The oil content and fatty acid composition remained unaffected by the salinity of the irrigation waters. The use of saline drainage water for cotton irrigation and growing safflower as a rotation crop with cotton in the San Joaquin Valley of California appears feasible. However, the salinity levels of the waters should be carefully monitored not to exceed levels of 3,000 ppm, especially if a long-term use of these waters is envisaged. A pre-plant irrigation with nonsaline water probably provided a zone of low salt concentration in the topsoil which may have aided in seed germination. Hence, this may be an important part of the management package while considering the use of saline water for crop irrigation. Overall, similar conclusions were supported by the study on cotton crop as reported in the previous paper (Goyal et al., 1999).

FIGURE 4. Electrical conductivity of soils as affected by the salinity of irrigation waters applied to two previously grown cotton crops. Soil samples were taken in mid-season safflower growing season (July 1986).

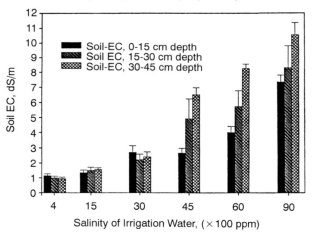

REFERENCES

Goyal, S.S., S.K. Sharma, D.W. Rains, and A. Lauchli. 1999. Long-term reuse of saline drainage water for crop irrigation in a cotton-safflower rotation system in the San Joaquin Valley of California–A nine year study: I. Cotton (*Gossypium hirsutum L.*). *J Crop Production* 2(2#4): 181-213.

Lauchli, A. and E. Epstein. 1990. Plant responses to saline and sodic conditions. pp. 113-137. *In*: K.K. Tanji (ed.). Agricultural Salinity Assessment and Management. ASCE, New York.

Maas, E.V. and G.J. Hoffman. 1977. Crop salt tolerance–current assessment. *J. Irrig. Drain. Div.* ASCE. 103 (IR2): 115-134.

Rains, D.W., S. Goyal, R. Weyrauch, and A. Lauchli. 1987. Saline drainage water reuse in a cotton rotation system. *Calif. Agric.* 41:24-26.

Yaron, B. and H. Frenkel. 1994. Water suitability for agriculture. pp. 25-41. *In*: K.K. Tanji and B. Yaron (eds.). Management of Water Use in Agriculture. Springer Verlag, Berlin.

Yermanos, D.M., L.E. Francois, and L. Bernstein. 1964. Soil salinity effects on the chemical composition of the oil and the oil content of safflower seed. *Agron. J.* 56:35-37.

Physiological Parameters, Harvest Index and Yield of Deficient Irrigated Cotton

Bruria Heuer
A. Nadler

SUMMARY. The agronomic and physiological response of cotton to deficient irrigation as well as its influence on yield quality was determined. Plant growth expressed as plant height and accumulation of fresh weight was significantly affected by water deficit. Nevertheless, neither seed cotton yield nor lint quality were decreased. Turgor pressure was maintained positive for the entire growth cycle. The relationship between growth and yield parameters is discussed. *[Article copies available for a fee from The Haworth Document Delivery Service: 1-800-342-9678. E-mail address: getinfo@haworthpressinc.com <Website: http://www.haworthpressinc.com>]*

KEYWORDS. Cotton, growth, harvest index, leaf water potential, yield, yield quality

INTRODUCTION

In arid and semiarid regions, characterized by water shortage, water availability is a major limitation in crop production. Cotton development and

Bruria Heuer and A. Nadler are affiliated with the State of Israel/Ministry of Agriculture, Agriculture Research Organization, The Volcani Center, The Institute of Soils, Water and Environmental Sciences, Bet Dagan.

Address correspondence to: Bruria Heuer, Institute of Soil, Water and Environmental Sciences, ARO, The Volcani Center, POB 6, Bet Dagan 50250, Israel (E-mail: vwheuer@volcani.agri.gov.il).

Contribution from the Agricultural Research Organization, Institute of Soils and Water, Bet Dagan, Israel, No. 634-1999.

[Haworth co-indexing entry note]: "Physiological Parameters, Harvest Index and Yield of Deficient Irrigated Cotton." Heuer, Bruria, and A. Nadler. Co-published simultaneously in *Journal of Crop Production* (Food Products Press, an imprint of The Haworth Press, Inc.) Vol. 2, No. 2 (#4), 1999, pp. 229-239; and: *Water Use in Crop Production* (ed: M. B. Kirkham) Food Products Press, an imprint of The Haworth Press, Inc., 1999, pp. 229-239. Single or multiple copies of this article are available for a fee from The Haworth Document Delivery Service [1-800-342-9678, 9:00 a.m. - 5:00 p.m. (EST). E-mail address: getinfo@haworthpressinc.com].

productivity are strongly affected by its water status (Constable and Hearn, 1981). Water stress early in the growing season of cotton usually affects subsequent growth (Steger et al., 1998). Vegetative growth is linearly related to water supply over a wide input range (Grimes et al., 1969), while harvest index is generally higher at low water applications (Alvarez-Reyna, 1990). It is accepted that water stress prior to flowering stimulates excessive shedding of squares, resulting in yield decrease (Cull et al., 1981; Brar, 1986).

Unfavorable growth conditions including water stress may modify the genetic potential of certain fiber characteristics (Grimes and El-Zik, 1990). Fiber elongation and strength are only slightly influenced by water status, while fiber length and micronaire are significantly reduced (Grimes and Kerby, 1992).

Optimal cotton production requires large amounts of irrigation water. Consequently, in countries where cotton is a major crop but at the same time water is limited, there is a continuous tendency to decrease these amounts of water down to a level beyond which yield and lint quality are reduced. This can be achieved by proper irrigation management and by an improved application timing.

Because of the high agricultural value of cotton, research studies were targeted towards the determination of critical growth periods when irrigation is obligatory and water stress should be avoided (Turner et al., 1986; Kock et al., 1990). However, this information regarding the effect of water deficit on cotton is in general not derived from field trials (Devendra et al., 1992; Zhao et al., 1993).

The purpose of the present study was to provide additional information on field cotton growth under restricted water supply and on irrigation scheduling for a better prediction of plant response to stress conditions. It was also aimed to determine the relationship between physiological parameters of growth and components of yield.

MATERIALS AND METHODS

Plant Material

Cotton (*Gossypium hirsutum* cv. Vered) was seeded on 16 April at Kibbutz Nirim, in the Northern Negev, Israel. The soil was a loam (Calcic Palexeralf), 14% clay, 38% silt, 48% sand, CEC = 11.9 mmol 100 g^{-1}, 11% $CaCO_3$, and a field capacity of 14.6% w/w. The experimental plot had been irrigated with brackish water (EC = 1.4-5.0 dS m^{-1}) for the last 10 years, alternating commercial growth of cotton or wheat. The cotton was seeded in 2 beds of paired, ridged, 16-m-long rows, 0.96 m apart with approximately 11 seeds per meter of row. Before seeding, the field was plowed, disked, ridged twice and fertilized with NirAmon supplying 230 kg total N ha^{-1} and 70 kg P_2O_5 ha^{-1} (no manure or gypsum was applied during the last 3 years). Each plot had 6 rows of which the 2 external ones on both sides were considered buffer rows. Drip laterals were laid along the centre of alternating rows with emitters every meter, delivering 3.5 L h^{-1}. Available water in the soil profile at the beginning of the experiment was estimated at 70 mm. This amount was sufficient to ensure good emergence and to support plant needs for 51 days.

Treatments

Irrigation treatments consisted of 3 applications of local saline well waters (EC = 3.6-4.0 dS m^{-1}, SAR 19.7). The highest level (Treatment C) received 4590 m^3 ha^{-1}, slightly below the locally common practice. The medium treatment (Treatment B) received 3861 m^3 ha^{-1} and the lowest one (Treatment A) supplied 3019 m^3 ha^{-1} (84% and 66% of Treatment C, respectively). The experiment consisted of one additional treatment (Treatment C_7) in which irrigation was withheld for 2 weeks, between day 101 and 115 after planting. In contrast to the common irrigation practice, we scheduled it based on quantity and timing of application, taking into consideration evaporation, plant size and developmental stages in the crop cycle. Irrigation water was applied twice weekly at a rate dependent on evaporation from a screened USWB Class A evaporation pan, and a crop coefficient which varied between 0.21 to 0.80, according to the soil coverage by the canopy. The extreme differences in applied irrigation amounts were between Treatments C and A (having consistently higher and lower returns as percentage of evaporation, 0.35-0.80 and 0.21-0.56, respectively), while Treatment B received intermediate quantities of water by a switch in its irrigation scheduled from that of Treatment A to that of Treatment C on day 94. The cumulative amounts of water applied with the daily evaporation rates are presented in Figure 1. The experiment was a completely randomized design with 4 replications and was part of a larger experiment testing the response of 7 cotton varieties to drought.

Measurements

Plant growth was estimated by measuring the following parameters: plant height, accumulation of fresh and dry weight, seed cotton yield and harvest index. Leaf area was determined from the relative fresh weights of leaves and of 25 discs randomly removed from them. Ten measurements were made for each sample. Eight plants from each treatment were sampled at different time intervals and weighed individually. The height estimate for each plot was

FIGURE 1. Cumulative amounts of water for irrigation and daily evaporation rate for the entire irrigation period. The arrow indicated the day when crop coefficient of treatment B was changed (day 94).

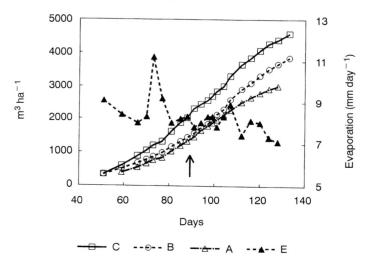

obtained by measuring 30 marked plants. Harvest index is defined as the ratio of harvested dry matter to total plant dry weight. As under field conditions it is difficult to measure root dry weight, shoot harvest index is commonly used in agronomic studies. Leaf water potential was determined with a pressure chamber (Arimad, Israel) and the leaf osmotic potential was determined by measuring the leaf sap, following freezing and thawing with a Precision 5004 osmometer. The data are presented with the least significant difference ($LSD_{0.05}$) between treatments, derived from an analysis of variance.

RESULTS AND DISCUSSION

Water restriction reduced plant height by about 18% in treatments B and A as compared with the control treatment (C), while withholding irrigation for 2 weeks (Treatment C_7) had no effect (Figure 2).

Accumulation of fresh weight in cotton plants was also significantly reduced by water deficit throughout the entire growth period (Figure 3). The same is true for accumulation of dry weight (data not shown).

The number of branches, leaves and bolls was also significantly reduced by water stress (Table 1). The maximal effect on all the growth parameters was observed after 100 days from seeding, and especially on leaf area which

FIGURE 2. Effect of irrigation regimes on plant height. Vertical bars indicate standard errors of the means (n = 30).

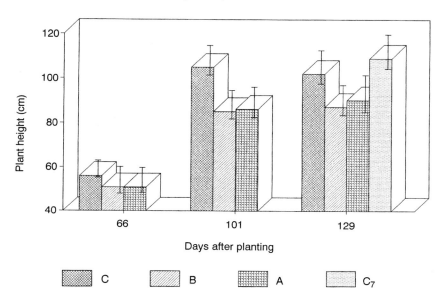

decreased by 30 and 44% in treatment B and A, respectively (Table 1). A short period of 2 weeks without irrigation was enough to reduce leaf number and area by 26 and 60%, respectively. Similar results were already reported, but mainly from pot experiments (Jagmail-Singh et al., 1990; Devendra-Singh et al. 1992).

Loss of only a few leaves does not affect plant growth, but if the rate of death approaches the rate of new production, a substantial drop in the supply of assimilates to the growing leaves may occur, leading to a significant suppression in plant growth. In our experiment, reduction of leaf number as a result of leaf death was only partially responsible for the decrease in leaf area. Thus, during the last month of crop irrigation, a reduction of 16 and 25% in leaf number led to a reduction of 53 and 47% in leaf area in treatments B and A, respectively (Table 1). Moreover, a loss of 43% of the leaves in treatment C did not affect leaf area. In general, plant development implies closure of the canopy followed by increased shading of lower leaves. We assume that well irrigated plants prefer to shed a large part of their leaves at the expense of large, fully expanded ones with a great photosynthetic activity. Following, the same leaf area was maintained. Restricted irrigation affected both number and size of the leaves, resulting in a significant decrease in leaf area.

FIGURE 3. Total accumulation of fresh weight in cotton plants as influenced by the treatments imposed. Vertical bars express standard errors of the means (n = 8).

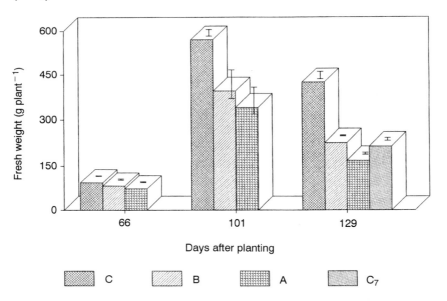

TABLE 1. The effect of irrigation regimes on leaf area, and the number of leaves, branches and bolls of cotton plants.

Treatment	Leaf area (m²)		Number per plant					
			Leaves		Branches		Bolls	
	26.7	23.8	26.7	23.8	26.7	23.8	26.7	23.8
C	2.07	2.00	68	39	16	16	16.6	8.6
B	1.45	0.68	42	35	10	10	13.0	8.3
A	1.15	0.54	40	30	10	10	10.5	7.3
C_7	—	0.81	—	29	—	12	—	7.6
$LSD_{0.05}$	0.23	0.48	14.8	5.8	3	3	3.7	0.9

It is generally accepted that late in the fruiting period, no new branches or squares are formed. Moreover, cotton normally produces more fruiting forms than can be carried through to maturity, even in the absence of stress (Grimes and El-Zik, 1990). Therefore, shedding of fruiting forms is considered normal. This explains the fact that fewer bolls could be counted at the last sampling, before irrigation was stopped. Water stress promotes the process of abscission; thus the number of bolls was reduced significantly only in treatments A and C_7, as at this stage treatment B received the same amount of water as the control (Table 1).

Generally, turgor pressure is accepted to be the best indicator of water stress in plants (Nielsen and Orcutt 1996), although it is not always associated with changes in physiological functions induced by water limitations (Boyer 1970). Plant water status, determined by measuring leaf water and osmotic potentials, was affected by the different irrigation regimes. Leaf water potential corresponded entirely to the amounts of water applied. It significantly declined in treatment B before the water return coefficent was changed to that of the control and fully recovered afterwards (Figure 4a). Similarly, a drastic drop in leaf water potential was observed in treatment C_7 with full recovery when irrigation was restored.

The leaf water potential of plants from treatment A was consistently lower.

FIGURE 4. Effect of water deficit on leaf water potential (a) and leaf osmotic potential (b). Vertical bars express standard errors of the means (n = 7).

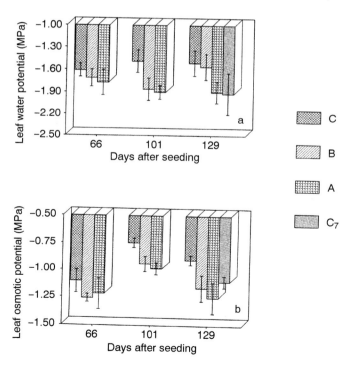

The treatments imposed led to a similar decrease in leaf osmotic potential (Figure 4b). As a result, turgor pressure, calculated as the arithmetic difference between leaf water and leaf osmotic potentials in stressed and non-stressed plants, was maintained positive. The ability of a crop to maintain turgor pressure as water potentials decline is an important mechanism in stress tolerance.

Seed cotton yield was not affected by the treatments imposed (Table 2), probably because an enormous increase in total dry matter production per unit leaf area (Table 3). However, withholding irrigation for only 2 weeks was enough to reduce yield by 36% (treatment C_7). According to Kock et al. (1990), the most sensitive growth period to water stress is during peak flowering (90-120 days after emergence). This period fully corresponds with treatment C_7 where irrigation was stopped between days 101 and 115 after seeding. Turner et al. (1986) claimed that water deficits reduced the fruiting capacity of cotton as a result of lower rates of photosynthesis. Our results did not confirm this finding (Table 3).

A significant increase in the harvest index was obtained in all the treatments imposed, suggesting a better water use efficiency (Table 3). Similar results were reported by Alvarez-Reyna (1990). Harvest index is generally higher at low water supply, reflecting a greater proportion of total biomass in reproductive growth as water stress severity is increased (Grimes and El-Zik 1990). The maintenance or the improvement of the harvest index is of critical importance in breeding programs when water is in short supply. Water use efficiency (WUE) is defined as the ratio between dry matter production and the amount of water transpired by a crop. Conditions of water deficit are responsible for temporary closure of stomata leading to an increased WUE (Fischer and Turner 1978). If harvest index is increased, the ability of the reduced photosynthetic surface per sink organ to supply assimilates is improved and so is net photosynthesis rate per unit leaf area (Table 3).

Lint quality, determined by fiber strength, elongation and micronaire, was

TABLE 2. Fiber yield and harvest index as a function of irrigation regimes.

Treatment	Seed cotton yield (t ha^{-1})	Harvest index (%)
C	5.00	35.21
B	5.34	66.65
A	4.55	72.07
C_7	3.17	37.10
$LSD_{0.05}$	0.77	8.69

very high and was not affected by the irrigation regimes (Table 4). Withholding irrigation for 2 weeks resulted in a higher percentage of short fibers and a lower micronaire. Similar results were obtained with sprinkle irrigation under various soil water regimes (Amir and Bielorai 1969; Marani and Amirav 1971). Fiber elongation and strength are slightly influenced by water status while fiber length and micronaire are decreased by severe water stress (Grimes and Kirby 1992). Micronaire is generally reduced by water management that delays maturity (Grimes and El-Zik 1990).

It can be concluded that if irrigation is scheduled based on quantity and especially on timing of application, taking into consideration developmental stages in the crop cycle and avoiding stress during critical growth stages, cotton plants are able to adapt to conditions of restricted irrigation without yield loss. The results obtained also show that there is a mismatch between parameters of plant growth, plant water status and crop yield.

TABLE 3. Effect of irrigation regimes on total dry matter production and total CO_2 per unit leaf area. Total CO_2 uptake was calculated using a conversion factor of 0.645 g dry matter per g CO_2 (Evans 1972).

Treatment	Total dry matter per leaf area $(g\ m^{-2})$	CO_2 uptake per leaf area $(g\ CO_2\ m^{-2})$
C	55.51	15.96
B	92.11	25.49
A	91.41	21.04
C_7	82.77	20.25
$LSD_{0.05}$	3.47	2.80

TABLE 4. The effect of irrigation regimes on lint quality.

Treatment	Fiber length (inch)	Elongation (%)	Fiber strength[+]	Micronaire (microwave units)	Fiber length uniformity (%)	Percentage of short fibers (%)
C	1.17	5.25	31.0	4.3	83.8	5.35
B	1.18	5.25	31.9	4.3	85.4	3.85
A	1.15	5.45	32.5	4.2	84.9	4.4
C_7	1.17	5.50	37.4	3.4	82.5	7

[+] Fiber strength is measured in (grams-force)/tex where "tex" is the weight in grams of 1,000 m of fiber. To convert into "KN m kg^{-1}" multiply by 9.81.

REFERENCES

Alvarez-Reyna, V. deP. (1990). Growth and development of three cotton cultivars of contrasting plant types differentially irrigated through a drip irrigation system. *Dissertation Abstracts International* 51: 4110B.

Amir, Y. and H. Bielorai. (1969). The influence of various soil moisture regimes on the yield and quality of cotton in an arid zone. *Journal of Agricultural Science* 73: 425-429.

Boyer, J.S. (1970). Leaf enlargement and metabolic rates in corn, soybean and sunflower at various leaf water potentials. *Plant Physiology* 46: 233-235.

Brar, A.S. (1986). Response of upland cotton (*Gossypium hirsutum* L.) to deficit irrigation management. *Dissertation Abstracts International* 46: 2129B.

Constable, G.A. and A.B. Hearn. (1981). Irrigation for crops in a subhumid environment. VI. Effect of irrigation and nitrogen fertilizer on growth, yield and quality of cotton. *Irrigation Science* 3: 17-20.

Cull, P.O., A.B. Hearn, and R.C. Smith. (1981). Irrigation scheduling of cotton in a climate with uncertain rainfall. I. Crop water requirements and response to irrigation. *Irrigation Science* 2: 127-130.

Devendra-Singh, R.K. Sahay, and D. Singh. (1992). Drought stress in cotton: identification of critical growth stages. *Plant Physiology and Biochemistry–New Delhi* 19: 55-57.

Evans, G.C. (1972). Growth studies and some problems of the physiology of gaseous exchange. In *The quantitive analysis of plant growth,* ed. G.C. Evans University of California Press, pp. 29.4.

Fischer, R.A. and N.C. Turner. (1978). Plant productivity in the arid and semiarid zones. *Annual Review of Plant Physiology* 28: 277-317.

Grimes, D.W., H. Yamada, and W.L. Dickens. (1969). Functions for cotton (*Gossypium hirsutum* L.) production from irrigation and nitrogen fertilization variables: II. Yield components and quality characteristics. *Agronomy Journal* 61: 773-776.

Grimes, D.W. and K.M. El-Zik. (1990). Cotton. *Agronomy* 30: 741-773.

Grimes, D.W and T.A. Kerby. (1992). Plant water relations and irrigation scheduling for Pima cotton. *Proceedings Beltwide Cotton Conferences* 3: 1041-1044.

Jagmail-Singh, S.N., Bhardwaj, Munshi-Singh, J. Singh, and M. Singh. (1990). Leaf size and specific leaf weight in relation to its water potential and relative water content in upland cotton (*Gossypium hirsutum* L.). *Indian Journal of Agricultural Sciences* 60: 215-216.

Kock, J.de, L.P. de Bryun, and J.J. Human. (1990). The relative sensitivity to plant water stress during the reproductive phase of upland cotton (*Gossypium hirsutum* L.). *Irrigation Science* 11: 239-244.

Marani, A. and A. Amirav. (1971). Effects of soil moisture on two varieties of upland cotton in Israel. III. The Bet Shean Valley. *Journal of Experimental Agriculture* 7: 289-301.

Nielsen, E.T. and D.M. Orcutt. (1996). Water limitation. In *Physiology of Plants Under Stress, Abiotic Factors*, eds. E.T. Nielsen and D.M. Orcutt. New York Wiley, pp. 322-361.

Stegger, A.J., J.C. Silvertooth, and P.W. Brown. (1998). Upland cotton growth and yield response to timing the initial postplant irrigation. *Agronomy Journal* 90: 455-461.

Turner, N.C., A.B. Hearn, J.E. Begg, and G.A. Constable. (1986). Cotton (*Gossypium hirsutum* L.): physiological and morphological responses to water deficits and their relationship to yield. *Field Crops Research* 14: 153-170.

Zhao, D.L., Y.Z. Xu, Y.F. Huang, and X. Xu. (1993). The effects of water deficiency on the development of cotton seeds during flowering and fruiting stages. *Acta Agronomica Sinica.* 19: 546-552.

Productive Water Use in Rice Production: Opportunities and Limitations

To Phuc Tuong

SUMMARY. It is crucial for food security to produce more rice with decreasing availability of water for rice production. This article seeks to answer the question " Does rice really require an abundance of water?" and discusses strategies and options to make rice production more water-efficient. It analyzes the effectiveness of prospective technological innovations in increasing the rice yield per unit water evapotranspired and reducing the water inflow to the rice fields. Possible impacts of these innovations on fertilizer-use efficiency, labor input, weed control, and water availability at locations downstream of the site of interest are discussed. Constraints to implementing these innovations include our inability to evaluate their on-farm and off-site effects and lacks of incentives for farmers and system managers to improve irrigation performance. *[Article copies available for a fee from The Haworth Document Delivery Service: 1-800-342-9678. E-mail address: getinfo@ haworthpressinc.com <Website: http://www.haworthpressinc.com>]*

KEYWORDS. Water productivity, water use efficiency, evapotranspiration, irrigation, percolation

INTRODUCTION

In recent years we have witnessed a growing scarcity and competition for water worldwide. As demand for water for domestic, municipal, industrial,

To Phuc Tuong is Water Management Engineer, Soil and Water Sciences Division, International Rice Research Institute (IRRI), MCPO Box 3127, 1271 Makati City, Philippines (E-mail: T.Tuong@CGIAR.ORG).

[Haworth co-indexing entry note]: "Productive Water Use in Rice Production: Opportunities and Limitations." Tuong, To Phuc. Co-published simultaneously in *Journal of Crop Production* (Food Products Press, an imprint of The Haworth Press, Inc.) Vol. 2, No. 2 (#4), 1999, pp. 241-264; and: *Water Use in Crop Production* (ed: M. B. Kirkham) Food Products Press, an imprint of The Haworth Press, Inc., 1999, pp. 241-264. Single or multiple copies of this article are available for a fee from The Haworth Document Delivery Service [1-800-342-9678, 9:00 a.m. - 5:00 p.m. (EST). E-mail address: getinfo@haworthpressinc. com].

and environmental purposes rises, less water will be available for agriculture. The potential for developing new projects for water resources and expanding irrigated area is limited. We therefore need to conserve water in agriculture so that it can be used to grow more food or be used in other sectors.

Rice is an obvious target for water conservation in Asia because it is the most widely grown of all crops under irrigation. More than 80% of the developed freshwater resources in Asia are used for irrigation purposes and about half of the total irrigation water is used for rice production (Bhuiyan, 1992; Dawe et al., 1998). The quantity of water is applied to the field to produce 1 kg of rice or each calorie equivalent of rice is significantly greater than that for any other major food crop (Bhuiyan, 1992). Food security in Asia depends on our ability to increase rice production with decreasing availability of water to grow rice.

This article discusses strategies and options to make rice production more water-efficient. It begins with a discussion of different definitions of common terms such as "water use, water use efficiency, and water productivity." The growing use of these concepts by different disciplines (agronomy, hydrology, plant physiology, engineering, etc.) requires careful and agreed definitions for common communication and progress (Stanhill, 1986). The second part of the article examines components of water use in rice production, seeking to answer the question, "Does rice production really require an abundance of water?" Finally, the article explores strategies and practices that produce more rice with less water by increasing rice yield per unit water evapotranspired, improving the efficiency of water application to rice fields, and making more effective use of rainfall. The article will also discuss constraints and limitations to implementing of these strategies.

WATER USE EFFICIENCY AND WATER PRODUCTIVITY: REVIEWING THE DEFINITIONS

Historically, the term "water use efficiency" has been used in two very different senses, hydrological and physiological (Stanhill, 1986). In the purely hydrological context, water use efficiency has been defined as the ratio of the volume of water used productively, i.e., evapotranspired (strictly speaking, it should only be transpired), to the volume of water supplied for that purpose. Engineers also use this context of water use efficiency in assessing the performance of irrigation projects (Jensen, 1980; Wolters and Bos, 1989). Overall irrigation efficiency has been divided into constituent components including conveyance, distribution, and field application efficiency. The latter is defined as the ratio of water evapotranspired to water received at the field inlet (Doorenbos and Pruitt, 1992). The hydrological concept of water use efficiency is useful in identifying components of water that is delivered to the field but that does not reach the crop. But this concept does not give information on the amount of food that can be produced with an amount of available water.

In the physiological context, often used by agronomists, water use efficiency refers to the biomass (or final grain yield) per unit volume of water evapotranspired (de Wit, 1958). Some authors used "water-use efficiency" to make a distinction from the hydrological meaning of "water use efficiency" (Viets, 1962; Tanner and Sinclair, 1983). The physiological concept of "water use efficiency" does not give information on components of water that is delivered to the field but that does not reach the crop.

To avoid ambiguity in the term "water use efficiency," this paper will use the term "application efficiency" to express the ratio of the weight of water evapotranspired to that of the water input to the field. Information on the amount of food that can be produced with an amount of available water will be expressed in terms of "water productivity," defined as the weight of biomass or yield produced per unit weight of water used (Viets, 1962; Tabbal et al., 1992; Molden, 1997). As it will be shown in the next sections, water used in rice production may have various components. It is important to specify which components are included when calculating water productivity (Molden, 1997). Water productivity per unit evapotranspiration (ET) will thus be equivalent to the physiological context of water use efficiency discussed above.

COMPONENTS OF WATER USED IN RICE PRODUCTION

Rice is unique among grain crops because land preparation is often done in flooded conditions and a layer of water is maintained over the soil surface during the crop growth period.

Water Balance Components During Land Preparation

The first step in land preparation for lowland rice is land soaking. Rice fields are irrigated until the topsoil (0-30-cm depth) is saturated and a water layer of 10-50 mm depth is ponded on the field. Land soaking is followed by plowing and harrowing operations under water-saturated conditions to "puddle" the topsoil so that rice seedlings can be transplanted easily. Theoretically, depending on the initial soil water condition and soil type, land preparation requires 150-250 mm of water to saturate the topsoil and to establish the ponded water layer. In practice, water also evaporates, recharges the groundwater, and flows out of the field during land preparation.

The amount of evaporation during land preparation can become very high when farmers take a long time to prepare the land. Long land preparation can be caused by inadequate canal discharged, and by the farmers' practice of soaking the field while they prepare the seedbed where seeds are germinated and nurtured for about 1 month until transplanting. It can also be caused by socioeconomic problems such as non-availability of labor and use of animals for draft power. If the canal that serves a block of farms has slow-flowing water, 2 months or more are taken before all farmers in the canal service area can complete land preparation (Valera, 1977). Taking an average evaporation rate of 6 mm d^{-1} in tropical rice-growing conditions, a 30-day land preparation period results in 180 mm of water depleted by evaporation.

Land soaking often involves applying water on cracked soils that resulted from soil drying during the fallow period after the harvest of the previous crop. Tuong et al. (1996b) reported that, in fields with relatively permeable subsoils, 45% of the water applied for land soaking moved through the cracks, bypassing the topsoil matrix, and flowed to the surroundings through lateral drainage. The bypass flow dominates water distribution during land soaking and contributes to a large extend to the high percolation (P) and seepage (S) rates reported by Wickham and Sen (1978). Cabangon and Tuong (1999) reported a percolation rate of 25 mm d^{-1} during the land-soaking period in a cracked clay soil. Percolation in the same soil after soil puddling was only about 5 mm d^{-1}. A large amount of water can flow out of the field when farmers take a long time to complete land preparation.

Because of high evaporation and outflow, and a long land preparation period, the amount of water diverted to the field for land preparation can be very high. Reported values vary from 350 mm in a field study in the Philippines (Cabangon and Tuong, 1999) to 1500 mm in the Ganges-Kobadak irrigation project in Bangladesh (Ghani et al., 1989).

Water Balance Components During Crop Growth Period

During the crop growth period, water is diverted to the field to meet the crop ET demand and to compensate for the unavoidable S and P in maintaining a saturated root zone.

The daily ET demand is very much driven by climatic condition, and thus depends on location and crop season. In temperate and subtropical areas, such as in China, ET varies from 2-3 mm d^{-1} at the beginning of the rice crop season (May) to 7-8 mm d^{-1} in July-October, giving an average ET of about 4-5 mm d^{-1} (Mao Zhi, 1982). In the Asian tropics, ET ranges from 4 to 12 mm d^{-1}, averaging 6-7 mm d^{-1} during the crop season. Thus, if the rice crop is in the field for 100 days, the seasonal ET can range from 500 to 700 mm of water.

Strictly speaking, ET represents transpiration from the crop and evaporation from the soil or from the surface of the ponded water layer in the field.

Though seldom separated, the two components of ET can be measured (Sugimoto, 1971) or simulated (Wopereis et al., 1996a). In the first 3 weeks after transplanting, E accounts for more than 85% of ET; as the plants grow, the shading effect of the expanding leaves reduces E rapidly to about 20% of ET at about 40 days after transplanting. After the canopy has closed (at leaf area index, LAI, of about 3), E is only about 10% of ET. On a seasonal basis, E accounts for nearly 30% of ET.

About 50-80% of the total water input leaves the field in the form of S and P (Sharma, 1989). In large irrigated areas, S occurs only in peripheries, but P occurs over the whole area. Percolation rate varies widely depending on soil texture and other factors, but usually increases as soil texture becomes lighter. Although values of 1-5 mm d^{-1} are often reported for puddled clay soils, percolation rates can be as high as 24-29 mm d^{-1} in sandy loam or loamy sand soils (Gunawardena, 1992; Khan, 1992). For sandy loam soils in Punjab, India, Sandhu et al. (1980) reported P of about 3 to 5 times higher than ET.

When the water supply within the irrigation system is unreliable, farmers try to store much more water in the field than is needed as insurance against a possible shortage in the future. In traditional transplanted rice, farmers prefer to maintain a relatively high depth of water to control weeds and reduce the frequency of irrigation (and hence labor cost). In rice irrigation systems where the plot-to-plot method of water distribution predominates, farmers have to build up the water head at the upper end of the farm to ensure the flow of water. These practices increase percolation because the percolation rate increases as the depth of water standing in the field increases.

In summary, though the ET demand of a rice crop is about 500-700 mm, the reported amount of inflow to produce one rice crop in the field ranges from 900 to 3000 mm. The low water application efficiency of the rice field is attributed to evaporation (100-180 mm) and bypass flow during land preparation (350-1500 mm), and S and P during the crop growth period (300-1500 mm).

STRATEGIES AND PRACTICES FOR INCREASING WATER PRODUCTIVITY IN RICE PRODUCTION

Water Productivity in Rice Production

Table 1 compares the productivity of water in rice production with that of other food crops. Rice yield per unit evapotranspiration (WP_{ET}) varies widely, from 0.5 to 1.6 g kg^{-1}. Wide variation in WP_{ET} also occurs in other crops and is due to differences in crop management which may affect crop yield and/or ET, such as soil nutrient, plant spacing, and planting time (Turner,

TABLE 1. Water productivity of rice in terms of grain yield (g) per kg of water evapotranspired (WP_{ET}), per kg of water inflow to the field during crop growth period (WP_{IF}), and per kg of water inflow to the field during land preparation and crop growth period (WP_{IFT}).

WP_{ET}	WP_{IF}	WP_{IFT}	Source of data used in calculating water productivity	Location
		Rice		
		0.05-0.25	Bhatti and Kijne (1992), rainwater not included	Pakistan
1.61	0.68 (0.42)[†]	0.39 (0.24)	Bhuiyan et al. (1995), wet-seeded rice	Philippines
1.39	0.48 (0.35)	0.29 (0.22)	Bhuiyan et al. (1995), transplanted rice	Philippines
1.1	0.45 (0.41)		Sandhu et al. (1980)	India
0.95	0.66 (0.69)	0.58 (0.61)	Kitamura (1990), dry season	Malaysia
0.88	0.48 (0.50)	0.33 (0.35)	Kitamura (1990), wet season	Malaysia
0.89	0.34 (0.36)		Mishra et al. (1990), continuous flooding	India
0.89	0.37 (0.42)		Mishra et al. (1990), alternate wetting and drying	India
0.4-0.5			Khepar et al. (1997)	India
		Wheat		
1.0-2.0			Turner (1997)	Australia
1.0-1.5	1.0-1.6 (1.1)		Deju and Lu Jingwen (1993), winter wheat	China
0.65	0.8 (1.2)		Sharma et al. (1990)	India
0.87	0.79 (0.91)		Pinter et al. (1990)	
		Corn		
2.8	2.2-3.9 (0.8-1.4)		Stegman (1982)	USA
1.9-2.8	1.9-2.5 (0.9)		Moridis and Alagcan (1992)	Philippines
1.7-2.1	1.6-1.7 (0.9)		Stockle et al. (1990)	USA

[†]Numbers in parentheses are water ratio of evapotranspiration to inflow water. When this ration is greater than 1, crops could use water stored in the root zone.

1997). The average yield of irrigated rice in tropical Asia is about 4-6 t ha^{-1}, but good crop management can achieve a yield of nearly 10 t ha^{-1} (Peng et al., 1999). In Table 1, rice and wheat have a considerably lower WP_{ET} than corn because of the inherent different metabolism between a C_4 crop (corn) and C_3 crops (rice and wheat). The lower value of WP_{ET} in rice than in wheat probably comes about because evaporation is higher in the rice field, and therefore a smaller portion of ET is productively transpired. Rice also has higher stomatal conductance and therefore transpires more than wheat (S.B. Peng, personal communication, 1999).

The difference in water productivity between rice and other food crops increases when water productivity is calculated with respect to the amount of water inflow to the field during the crop growth period. In rice, water productivity per unit inflow (WP_{IF}) is about 0.3-0.7 g kg^{-1}, while it is 0.8-1.6 g

kg^{-1} in wheat and 1.6-3.9 g kg^{-1} in corn (Table 1). Lower WP_{IF} in rice conforms to higher S and P in rice than in other food crops. The ET/inflow ratio in other crops is sometimes greater than 1, indicating that other crops could make use of water stored in the root zone. This is not possible in rice because of its shallow root systems. In addition, rice suffers from water stress at soil moisture just below saturation.

If the amount of water applied during land preparation is included in the computation, water productivity of rice becomes dismally small, from 0.25 to 0.6 g kg^{-1} (Table 1). The very low water productivity of rice with respect to the total inflow conforms to previous researchers' finding, such as Revelle (1963), that rice uses as much as 4000 kg of water to produce 1 kg of grain, whereas wheat uses only about 1000 kg of water. The word "use" can however be misleading, because only a small fraction of 4000 kg of water is evapotranspired by the crop, the rest does not reach the crop and is not "used."

Strategies for Increasing Water Productivity in Rice Production

Producing more rice with less water involves (i) increasing yield per unit ET and (ii) reducing the portion of water input to the field that is not available for crop ET. From the previous discussion, this portion includes the non-beneficial depletion by evaporation during land preparation and the outflows (bypass flow during land preparation and S and P during crop growth). Water input to the field consists of irrigation water and rainfall. For economic reasons, we are more interested in reducing the amount of irrigation water. This can be achieved by more effective use of rainfall (Khepar et al., 1997).

The following sections discuss practices that can fulfill the above strategies. The discussion does not include the role of water storage in the soil profile. Most of lowland rice roots are in the 0-20 cm surface layer (Samson et al., 1995). Transpiration of rice declines quickly as soil moisture gets just below saturation point (Tuong et al., 1996a). The amount of water that rice can extract from the soil profile is very small compared to other components of water balance.

Increasing Yield Per Unit ET

Changing Crop Schedule

For a given crop and soil fertility, the potential biomass accumulation largely depends on the amount of radiation that the canopy receives. Evapotranspiration on the other hand is driven by other climatic factors, such as temperature, wind speed, and especially, the saturation vapor pressure deficit of the atmosphere. These climatic parameters vary widely from region to region and from season to season. It is therefore possible to change the potential biomass production per unit ET (strictly speaking, per unit transpiration) by adjusting the growing season (Stanhill, 1986). Untimely planting may also lead to extremely high yield loss due to high spikelet sterility caused by chilling or high temperature. In the Sahel and savanna environment of West Africa, yield per unit ET declines sharply as the planting date is delayed from early August to mid September (Wopereis, 1998). In Punjab, India, the highest yield per unit ET is obtained if rice is transplanted in June, though rice transplanted in May would yield more (Table 2). The variation in yield per unit ET due to changing the planting date is larger in a temperate climate than in a tropical, equatorial climate (Stanhill, 1986).

Selecting Suitable Variety

In well-watered conditions, daily ET of a crop is very much driven by weather conditions. Under the same growing conditions, rice varieties (at least for those of the same type, japonica, indica) do not differ much in their daily ET; total ET is proportional to crop duration. The adoption of improved, early maturing, high-yielding indica varieties of rice during the past 25 years in Asia has increased the average yield of irrigated rice from 2-3 t ha^{-1} to 5-6 t ha^{-1} and reduced crop duration from about 140 days to about 110 days. Compared with traditional varieties, modern indica varieties have a 2.5-3.5-fold increase in WP_{ET}.

Since the release of the first semi-dwarf indica variety, IR8, in 1966, breeders have continued to increase yields of modern varieties and have achieved an annual gain in rice yield of 75 to 81 kg ha^{-1}, equivalent to 1% per year (Peng et al., 1999). Over the same period, there has been a tendency to increase WP_{ET} (Figure 1). Most of the increase in WP_{ET}, however, occurs in cultivars released before 1980. This is because the increase in yield from

TABLE 2. Effect of transplanting date on evapotranspiration (ET), yield and yield per unit ET of rice (WP_{ET}) in Punjab.

Date of transplanting[†]	ET[†] (mm)	Yield[†] (t ha^{-1})	WP_{ET} (g rice kg^{-1} water)
May 1	770	3.53	0.46
May 16	720	3.54	0.49
June 1	660	3.58	0.54
June 16	610	3.41	0.56
July 1	550	2.96	0.54
July 16	530	2.54	0.48

[†] Adapted from Khepar et al. (1997).

FIGURE 1. Yield and water productivity (in yield, g, per kg of evapotranspiration, ET) of cultivars/lines developed since 1996. Data on yield and crop duration are taken from Peng et al. (1999). ET was calculated by ET = $K_c.K_p.K_p$. K_c and K_p are crop- and pan-factor (Doorenbos and Pruitt, 1992). E_p is the total class-A pan evaporation during crop duration, measured at the site of the experiment.

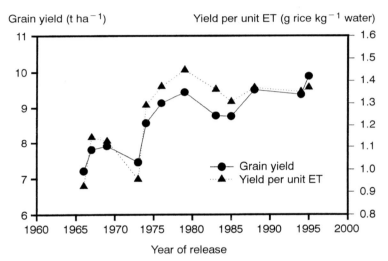

1966 to the early 1980s is coupled with a decrease in growth duration, whereas cultivars released after the mid-1980s have a longer duration than those released before 1980 (Peng et al., 1999).

Since 1989, IRRI has worked on tropical japonicas (also called "new plant type," IRRI, 1998) to improve yield potential and increase genetic diversity of rice (Khush, 1990). Peng et al. (1998) reported that the transpiration efficiency (A/T: ratio of photosynthesis, A, to transpiration, T) was 25-30% higher for the tropical japonica than the indica type. Most of the increase in A/T comes from lower T, suggesting that, in the field, the tropical japonicas may use less water than the indicas.

Manipulating Soil Fertility

Soil fertility can affect WP_{ET} by altering plant growth and yield. In normal rice-growing conditions, each kg of nitrogen fertilizer applied to the field can produce 10-15 kg more rice (S.B. Peng, 1999, personal communication). An improved supply of nutrient will result in rapid expansion of leaf area, increasing the potential for water use for transpiration. But most of the increase in canopy transpiration due to improved nutrient status occurs before the canopy closes (at leaf area index, LAI, of about 3), at about 40 days after transplanting. Because of mutual shading, light intensity diminishes exponentially with cumulative leaf area from the top of the canopy; therefore, a further increase in LAI does not increase transpiration. A more rapid expansion of leaf area reduces the penetration of radiant to the soil surface, or to the water layer if the rice field is flooded. The two processes–increased potential for transpiration and reduced evaporation–may offset each other, providing little change in total ET (Table 3). Better nutrient management increases yield but not ET and can thus increase WP_{ET} several-fold.

Weed Management

Weeds are a common problem in rice fields. Besides nutrient and light, weeds also compete with rice for water. Few data are available on the exact partitioning of the amount of water transpired by rice and that by weeds in a rice-weed community (Kropff, 1993). This information would require data on the LAI of rice and weeds and their distribution at different heights of the crop-community canopy. It is likely, however, that the total ET of the community would not be much different from the weed-free rice crop. On the other hand, weed infestation can cause losses ranging from 30% to 90% of the rice yield (Moody, 1993), and therefore a corresponding reduction WP_{ET}. Weed control is a practical way to increase the water productivity of water in rice production. Weed control is now facilitated by selective and effective herbicides. Unfortunately, associated crop management practices are generally far from optimal for weed control. For example, poor leveling resulting in wide variation in the ponded water depth or soil moisture status in the field reduces the efficiency and effectiveness of herbicide application (Moody, 1993). Farmers' knowledge of safety, timing, and dosage of herbicide use is usually poor.

TABLE 3. Effect of nitrogen rate on evaporation (E), transpiration (T), evapotranspiration (ET), yield, and yield per unit ET of rice cv. IR8.

	N-fertilizer rate (kg ha^{-1})		
	0	80	100
E[†] (mm)	212	157	149
T (mm)	302	352	362
ET (mm)	514	509	511
Yield (t ha^{-1})	2.06	4.55	4.84
Yield per unit ET (g rice kg^{-1} water)	0.40	0.89	0.95

[†]E, T, and ET were computed using model Oryza_W (Wopereis et al. 1996a)

Reducing Evaporation from Soil/Water Surface During Crop Growth

Evaporation from the soil or from the surface of the ponded water layer in the field accounts for about 30% of ET. Reducing evaporation may contribute greatly to increasing WP_{ET}. Unfortunately, it has been taken for granted that, from the management point of view, evaporation can not be isolated from ET and no systematic attempt has been made to reduce the evaporation component during crop growth.

One obvious option to reduce evaporation is to reduce spacing between plants. In rice, while this leads to increase of total biomass, it may not result in yield increase. Due to source limitation, harvest index will decline when the plants are too close. This is best illustrated by comparing transplanted (20 to 25 hills m^{-2}, 2-4 plants per hill) and direct-seeded rice (seeds are sown directly on rice fields, 300 to 500 plants m^{-2}). Inter- and intra-plant competition due to high plant density and excessive vegetative growth result in a high rate of tiller abortion, low percentage of filled grains, and lower harvest index in direct-seeded rice compared to transplanted rice. Yield of direct-seeded rice is often less than transplanted rice, especially in the wet season (Dingkuhn et al., 1990). In transplanted rice, reducing plant spacing is also constrained by transplanting labor cost and availability. In China, where hybrid rice is grown, the cost of seed also restrains farmers from increasing plant density (Wang et al., 1998).

Reducing Non-Beneficial Depletion of Water

Besides evaporation from soil or surface water during the crop growth period non-beneficial depletion of water in rice culture also includes evaporation from the field during land preparation. The amount of evaporation during land preparation can be reduced by changing the farming schedule or by reducing the duration of land preparation.

Changing Crop Schedule

Avoiding land preparation during periods of high evaporation rate can reduce the amount of water depleted into the atmosphere. In Punjab, by delaying land preparation until late June or early July, farmers can reduce evaporation during land preparation by about 60-100 mm. May and June have the highest evaporation (Hira, 1996). Delaying land preparation, however, may affect yield of the wheat crop, which is established when rice is harvested. Furthermore, the later transplanted rice is more susceptible to stem borer incidence.

Reducing the Duration of Land Preparation

In the transplanted-rice (TPR) system, seedlings are usually nurtured in the seedbed for about 1 month. In many irrigation systems, farmers set aside a small area of their farm to make the seedbeds. Because of a lack of tertiary and field channels, and with field-to-field irrigation, all the surrounding fields have to be flooded before farmers can get water to prepare the seedbeds and irrigate the seedlings. Farmers thus begin land soaking (of the whole farm) as they start seedbed preparation and have no reason to complete plowing and harrowing activities until the seedlings are ready. In some cases, land soaking starts even before farmers are ready for their seedbeds because neighboring farmers like to prepare their seedbeds earlier. The following sections discuss several options available avoiding the unnecessary prolongation of land preparation.

Provision of tertiary infrastructure. The provision of tertiary infrastructure to supply irrigation water directly to individual farm lots will allow farmers to saturate only the nursery field without submerging the main field. It would also allow farming activities to be independent of those of neighboring farms. In 1979, the Muda irrigation scheme, Malaysia launched a tertiary infrastructure development project in a portion of the system, increasing canal and drainage intensities from 10 m ha^{-1} to 30 m ha^{-1}. This enabled farmers to shorten their land preparation by 25 days, resulting in an annual water savings of 375 mm in two rice-cropping seasons (Abdullah, 1998). This author indicated, however, that tertiary infrastructure was costly; the economic rate of return of areas with tertiary facilities was only 6% compared with 12% in areas without tertiary channels.

Community seedbeds. In countries such as Vietnam and China, specific land was set aside for community seedbeds where seedlings were raised for the whole cooperative. Seedlings were then distributed to farmers in their allotted fields. The community seedbeds could be irrigated independently. Land preparation of the main fields could be carried out independently of the seedling age. But the arrangement needs a strong institutional set-up and cannot be applied to areas where farmers cultivate their land individually. In fact, the system of community seedbeds disintegrated with the weakening of the cooperatives when the "free market systems" were applied to China and Vietnam (Mao Zhi, 1998, personal communication).

Adoption of a water-efficient method of rice establishment. In recent years, there has been a shift from transplanted rice to the direct-seeded method of crop establishment in several countries in Southeast Asia (Khan et al., 1992; Sattar and Bhuiyan, 1993; Khoo, 1994). This change to direct seeding was brought about largely by increased wages that had to be paid for the transplanting operation because of the acute shortage of farm labor (De Datta,

1986; Chan and Nor, 1993). The shift to direct seeding does offer opportunities to reduce non-beneficial evaporation by shortening land preparation.

There are two forms of direct-seeded rice: wet seeding and dry seeding. In wet-seeded rice (WSR), pre-germinated seeds are broadcast on saturated and usually puddled soil. In contrast, dry-seeded rice (DSR) is grown by sowing ungerminated seeds on dry or moist but unpuddled soil.

In WSR, seeds requires only 24-36 hours of soaking and incubation to be ready for sowing in the field. Land preparation is not tied to seedling raising. In Central Luzon, Philippines, the average time taken for farmers with farm labor for WSR to complete land preparation is 6 days, whereas farmers who transplant need 24 days (Bhuiyan et al., 1995). In the Muda irrigation scheme, shortening the land preparation period by shifting from transplanted rice to WRS resulted in a reduction in irrigation duration from 140 to 105 days (Fujii and Cho, 1996).

Wider adoption of wet-seeded rice is faced with many constraints, however, the major ones being inconsistent seedling establishment and high weed infestation. Because weeds emerge almost at the same time or at the same time as the crop, weed competition is greater and weed control by herbicides is more critical in wet-seeded rice (Moody, 1993). Rice farmers who practice wet seeding have to resort to using higher quantities of herbicides or more labor-intensive hand weeding than farmers of transplanted rice. Increased use of herbicide may raise the economic, social, and environmental costs of growing rice (Williams et al., 1990). Though technologies are available for minimizing weed infestation in WSR (Moody, 1993; Tuong et al., 1999), further research is needed on weed dynamics and alternative environmentally friendly weed management strategies for WSR systems. Water management for direct seeding is different from that for transplanting, particularly in crop establishment and early growth. Because the drainage requirement is more stringent with WSR for satisfactory crop establishment, a change in water management may be necessary. We also need to develop an effective and affordable method of land leveling, which is crucial for good crop establishment of WSR.

Reducing Outflows

Minimizing Bypass Flow Through Cracks During Land Preparation

The amount of bypass flow can be reduced by measures that restrict the formation of soil cracks or impede the flow of water through the cracks. Shallow, dry tillage soon after harvesting the previous rice crop is an effective strategy for minimizing the formation of soil cracks and occurrence of bypass flow. The tilled layer acts as mulch and therefore reduces soil drying and subsequent cracking. In soils that already have cracks, dry tillage produces small soil aggregates that block the cracks, thereby reducing bypass flow. Cabangon and Tuong (1999) found that in farmers' fields in Bulacan and Nueva Ecija, Philippines, shallow tillage reduced the total water input for land preparation by 31-34%, which corresponds to 108-117 mm of water. Dry tillage is now widely practiced in the Muda irrigation scheme in Malaysia and is responsible for reduced water released from the reservoir and timely crop establishment in the area (Ho Nai Kin et al., 1993). Increasing access to high-powered tractors makes dry tillage possible in many irrigated rice systems in Asia.

Reducing Soil Permeability

Puddling the soil during land preparation is an effective way to reduce percolation during crop growth. Puddling causes the formation of a semi-impermeable layer with a very low hydraulic conductivity beneath the puddled topsoil (Sanchez, 1973; De Datta and Kerim, 1974; Tuong et al., 1994). Dayan and Singh (1980) reported that puddling can reduce input water by 40-60% during crop growth because of the reduced percolation rate. In permeable subsoil conditions, even a small area of unpuddled soil (on the order of 1% of the area of puddled soil) could increase the percolation rate in the field by a factor of five (Tuong et al., 1994). In most cases, however, a semipermeable soil layer or hardpan develops through years of puddling the soil, which substantially reduces percolation loss (De Datta, 1981). Hence, in soils with a developed hardpan, puddling is not needed every year to reduce percolation.

Reducing Water Depth and Duration of Flooding:
Water-Saving Irrigation Techniques

Numerous studies conducted on the manipulation of depth and interval of irrigation have demonstrated that continuous submergence is not essential for obtaining high rice yields. Hatta (1967), Tabbal et al. (1992), and Singh et al. (1996) reported that maintaining a very thin water layer or saturated soil conditions or alternate wetting and drying could reduce water applied to the field by about 40-70% compared with the traditional practice of continuous shallow submergence, without a significant yield loss. The reduction comes from lowering seepage and percolation. In general, the lighter the soil, the greater the reduction in water needed for the rice field when these water-saving irrigation (WSI) techniques are used. The dry period after the disappearance of ponded water depends on the depth of the groundwater table. The shallower the groundwater table, the longer the interval between irrigations (Mishra et al., 1990, 1997). Though experimented in many countries, WSI techniques are widely practiced by rice farmers only in China (Wei and Song, 1989). One of these techniques practiced in China also maintains soil moisture at 50-100% of the soil saturation value throughout the period following the start of the booting stage.

Water-saving irrigation techniques do not increase WP_{ET} in rice (Table 1; Mishra et al., 1990). The drying periods in some WSI techniques may induce water stress in the rice plant, because transpiration of the plant starts declining at soil moisture just below saturation (Tuong et al., 1996a). But stresses during the vegetative stage of the crop may not lead to a reduction in total transpiration and yield loss because the stresses delay anthesis and prolong the crop duration. The crop recovers when the stress is relieved. Stresses during the productive stage of the crop may lead to a reduction in rice yield (Wopereis et al., 1996b) and WP_{ET}.

Alternate wetting and drying cycles in some WSI techniques may have enormous consequences for nutrient availability. Ammonium (NH_4^+) is the predominant soil mineral N form in continuously flooded or saturated rice systems, while nitrate (NO_3^-) is the major form in non-flooded systems. Nitrate can be formed during the drying period of the alternate wetting and drying cycles of some SWI techniques. Upon subsequent flooding, NO_3^- can be leached or subjected to denitrification. Wang et al. (1998) reported low N recovery efficiency (29% in early rice and 5% in the late rice crop) in Zheijiang province, China, and attributed the high nitrogen loss partly to the midseason drainage and subsequent wet and dry cycles at the later growth stages of the WSI. Khunthasuvon et al. (1998) also indicated that the loss of standing water during some stages of growth in the rice field may not cause severe water stress, but may reduce yield through reduced availability of certain nutrients. The results indicate a possible trade-off between water savings in WSI and nutrient-use efficiency. This trade-off has to be evaluated to identify the optimum combination of water and agronomic management.

Keeping a shallow water depth, maintaining saturated soil conditions and alternating wetting and drying cycles also increase the risk of weed infestation (Moody, 1993). Rice farmers who practice WSI have to resort to using higher quantities of herbicides or more labor-intensive hand weeding than those who practice continuous submergence to reduce weed problems. WSI techniques also require a high degree of management and infrastructure at both the farm and system levels. Because of smaller quantities of irrigation water and more frequent applications, more supervision and labor are required than in the traditional shallow-flooding system. Adoption may also be hampered by farmers' concerns about not having access to water when they need it because of the lack of reliability in the system's water supply performance. The lack of field channels, which are necessary for effective water distribution, is another constraint to the adoption of WSI regimes. We need to understand more about the costs and benefits of WSI techniques, including the requirement for labor (for water management and weed control) and other inputs such as fertilizer and herbicides. The environmental impact of WSI caused by higher losses of fertilizer, runoff of herbicides, etc., also needs investigation.

By minimizing seepage and percolation, WSI techniques reduce the amount of water diverted to the field and are therefore important for farmers when water applied to the field is costly. Their effects on overall irrigation system efficiency and regional water productivity are much less understood and defined. The effects depend heavily on the consequences of runoff and S and P after water leaves the farm. Some authors, such as Keller et al. (1996), argued that reducing S and P of upstream farms may not improve overall efficiency if S and P water is recycled (reused) downstream. In this case, WSI does not result in more water to irrigate more land nor does food production increase because WSI does not increase WP_{ET}. The issue becomes comparing the cost-benefit of WSI and recycling options.

On the other hand, if S and P flow to sinks (such as saline groundwater or the sea), reducing S and P will result in more fresh water available to irrigate more land or for other uses. Systematic analyses of scale effects in moving the analysis from the farm to the irrigation system to the river basin are lacking.

Making Use of Rainfall More Effectively

Changing the Crop and Irrigation Schedule

Normally, no water or only a small amount of water is available for release from the reservoir at the beginning of the rainy season. Farmers do not often start their rainy season crop until irrigation water is released from the canal, that is, when enough water is collected in the reservoir. Complete dependence on the irrigation water supply at that time leads to a delayed start of the rice crop, which cannot make use of early rainfall. Developing and adopting new irrigation schedules for preparing land using early season rainfall could make it possible to conserve water in the reservoir, allowing more opportunity for increasing irrigated area in the dry season.

Projects such as the Kadulla irrigation scheme (Bird et al., 1991) and the Walagambahuwa minor-tank settlement scheme (Upasena et al., 1980) reported initial success. But considerable coordination is needed between farmers who must adjust their planting schedules and irrigation administrators who must provide the timely release of water for farmers to adopt this system. The failure on the part of farmers may be related to their own economic situation (e.g., lack of money to finance inputs for early planting) and/or risk-averting decision-making, whereas the failure on the part of irrigation administrators may reflect a lack of motivation and incentives.

Adopting Dry-Seeded Rice System

Dry-seeded rice technology offers a significant opportunity for conserving irrigation water by using rainfall more effectively. In transplanted and wet-seeded rice systems, early rainfalls are not adequate to saturate the soil for wet land preparation. Farmers normally wait for delivery of canal water before they start soaking land for plowing. In DSR, early premonsoon rainfall is used effectively for crop establishment and the early stage of crop growth. Later in the season, when the reservoir has been filled and irrigation has begun, the crop can be irrigated as needed. Early crop establishment results in early harvest of the first crop. This permits a reduction in irrigation inflow requirements from reservoirs in the wet season, leading to an increase in the availability of water in the dry season.

In the Muda irrigation scheme, Malaysia, DSR requires about 300 mm of irrigation water, less than TPR and WSR (Table 4). Rice production per unit volume of irrigation water of DSR doubled that of WSR and TPR. The average yield per hectare of land in DSR, however, was slightly less than that of WSR and TPR, leading to a slightly lower WP_{ET} in DSR (Table 4). Lower yield in DSR was caused by higher weed infestation.

Dry seeding of rice may expose the crop to several risks during crop establishment. A rain break after sowing could result in seed loss. The stand may be thinned or lost to seedling drought, or seedling vigor may be impaired. Weeds may emerge before or with the rice seedlings. Similar to wet-seeded rice, seedlings of dry-seeded rice lack the early size advantage of transplants. Changing from transplanting to dry seeding also changes the dominance of weed infestation, which is highly dependent on the hydrology of the field at crop establishment. Dry seeding in the Muda irrigation scheme has led to heavy infestation of "weedy rice" (off type rice which has undesir-

TABLE 4. Effect of crop establishment by transplanting (TP), wet seeding (WS) and dry seeding (DS) on water productivity of rice in Muda irrigation scheme, Malaysia.

	TP and WS	DS
Irrigation (mm)	589 ± 47	261 ± 83
Rainfall (mm)	1371 ± 94	1380 ± 132
ET (mm)	850 ± 17	866 ± 24
Yield (t ha^{-1})	4.43 ± 0.28	4.28 ± 0.1
Yield per unit irrigation water (g rice kg^{-1} water)	0.75 ± 0.07	1.64 ± 0.32
Yield per unit ET (g rice kg^{-1} water)	0.52 ± 0.03	0.49 ± 0.01

able characteristics of early and easy shattering), a threat to the rice industry in Malaysia (Zainal and Azmi, 1996). We do not fully understand the mechanism and the consequences of the shifting weed communities.

Dry seeding may offer the prospect of reducing the amount of irrigation water as long as a suitable plant stand relatively free from weeds can be established. To capitalize fully on the potential advantages of direct dry seeding, research is needed to develop integrated strategies for reliable establishment of direct dry-seeded crops with adequate management or suppression of weeds. The potential contribution of short-residual, postemergence herbicides should be fully explored. Although crop management is the basis of any effective strategy for establishing a good stand and competing with weeds, selection of improved cultivars may also be helpful. Most cultivars, whether traditional or improved, have been selected for performance in transplanted conditions. Breeding lines are now evaluated under dry-seeding conditions, and selections are made for quality of plant stand and seedling vigor (Sarkarung et al., 1995).

Dry seeding excludes the benefit of soil puddling to reduce percolation. In some soils, the gain in capturing early rainfall may be offset by higher outflow by percolation. Soil hydraulic properties have to be investigated before adopting dry seeding as an option for saving water.

CONCLUSIONS

Compared with other grain crops, rice has a lower WP_{ET}. The fraction of water supply that is available for crop ET is also smaller in rice than in other grain crops. A large proportion of water inflow to the rice field is depleted by evaporation during land preparation and flow out of the field in the forms of bypass flow, seepage, and percolation. These non-beneficial depletions and outflows are inevitable in traditional rice cultivation, which favors preparing land in a flooded condition and maintaining a water layer on the soil surface during crop growth.

This paper has presented a range of strategies and options to increase water productivity at the farm level. Implementing options that increase yield per unit ET, reduce evaporation during land preparation, or make more effective use of rainfall will always result in more rice with less water needed from the irrigation system or river basin. The effects of options that minimize outflows (bypass flow, seepage, and percolation) are scale-dependent and site-specific. These options do not increase WP_{ET} of rice and therefore may not result in more rice production at a regional scale if the outflow is recycled (reused) at a point downstream in the basin. Some options have negative impact on fertilizer-use efficiency, labor use, weed management, and may not be cost effective. These illustrate the strong interactions that exist among different components and scales of an irrigation system or a basin and among different inputs in rice farming. Farm- and system-level options for increasing irrigation efficiency and water productivity have to be analyzed interactively. They have to take into account their on-farm impact on other inputs. Unfortunately, we do not now have adequate information on the off-site effects of the farm level options, to judge whether their implementation will be translated into net gains for the irrigation system and the entire basin. We have few data on the productivity of water over a large scale and on the cost of various options for increasing productivity to evaluate the trade-off among options. Implementing these options may be constrained by the continuing lack of incentives for farmers and irrigation systems managers to improve performance, and by poorly defined land and water rights and inadequate support systems that discourage farmer participation in management. The challenge to improve water management and control on-farm water and water in the irrigation system and to grow more rice with less water is formidable.

REFERENCES

Abdullah, N. (1998). Some approaches towards increasing water use efficiency in the Muda irrigation scheme. In *Proceedings of the Workshop on Increasing Water Productivity and Efficiency in Rice-Based Systems. July 1998*, Los Baños, Philippines: IRRI.

Bhatti, M.A. and J.W. Kijne. (1992). Irrigation management potential of paddy/rice production in Punjab of Pakistan. In *Soil and Water Engineering for Paddy Field Management. Proceedings of the International Workshop on Soil and Water Engineering for Paddy Field Management, January 28-30, 1992*, eds. V.V.N. Murty and K. Koga, Bangkok, Thailand: Asian Institute of Technology, pp. 355-366.

Bhuiyan, S.I. (1992). Water management in relation to crop production: case study on rice. *Outlook on Agriculture* 21(4):293-299.

Bhuiyan, S.I., M.A. Sattar, and M.A.K. Khan. (1995). Improving water use efficiency in rice irrigation through wet seeding. *Irrigation Science* 16(1):1-8.

Bird, J.D., M.R.H. Francis, I.W. Makin, and J.A. Weller. (1991). Monitoring and evaluation of water distribution: an integral part of irrigation management. In *Improved Irrigation Systgem Performance for Sustainable Agriculture. Proceedings of the Regional Workshop on Improved Irrigation System Performance for Sustainable Agriculture, 22-26 October 1990, Bangkok*. Rome, Italy: Food and Agriculture Organization of the United Nations, pp.112-131.

Cabangon, R.J. and T.P. Tuong. (1999). Management of cracked soils for water saving during land preparation for rice cultivation. *Soil and Tillage Research* (Accepted.)

Chan, C.C. and M.A.M. Nor. (1993). Impacts and implications of direct seeding on irrigation requirement and systems management. In *Workshop on Water and Direct Seeding for Rice, 14-16 June 1993*, Ampang Jajar, Alor Setar, Malaysia: Muda Agricultural Development Authority.

Dawe, D., R. Barker, and D. Seckler. (1998). Water supply and research for food security in Asia. In *Proceedings of the Workshop on Increasing Water Productivity and Efficiency in Rice-Based Systems. July, 1998*. Los Baños, Philippines: International Rice Research Institute.

Dayan and A.K. Singh. (1980). Puddled vs. unpuddled paddy. *Intensive Agriculture* 8(3):7.

De Datta, S.K. (1981). *Principles and Practices of Rice Production*. New York, N.Y. (USA): John Wiley and Sons.

De Datta, S.K. (1986). Technology development and the spread of direct-seeded flooded rice in Southeast Asia. *Experimental Agriculture* 22:417-426.

De Datta, S.K. and M.S. Kerim. (1974). Water and nitrogen economy of rainfed rice as affected by soil puddling. *Soil Science Society of America Proceedings* 38(3):515-518.

de Wit, C.T. (1958). Transpiration and crop yields. *Versalg Landouw-kundig Onderzoek*, Wageningen, Netherlands: Wageningen University.

Dingkuhn, M., H.F. Schnier, S.K. De Datta, E. Wijangco, and K. Doerffling. (1990). Diurnal and developmental changes in canopy gas exchange in relation to growth in transplanted and direct-seeded flooded rice. *Australian Journal of Plant Physiology* 17:119-134.

Doorenbos, J. and W.O. Pruitt. (1992). Crop water requirements. *FAO Irrigation and Drainage Paper 24*. Rome, Italy: Food and Agriculture Organization of the United Nations.

Deju, Z. and L. Jingwen. (1993). The water-use efficiency of winter wheat and maize on a salt-affected soil in the Huang Huai river plain of China. *Agricultural Water Management* 23: 67-82.

Fujii, H. and M.C. Cho. (1996). Water management under direct seeding. In *Recent Advances in Malaysian Rice Production: Direct Seeding Culture in the Muda Area*, eds. Y. Morooka, S. Jegatheesan, and K. Yasunobu, Ampang Jajar, Alor Setar, Malaysia: Muda Agricultural Development Authority and Japan International Research Center for Agricultural Sciences, pp. 113-129.

Ghani, Md. A., S.I. Bhuiyan, and R.W. Hill. (1989). Gravity irrigation management in Bangladesh. *Journal of Irrigation and Drainage Engineering* 115(4):642-655.

Gunawardena, E.R.N. (1992). Improved water management for paddy irrigation: a case study from the Mahaweli System "B" in Sri Lanka. In *Soil and Water Engineering for Paddy Field Management. Proceedings of the International Workshop on Soil and Water Engineering for Paddy Field Management, 28-30 January*, eds. V.V.N. Murty and K. Koga, Bangkok, Thailand: Asian Institute of Technology, pp. 307-316.

Hatta, S. (1967). Water consumption in paddy field and water saving rice culture in the tropical zone. *Japanese Tropical Agriculture* 11(3):106-112.

Hira, G.S. (1996). Evapotranspiration of rice and falling water table in Punjab. In *Evapotranspirtion and Irrigation Scheduling. Proceedings of a Meeting 3-6 November 1996, San Antonio Convention Center, San Antonio, Texas*, eds. C.R. Camp, E.J. Sadler, and R.E. Yoder, St. Joseph, Michigan: American Society of Agricultural Engineers, pp. 579-584.

Ho Nai Kin, C.M. Chang, M. Murat, M.Z. Ismail. (1993). MADA's experiences in direct seeding. In *Workshop on Water and Direct Seeding for Rice, 14-16 June 1993*. Ampang Jajar, Alor Setar, Malaysia: Muda Agricultural Development Authority.

IRRI. (1998). *Sustaining Food Security Beyond the Year 2000: A Global Partnership for Rice Research. IRRI Rolling Medium-Term Plant 1999-2001*. Los Banos (Philippines): International Rice Research Institute.

Jensen, M.E. (1980). The role of irrigation in food and fiber production. In *Design and Operation of Farm Irrigation Systems*, ed. M.E. Jensen, St. Joseph, Michigan: American Society of Agricultural Engineers, pp. 15-41.

Keller, A., J. Keller, and D. Seckler. (1996). *Integrated Water Resources Systems: Theory and Policy Implications. Research Report 3*. Colombo (Sri Lanka): International Irrigation Management Institute.

Khan, L.R. (1992). Water management issues for paddy fields in the low-lying areas of Bangladesh. In *Soil and Water Engineering for Paddy Field Management: Proceedings of the International Workshop on Soil and Water Engineering for Paddy Field Management, 28-30 January 1992*. eds. V.V.N. Murty and K. Koga, Bangkok, Thailand: Asian Institute of Technology, pp. 339-345.

Khan, M.A.K., S.I. Bhuiyan, and R.C. Undan. (1992). Assessment of direct-seeded rice in an irrigated system in the Philippines. *Bangaldesh Rice Journal* 3(1&2): 14-20.

Khepar, S.D., S.K. Sondi, S. Kumar, and K. Singh. (1997). Modelling effects of cultural practices on water use in paddy fields—a case study. *Research Bulletin, Publication No. NP/SWE-1*. Ludhiana, India: Punjab Agricultural University, Ludhiana, India.

Khoo, C.N. (1994). Improved irrigation/water management for sustainable agricultural development in Malaysia. In: *Irrigation Performance and Evaluation for Sustainable Agricultural Development. Report of the Expert Consultation of the Asian Network on Irrigation/Water Management, 16-20 May 1994*, Bangkok, Thailand: RAPA Publication 1994/17. pp. 151-160.

Khunthasuvon, S., S. Rajastasereekul, P. Hanviriyapant, P. Romyen, S. Fukai, J. Basnayake, and E. Skulkhu. (1998). Lowland rice improvement in northern and northeast Thailand. 1. Effects of fertiliser application and irrigation. *Field Crops Research* 59:99-108.

Khush, G.S. (1990). Varietal needs for different environments and breeding strategies. In *New Frontiers in Rice Research*, eds. K. Muralidharan and E.A. Siddiq, Hyderabad, India: Directorate of Rice Research, pp. 68-75.

Kitamura, Y. (1990). Management of an irrigation system for double cropping culture in the tropical monsoon area. *Technical Bulletin 27*. Tropical Agriculture Research Center, Ministry of Agriculture Forestry and Fisheries, Tsukuba, Ibaraki 305, Japan.

Kropff, M.J. (1993). Mechanisms of competition for water. In *Modelling Crop-Weed Interactions*, eds. M.J. Kropff and H.H. van Laar, Wallingford, Oxon, United Kingdom: CAB International, pp. 63-76.

Mao Zhi. (1982). Calculation of evapotranspiration of rice in China. In *Soil and Water Engineering for Paddy Field Management. Proceedings of the International Workshop on Soil and Water Engineering for Paddy Field Management, 28-30*

January, eds. V.V.N. Murty and K. Koga, Bangkok, Thailand: Asian Institute of Technology, Bangkok, Thailand. pp. 21- 31.

Mishra, H.S., T.R. Rathore, and R.C. Pant. (1990). Effect of intermittent irrigation on groundwater table contribution, irrigation requirement and yield of rice in Mollisols of the tarai region. *Agricultural Water Management* 18:231-241.

Mishra, H.S., T.R. Rathore, and R.C. Pant. (1997). Root growth, water potential, and yield of irrigated rice. *Irrigation Science* 17:69-75.

Molden, D. (1997). Accounting for water use and productivity. *SWIM Paper 1.* Colombo (Sri Lanka): International Irrigation Management Institute.

Moody, K. (1993). The role of water for weed control in direct-seeded rice. In *Workshop on Water and Direct Seeding for Rice, 14-16 June 1993*, Ampang Jajar, Alor Setar, Malaysia: Muda Agricultural Development Authority.

Moridis, G.J. and M. Alagcan. (1992). High-frequency basin irrigation design for upland crops in rice lands. *Irrigation and Drainage Engineering* 118:564-583.

Peng, S., R.C. Laza, G.S. Khush, A.L. Sanico, R.M. Visperas, and F.V. Garcia. (1998). Transpiration efficiencies of indica and improved tropical japonica rice grown under irrigated conditions. *Euphytica* 103:103-108.

Peng, S., R.C. Laza, R.M. Visperas, A.L. Sanico, K.G. Cassman, and G.S. Khush. (1999). Grain yield of rice cultivars and lines developed in the Philippines since 1966. *Crop Science* (Submitted.)

Pinter, P.J., G. Zipoli, R.J. Reginato, R.D. Jackson, S.B. Idso, and J.P. Hohman (1990). Canopy temperature as an indicator of differential water use and yield performance among wheat cultivars. *Agricultural Water Management* 18:35-48.

Revelle, R. (1963). Water. *Scientific American* 209(3):92-109.

Samson, B.K., L.J. Wade, S. Sarkarung, M. Hasan, R. Amin, D. Harnpichitvitaya, G. Pantuwan, R. Rodriguez, T. Sigari and A.N. Calendacion. (1995). In *Fragile Lives in Fragile Ecosystems. Proceedings of the International Rice Research Conference, 13-17 February 1995F*, ed. T.P. Tuong, Los Banos, Philippines: International Rice Research Institute, pp. 521-534.

Sanchez, P.A. (1973). Puddling tropical rice soils. 2. Effects on water losses. *Soil Science* 115(4):303-308.

Sandhu, B.S., K.L. Khera, S.S. Prihar, and B. Singh. (1980). Irrigation need and yield of rice on a sandy-loam soil as effected by continuous and intermittent submergence. *Indian Journal of Agricultural Science* 50(6):492-496.

Sarkarung, S., O.N. Singh, J.K. Roy, A. Vanavichit, and P. Bhekasut. (1995). Breeding strategies for rainfed lowland ecosystem. In *Fragile Lives in Fragile Ecosystems. Proceedings of the International Rice Research Conference, 13-17 February 1995*, ed. T.P. Tuong, Manila (Philippines): International Rice Research Institute, pp. 709-720.

Sattar, M.A. and S.I. Bhuiyan. (1993). A study on the adoption of direct-seeded rice in some selected areas of Philippines. Conference 93. *Farm Economics Journal* 9:128-138.

Sharma, P.K. (1989). Effect of period moisture stress on water-use efficiency in wetland rice. *Oryza* 26(3):252-257.

Sharma, D.K., A. Kumar, and K.N. Singh. (1990). Effect of irrigation scheduling on growth, yield and evapotranspiration of wheat in sodic soils. *Agricultural Water Management* 18: 267-276.

Singh, C.B., T.S. Aujla, B.S. Sandhu, and K.L. Khera. (1996). Effect of transplanting date and irrigation regime on growth, yield and water use in rice (*Oryza sativa*) in northern India. *Indian Journal of Agricultural Science* 66(3): 137-141.

Stanhill, G. (1986). Water use efficiency. *Advances in Agronomy* 39:53- 85.

Stegman, E.C. (1982). Corn grain yield as influenced by timing of evapotranspiration deficits. *Irrigation Science* 3:75-87.

Stockle, C.O., L.G. James, D.L. Bassett, and K.E. Saxton. (1990). Effect of evapo-transpiration underprediction on irrigation scheduling and yield of corn: a simulation study. *Agricultural Water Management* 19:167-179.

Sugimoto, K. (1971). Plant-water relationship of indica rice in Malaysia. *Technical Bulletin TARC No. 1*, Tokyo, Japan: Tropical Agricultural Research Center, Ministry of Agriculture and Forestry.

Tabbal, D.F., R.M. Lampayan, and S.I. Bhuiyan. (1992). Water-efficient irrigation technique for rice. In *Soil and Water Engineering for Paddy Field Management. Proceedings of the International Workshop on Soil and Water Engineering for Paddy Field Management, 28-30 January 1992*, eds. V.V.N. Murty and K. Koga, Bangkok, Thailand: Asian Institute of Technology, pp. 146-159.

Tanner, C.B. and T.R. Sinclair. (1983). Efficient water use in crop production: research or re-search. In *Limitations to Efficient Water Use in Crop Production*, eds. H.M. Taylor, W.R. Jordan, and T.R. Sinclair, Madison, Wisconsin: American Society of Agronomy, Crop Science Society of America, Soil Science Society of America. pp. 1- 27.

Tuong, T.P., A. Boling, A.K. Singh, and M.C.S. Wopereis. (1996a). Transpiration of lowland rice in response to drought. In *Evaporation and Irrigation Scheduling. Proceedings of the International Conference, 3-6 November, San Antonio, Texas*, eds. C.R. Camp, E.J. Sadler, and R.E. Yoder, St. Joseph, Michigan: American Society of Agricultural Engineers. pp. 1071-1077.

Tuong, T.P., R.J. Cabangon, and M.C.S. Wopereis. (1996b). Quantifying flow processes during land soaking of cracked rice soils. *Soil Science Society of America Journal* 60:872-879.

Tuong, T.P., P.P. Pablico, M. Yamauchi, R. Confesor, and K. Moody. (1999). Increasing water productivity and weed suppression of wet seeded rice: effect of water management and rice genotypes. *Experimental Agriculture* (In press).

Tuong, T.P., M.C.W. Wopereis, J.A. Marquez, and M.J. Kropff. (1994). Mechanisms and control of percolation losses in irrigated puddled rice fields. *Soil Science Society of America Journal* 58(6):1794-1803.

Turner, N.C. (1997). Further progress in crop water relations. *Advances in Agronomy* 58: 293-338.

Upasena, S.H., M. Sikurajapathy, and T. Seneviratna. (1980). A new cropping system strategy for the poorly irrigated ricelands of the dry zone. *Tropical Agriculture* 136: 51-58.

Valera, A. (1977). *Field Studies on Water Use and Duration for Land Preparation for Lowland Rice*. Unpublished M.Sc. thesis. Los Baños, Laguna, Philippines: University of the Philippines.

Viets, F.G. (1962). Fertilizer and efficient use of water. *Advances in Agronomy* 14:223-264.

Wang, G., A. Dobermann, F. Rongxing, D. Xianghai, G. Simbahan, M.A.A. Adviento, and H. Changyon. (1998). Indigenous nutrient supply and nutrient efficiency in intensive rice systems of Zhejiang province, China. *Paper Prepared for "Agronomy, Environment and Food Security for the 21st Century, November 22-27, 1998*, New Delhi, India: First International Agronomy Congress.

Wei, W. and S. Song. (1989). Irrigation model of water saving-high yield at lowland paddy field. *International Commission on Irrigation and Drainage, Seventh Afro-Asian Regional Conference, 15-25 October 1980, Delhi, India.* Tokyo, Japan. Vol. I-C:480-496.

Wickham, T.H. and L.N. Sen. (1978). Water management for lowland rice: water requirements and yield response. In *Soil and Rice*. Manila (Philippines): International Rice Research Institute. pp. 649-669.

Williams, J.F., R.R. Roberts, J.E. Hill, S.C. Scardaci, and G. Tibbits. (1990). Managing water for weed control in rice. *California Agriculture* 44(5):7-10.

Wolters, W. and M.G. Bos. (1989). Irrigation Performance Assessment and Irrigation Efficiency. *International Land Research Institute (ILRI) Annual Report*, Wagennjen, The Netherlands pp. 25-37

Wopereis, M.C.S., B.A.M. Bouman, T.P. Tuong, H.F.M. Berge, and M.J. Kropff. (1996a) ORYZA_W: Rice growth model for irrigated and rainfed environments. In *Simulation and System Analysis for Rice Production Proceedings*. Wageningen, Netherlands: Wageningen University and Manila, Philippines: International Rice Research Institute, pp. 159.

Wopereis, M.C.S., M.J. Kropff, and T.P. Tuong. (1996b). Drought responses of two lowland rice cultivars to soil water status. *Field Crops Research* 46: 21-39.

Wopereis, M.C.S. (1998). Issues relevant to water efficiency and productivity of rice-based irrigation systems in West Africa. *Proceedings of the Workshop on Increasing Water Productivity and Efficiency in Rice-Based Systems, July 1998*, Los Baños, Philippines: International Rice Research Institute.

Zainal, A.H. and M. Azmi. (1996). A new weed problem in rice: weedy rice. In *Integrating Science and People in Rice Pest Management. Proceedings of the Rice Integrated Pest Management (IPM) Conference, 18-21 November 1996 Kuala Lumpuv, Malaysia*, eds. A.A. Hamid, L.K. Yeang, and T. Sadi, Malaysian Agriculture Research and Development Institute, Penang, Malaysia pp. 166-174.

Genotype and Water Limitation Effects on Transpiration Efficiency in Sorghum

Miranda Y. Mortlock

Graeme L. Hammer

SUMMARY. Sorghum is grown in many parts of the semi-arid tropics in environments where water limitation is common. Recent studies have identified genetic variation in transpiration efficiency (TE) in sorghum under well-watered conditions. Crop simulation studies suggest that improvement in TE in sorghum could have considerable pay-off in many water-limited environments. The objectives of this study were to examine the variation in TE for a range of sorghum genotypes grown under well watered and water limited conditions, and to seek selection indices for this trait by measuring a range of associated physiological and morphological attributes. A glasshouse study was conducted with 17 genotypes grown under well-watered (WW) or water-limited (WL) conditions. Plants were grown in mini-lysimeters and water use and biomass production were measured. A range of other attributes were measured at plant and leaf level. Genotypes varied significantly in TE (highest about 50% greater than lowest) and TE was

Miranda Y. Mortlock is Research Fellow in the School of Land and Food, University of Queensland, St. Lucia, Qld 4072, Australia, and Graeme L. Hammer is affiliated with the Queensland Department of Primary Industries (QDPI), and The Commonwealth Scientific and Industrial Research Organization (CSIRO), Principle Scientist, Agricultural Production Systems Research Unit, P.O. Box 102, Toowoomba, Qld 4350, Australia.

This experiment was financially supported by the Australian Grains Research and Development Corporation. Seed was supplied by Dr. Bob Henzell (DPI, Warwick, Queensland) and the Australian Tropical Field Crops Genetic Resource Centre (DPI, Biloela, Queensland).

[Haworth co-indexing entry note]: "Genotype and Water Limitation Effects on Transpiration Efficiency in Sorghum." Mortlock, Miranda Y., and Graeme L. Hammer. Co-published simultaneously in *Journal of Crop Production* (Food Products Press, an imprint of The Haworth Press, Inc.) Vol. 2, No. 2 (#4), 1999, pp. 265-286; and: *Water Use in Crop Production* (ed: M. B. Kirkham) Food Products Press, an imprint of The Haworth Press, Inc., 1999, pp. 265-286. Single or multiple copies of this article are available for a fee from The Haworth Document Delivery Service [1-800-342-9678, 9:00 a.m. - 5:00 p.m. (EST). E-mail address: getinfo@haworthpressinc.com].

265

about 10% greater under WL. There was no interaction among genotype and water treatments. TE correlated well with transpiration per
unit leaf area, which is a plant scale index of conductance. Leaf level
measurements supported the association of TE with conductance. The
best indicators of variation in TE were leaf C concentration and leaf
ash, which offer promise as an avenue for development of a selection
index. The mechanism underlying this association, however, remains
unclear. Before TE can be used actively in a breeding program, field
studies are required to confirm these findings from glasshouse studies
and more robust selection indices are needed. *[Article copies available for
a fee from The Haworth Document Delivery Service: 1-800-342-9678. E-mail
address: getinfo@haworthpressinc.com <Website: http://www.haworthpressinc.
com>]*

KEYWORDS. Carbon isotope discrimination, ash content, leaf C concentration

INTRODUCTION

Sorghum (*Sorghum bicolor* [L.] Moench) originated in Africa and is now
widely grown in the semi-arid tropics, predominantly as a dryland crop.
Annual world production has averaged 57 million tonnes over the 1989-1991
period. In 1994 sixty-three million tonnes of sorghum was produced, twice
that of oats or rye (FAO, 1995). It is the dietary staple of 400 million people
in 30 countries, and is the fifth major crop feeding the human race (National
Academy of Sciences, 1996). Subsistence yields as low as 300 kg/ha from
marginal dryland agriculture are vital to rural communities in many developing countries. In USA, Argentina and Australia it is grown as a feed grain and
globally represents nearly 4% of world stockfeed grains. Israel and Japan are
the main importers of sorghum (Lawrence, 1996). The yields are higher in
developed countries but droughts are common in all regions where sorghum
is grown. Throughout these regions, efficient use of the water resource is
increasingly important and sorghum improvement programs are incorporating drought resistance traits into the germplasm to contribute to yield increases or yield stability.

Crop species differ in their water requirements and this was identified
early this century by Briggs and Shantz (1914). Typically, more biomass is
produced per unit of water used by crops with the C_4 photosynthetic pathway
than it is by crops with the C_3 pathway due to differences in carboxylation
pathways. In addition, there are differences in the energy required to produce
the diverse biomass composition of different species.

Definitions of water use efficiency may include both components of soil water evaporation (E) and plant transpiration (T). Soil evaporation under field conditions can be altered by crop management, for example, by attention to plant population, row spacing, and soil surface management. Soil evaporation is also influenced by soil type and frequency of precipitation or irrigation. Water use efficiency is also affected by climate, in particular the vapour pressure deficit (VPD), which affects the drying power of the aerial environment and thus the transpiration rate of plants. Confusion often arises in studies of crop water use efficiency due to lack of clarity in definitions and whether water use includes only transpiration or transpiration plus soil evaporation.

Transpiration efficiency (TE), which is defined here as the amount of biomass accumulated per unit of water transpired, is the preferred measure for examining potential genetic variation in crop water use efficiency. The use of TE avoids confounding due to differences in soil evaporation. TE is inversely proportional to the VPD of the environment, and crop specific transpiration efficiency coefficients (k) can be determined from equation (1)

$$TE = k/VPD \tag{1}$$

A k value of 9 Pa is accepted for sorghum and is used in crop modelling (Tanner and Sinclair 1983; Hammer and Muchow, 1994). Tanner and Sinclair (1983) were sceptical about the possibility to improve TE *per se*, and suggested that most improvements in crop water use efficiency have resulted from reducing soil evaporative losses as a proportion of total water use. Several recent studies, however, have identified significant genetic variation in transpiration efficiency within a species; examples include wheat (Farquhar and Richards, 1984; Matus, Slinkard, and van Kessel, 1996), cotton (Hubick and Farquhar, 1987), peanut (Hubick, Farquhar, and Shorter, 1986; Wright, Hubick and Farquhar, 1988), barley (Hubick and Farquhar, 1989), mustard, canola, pea (Knight, Livingstone and van Kessel., 1994) and sorghum (Hammer, Farquhar, and Broad, 1997; Henderson et al., 1998).

Sorghum is the only C_4 species to date with reported significant genetic variation in TE (Hammer, Farquhar and Broad, 1997; Henderson et al., 1998). In an experiment looking at sorghum genotypes under a range of water limitations, Donatelli, Hammer and Vanderlip (1992) found significant genotype and water effects for TE. When expressed relative to TE in well-watered treatments, TE in water-limited treatments was observed to increase by 28%. Hammer Farquhar, and Broad (1997) found substantial genetic variation in TE among a diverse range of 49 sorghum genotypes when screened under well-watered conditions. The extent of the variation, however, was similar to that found in an adapted range of genotypes by Henderson et al. (1998), suggesting that pursuit of enhanced variation in less adapted material may not be fruitful. Hammer, Farquhar, and Broad (1997) observed TE values up to 9% higher than the accepted standard of 9 Pa for sorghum.

The level of genetic variation in TE observed in sorghum could give significant yield increases in particular environments. Crop models can be used integrate our understanding of crop growth and development in response to environmental inputs (Muchow, Hammer and Carberry, 1991; Hammer and Muchow, 1992) and to examine the potential value of specific traits (Shorter, Lawn, and Hammer, 1991). Recent simulation studies have indicated that TE is a trait that has the potential to improve productivity of sorghum in moderately water-limited environments (Muchow, Hammer, and Carberry, 1991; Hammer et al., 1996). To estimate the value of the TE trait for sorghum yield it was included in a simulation study where parameters were changed to represent genotypes with and without a 10% increase in TE, in association with three levels of maturity. Cluster analysis of the results from the 30 environments tested revealed a significant yield benefit over a wide range of environments for genotypes having enhanced TE (Hammer et al., 1996). The modelling studies support the need to further explore and understand genetic variation in TE, particularly in water-limiting environments.

For the effective use of genetic variation in traits such as TE in crop improvement programs, simple selection indices or genetic markers are needed. This is because TE is difficult, time consuming and costly to measure, especially in the field. The selection index is used as a tool to allow simple, practical selection of improved genotypes. The index needs to correlate with the trait of interest and be relatively simple to measure. Studies in peanut provide an example of a successful hunt for selection indices. In peanut the significant correlation between TE and carbon leaf discrimination (Δ) found by Hubick, Farquhar, and Shorter (1986) and Wright, Hubick, and Farquhar, (1988) allowed high TE genotypes to be readily selected. Carbon isotope discrimination (Δ) occurs as ^{13}C is not incorporated to the same extent as ^{12}C at each level of CO_2 diffusion and metabolism; fractionation occurs at each biochemical step resulting in an integrated value for a plant that relates to the type of photosynthesis and growth conditions. The biomass results in its own 'signature' of Δ that relates to its assimilation history. In C_3 plants there is a theoretical relationship between TE and Δ that can be observed as a negative correlation between TE and Δ (Farquhar, O'Leary, and Berry, 1982; Farquhar and Richards, 1984). In peanut the carbon isotope discrimination (Δ) was in turn correlated to specific leaf area (SLA), which had an advantage in that it was a lot less costly to measure. Such indices can be measured and used to select genotypes in large breeding programs.

Selection indices for TE in sorghum, such as carbon isotope discrimination, may be more difficult to find. In theory, carbon isotope discrimination

(Δ) has a positive relationship with TE in sorghum, which is opposite to that in C_3 species (Farquhar and Richards, 1984; Farquhar, von Caemmerer, and Berry, 1980; Farquhar, O'Leary, and Berry, 1982; Farquhar, 1983). Twelve sorghum genotypes varied in carbon isotope discrimination (Δ) under adequately watered field conditions, and the Δ was negatively correlated with yield (Hubick et al., 1990). In subsequent controlled studies, however, either no relationship was found with TE (Hammer et al., 1997) or the theoretical positive relationship with TE was observed (Henderson et al., 1998). Variation in Δ has not been found in maize (O'Leary, 1981). Interspecific variation in Δ was observed in some *Panicum* species (Ohsugi et al., 1988).

There are three hypothesis put forward to explain the genetic variation of TE in sorghum:

  i. genetic differences in the ratio of carbon assimilation rate (A) to stomatal conductance *(g)* that would relate to transpiration efficiency, or
 ii. genetic differences in leakiness (ϕ) of the bundle sheath cells, which is related to the light-use efficiency of the leaf (Hubick et al., 1990; Farquhar, 1983), or
iii. post-photosynthetic fractionations may be more important in C_4 species (Henderson et al., 1998).

Prior to the discovery of TE variation in sorghum, genetic variation in a range of associated traits had been reported. These attributes include stomatal response to water deficits (Henzell et al., 1976) and differences in p_i/p_a (the ratio of intercellular to ambient partial pressure of CO_2) (O'Leary, 1981; Kreig, 1983). Henderson et al. (1998) quantified the relationship between Δ and p_i/p_a for a range of sorghum genotypes and found variation in Δ to be much less than in C_3 species. In C_3 species Δ varied 22.6% (22.6×10^{-3}) per unit change in p_i/p_a, whereas in C_4 species it was only 3.7% change in Δ per unit change in p_i/p_a (Henderson et al., 1998). Mean p_i/p_a for sorghum lines was 0.33 and it was positively correlated to leaf conductance to CO_2 transfer *(g)* but not to rate of CO_2 assimilation (A) (Henderson et al., 1998).

The evidence to date has shown useful genetic variation in TE of sorghum, although detailed studies under water limitation have not been conducted. There has been little progress, however, in finding selection indices and elucidating the mechanisms of improved TE. It appears that biochemical differences between C_3 and C_4 species will mean that selection indices other than Δ may be needed in C_4 species. Physiological attributes need to be measured along with TE to broaden the search for selection indices and to assist in elucidating mechanisms contributing to enhanced transpiration efficiency in sorghum.

Hence the objectives of this study were–

i. to examine the variation in TE for a range of sorghum genotypes grown under well watered and water limited conditions, and

ii. to seek selection indices for this trait by measuring a range of associated physiological and morphological attributes.

MATERIALS AND METHODS

Experimental Design

A selection of seventeen sorghum (*Sorghum bicolor* [L.] Moench) genotypes (Table 1) was grown under well watered (WW) and water limited (WL) conditions in a glasshouse experiment at the University of Queensland, St. Lucia, Brisbane in the 1997 summer. The experiment was laid out as a randomised complete block design with 34 treatments replicated 4 times. The 4 blocks were positioned to account for the effect of the likely temperature gradient across the glasshouse. Two plants in a mini-lysimeter represented an

TABLE 1. Sorghum (*Sorghum bicolor* [L.] Moench) genotypes used in the experiment.

Genotype	Previous TE Value[1]	Seed Source
M35-1	7.0-7.5	Hermitage
SC237-14E	7.5-8.0	Biloela
R9188	7.0-7.5	Hermitage
SC110-14e	7.0-7.5	Biloela
QL12	6.5-7.0	Hermitage
QL41	6.0-6.5	Hermitage
72389-1-2-3/QL12	-	Hermitage
Buster	-	Hermitage
QL12/69264[2]	-	Hermitage
QL41/69264	-	Hermitage
A35/69264	-	Hermitage
QL39/QL36	-	Hermitage
QL41/QL36	-	Hermitage
A35/QL36	-	Hermitage
QL39/QL12	-	Hermitage
QL41/QL12	-	Hermitage
A35/QL12	-	Hermitage

[1] These values were determined by Hammer et al. (1997).
[2] These genotypes were used in a stay-green experiment, at DPI, Warwick (Borrell, pers. comm.). The nine hybrids are formed by combining 3 male parents with high (QL12), intermediate (QL36), or low 69264 levels of the stay-green trait with 3 female parents with high (A35), intermediate (QL41), or low QL39 levels of stay-green.

experimental unit. The mini-lysimeters were pots made from PVC pipe (0.3 m internal diameter) cut to 0.65 m length and capped with a zincalum base, which was made watertight with a bead of silica.

Each mini-lysimeter was filled with a soil mix consisting of a (3:1:1) mixture of sandy loam soil, sand, and peat. Lime and dolomite neutralised the soil mixture. Slow release fertiliser (Osmocote®, 18% N, 4.8% P and 9.1% K) was added at 2 kg per 500 litres of soil mix. The soil mix was steam sterilised and pots were filled with the equivalent of 61 kg dry soil. Two pipes (50 mm diameter) were placed into the soil profile to allow sub-surface watering.

Pots were wetted to 90% of field capacity (17% gravimetric moisture content) and 8 germinated seeds were sown on October 15. Four pots (one in each block) were treated similarly, but were not planted, so that the amount of water lost as soil evaporation could be measured. Seeds had been rinsed in 70% ethanol, followed by 1% bleach and germinated for 30 h at room temperature. Temik (a.i. 100 g/kg aldicarb) was applied to the soil surface (0.6 g per pot) at 16 days after sowing. After 18 days, plants were thinned to 2 plants/pot (mean dry weight of 0.45 g plant^{-1}). The stress period was initiated at 26 days after sowing. At this time, plastic covers were fitted to the tops of the mini-lysimeters to minimise soil evaporation. The mini-lysimeters were weighed frequently with an electronic load cell attached to a mobile gantry and hand lift device. The WL treatment was not watered until water use was half that of the similar WW treatment (determined within a block). At that stage water was added to the WL treatment at the rate of half that given to the WW treatment. Donatelli, Hammer, ande Vanderlip (1992) had showed that phenology was not affected by water limitation until the relative transpiration ratio (transpiration of WL as fraction of transpiration of WW) was less than 0.5. Hence, this watering regime was designed to impose a significant degree of water limitation without introducing major differences in phenology between water treatments.

Measurements

Several measurements were taken on the plants before harvest. Leaf greenness of each fully expanded leaf was measured with a chlorophyll meter (SPAD [Soil-Plant-Analysis-Development] meter, Minolta). Five readings were taken on one side of the leaf midrib, and taken as a mean for the plot. As leaves (4, 6, 8, and 10) reached their full size, length and maximum width were measured. Plant height was determined on several occasions.

After the final pot weighing, plants were harvested at 64 days after sowing, which was near anthesis for many of the genotypes. Soil cores were removed from the pots and roots extracted by washing away the soil mix over a framed sieve. Main culms and tillers were separated and partitioned into leaf and stem fractions. For the main culm, leaves were harvested as upper (top 5-6 leaves) and lower layers. Plant leaf area was measured by passing all leaves through a belt planimeter. Fresh weight of plant parts was recorded and biomass was determined after oven drying at 80°C for at least 72 h. Dried leaf samples were ground on a 0.5 mm screen and leaf nitrogen concentration (%) and leaf C concentration (%) were determined. Half a gram of ground leaf was ashed for 5 h at 540°C in a muffle furnace to determine ash percent. Carbon isotope ratio was determined on finely ground upper leaf samples (ground on a 0.2 mm screen) using a VG Isogas SIRA 24 ratio mass spectrometer (with respect to a Pee Dee Belemnite [PDB] carbonate standard). Carbon isotope discrimination (Δ) was calculated using equation (2) with the carbon isotope ratio of the plant sample (δ_p), and air (δ_a), which was taken as 8.00 % o (Farquhar and Richards, 1984).

$$\Delta = (\delta_a - \delta_p)/(1 + \delta_p) \tag{2}$$

The standard deviation on a dry matter sample determined on 8 replicate leaf samples was 0.041%.

Data Logging

Relative humidity and air temperature were measured with a Vaisala Humitter sensor, which was placed in the centre of the experiment. In addition, air temperature was logged in each block of the experiment with calibrated thermocouples. All readings were logged hourly using a data logger (Campbell Scientific). These measurements allowed absolute vapour density to be determined for the glasshouse and vapour pressure deficit (VPD) to be estimated for each block of the experiment. Total incoming solar radiation was measured inside and outside the glasshouse.

Photosynthesis

Photosynthesis was measured using a portable LiCor 6200 (LI-Cor Inc. Lincoln, NE) at 24 days after sowing (prestress) and at 48 and 61 days using 14 cm^2 of lamina on the uppermost fully expanded leaf. Net photosynthesis (μmol m^{-2} s^{-1}), stomatal conductance (mol m^{-2} s^{-1}), leaf temperature, and internal CO_2 concentration (ppm CO_2) were recorded. Measurements were taken on sunny, cloudless days between 10.00 and 14.00 h when the irradiance in the glasshouse was greater than 1500 μmol quanta m^{-2} s^{-1}.

A transverse section of the uppermost fully expanded leaf lamina was taken in the late afternoon on day 63 and placed into fixative. Sections were cut in wax, stained and mounted for light microscopy.

Data Analysis

Transpiration was calculated as the total water applied adjusted for both differences in pot weight over the stress period, and the final plant dry weight. TE was calculated as the ratio of total biomass produced to water transpired in units of g/kg. Analyses of variance were conducted on all variables using SAS for analysis of variance (SAS, 1987). The associations among variables were examined using correlation and regression analysis in Excel 5.0 (Microsoft, 1994) and Sigmplot (Jandel Scientific). Six experimental units were treated as missing due to poor growth and were removed from all the analyses.

RESULTS AND DISCUSSION

The plants grew well, although growth in blocks differed due to effects of temperature and VPD gradients across the glasshouse. The water limitation caused a reduction in leaf area of the tenth leaf and above in WL compared to WW (data not shown). Negligible soil evaporation occurred from the bare soil lysimeters so all water use measured was regarded as transpiration.

Over the experimental period air temperatures were between 22°C and 34°C and relative humidity was between 42% and 75% at 9.00 am, with maximum daily air temperatures between 24° and 37°C. Daily vapour pressure deficit (VPD), taken as the mean of VPD calculated at 9 am and at time of maximum air temperature, ranged from 0.1 to 2.85 kPa, with a mean of 1.68 kPa. Radiation levels of 17 to 32 MJ m^{-2} d^{-1} were recorded outside, and 10 to 17 MJ m^{-2} d^{-1} inside, the glasshouse.

Plant biomass, leaf area, transpiration and TE were all affected significantly by water and genotype treatments but there were no significant interactions (Table 2). The associated mean values for genotypes and water treatments (Table 3) show the extent of the variation. As expected, the plants in WW treatment grew more leaf area, transpired more, and produced more biomass than plants in WL treatment (Table 3). Plants grown under water limited (WL) conditions used 65% of the water of those grown under WW conditions and produced 71% as much biomass. Genotypes varied more than twofold in their size and total biomass, which is similar to variation reported by Donatelli, Hammer, and Vanderlip (1992) in a similar experiment. Differences in biomass production were strongly correlated with transpiration across water treatments and genotypes (Figure 1).

Differences in plant size and transpiration were generated by differences in leaf area (Figure 2), which were associated with differences in maturity (leaf number and size) and tillering. At a particular leaf area, less water was

TABLE 2. Summary of the ANOVA for transpiration, leaf area, total biomass and transpiration efficiency for the main effects of water (W), sorghum genotype (G) and the interaction (W × G). Significance levels are denoted as *, **, and *** for P < 0.05, P < 0.01 and P < 0.001, respectively.

Variable	Water	Genotype	W × G
Transpiration (kg)	***	***	ns
Total Leaf area	***	***	ns
Total biomass (g)	***	***	ns
Transpiration Efficiency[1] (g/kg)	*	*	ns
Transpiration Efficiency[2] (g/kg)	*	*	ns

[1] TE calculated using total biomass
[2] TE calculated using above ground biomass

lost from the WL treatments. There was over a 5 kg range in transpiration for a particular value of leaf area, when both water levels are considered across all genotypes. In general, there was a greater effect of WL on transpiration, than on leaf area, although both were reduced significantly (Table 3). Leaf area of these genotypes in spaced pots varied from 2000 cm^2 for R9188 to over 8000 cm^2 for M35-1.

Transpiration efficiency calculated as g of total biomass per unit of water transpired is represented by the slope of the line from the origin to each point in Figure 1. Hence, the scatter of points about the regression line in Figure 1 is what is of interest in the current study. As the ratio of root to total biomass did not differ significantly among genotypes or water treatments (average 0.73; data not shown), and as root biomass was not measured as precisely as above ground biomass, in this study TE was calculated using above ground biomass only. TE varied significantly among genotypes and across water treatments, but there was no significant interaction (Table 2). This result and magnitude of the difference among genotypes in TE is similar to results found by Hammer et al. (1996) under well-watered conditions. The common genotypes in the two studies grouped similarly for TE level, with the exception of QL41, which gave an intermediate level of TE in this study. When analysed separately, the nine hybrids varying in levels of the drought resistance trait stay-green (Table 1) showed no significant differences either among parents or among the nine individual hybrids (data not shown). This suggests that genetic variation in TE is not associated with genetic sources of stay-green. It should be noted, however, that TE in this experiment was measured prior to anthesis and that stay-green is a trait that is expressed as maintenance of green leaf area under terminal water deficit after anthesis.

TABLE 3. The effect of (a) sorghum genotype and (b) water treatment on transpiration efficiency (TE), transpiration, above ground biomass, root biomass and total leaf area. TE means followed by the same letter indicate no significance difference at P < 0.05. Genotypes are ranked by TE. TE is calculated using above ground biomass.

(a)

Genotype	TE	Rank	Transp.	Above grd. biomass	Root biomass	Leaf area
	g/kg		kg	g	g	cm^2
M35-1	9.15 a	1	15.05	136.2	134	8701
SC237	8.84 ab	2	12.45	108.6	74	6219
Q41-36	8.18 abc	3	13.32	106.1	55	4899
A35-69	8.15 abc	4	13.58	108.5	74	5839
Q41-12	8.11 abc	5	11.39	89.8	56	4183
72389	8.10 abc	6	9.76	77.8	54	3289
A35-36	7.81 abc	7	12.29	94.7	71	6296
Q41-69	7.74 abc	8	13.69	106.4	74	5092
QL41	7.57 bcd	9	10.01	73.8	60	4219
Q39-69	7.55 bcd	10	11.74	87.4	52	4611
R9188	7.47 bcd	11	9.50	67.9	32	3027
Q39-12	7.47 bcd	12	9.31	69.0	19	2336
Q39-36	7.39 bcd	13	13.26	93.7	83	5393
Buster	7.33 cd	14	10.96	80.6	48	3935
A35-12	7.18 cd	15	11.67	83.1	58	4087
SC110	7.06 cd	16	12.93	93.3	73	4325
QL12	6.10 d	17	10.01	57.9	48	3184
LSD (P > 0.05)						
	1.48		2.55	22.97		

(b)

Water	TE	Transpiration	Above grd. biomass	Root biomass	Leaf area
	g/kg	g	g	g	cm^2
Water limited	8.09	9.34	75.3	47.9	4187
Well watered	7.33	14.29	105.2	69.2	5181
se	0.13	0.32	2.88	4.25	174

se = standard error

In this study, genetic differences in TE were maintained under water limitation but at an increased level (Tables 2 and 3), about 10% higher in WL treatments. This confirms the preliminary findings of Donatelli, Hammer, and Vanderlip (1992). The lack of significant interaction between genotype and water treatments indicates that water limited environments may not be

FIGURE 1. Total biomass versus transpiration of (a) water limited (WL) and well watered (WW) for 17 sorghum genotypes (points mean of 4 blocks), and (b) genotype means (points are mean of 8 points).

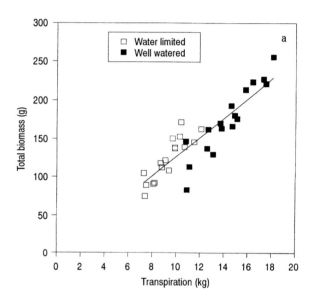

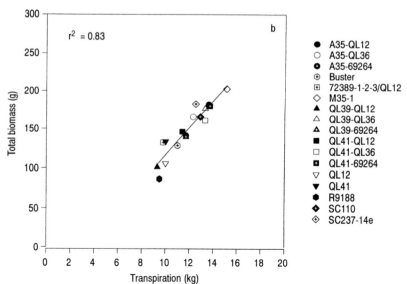

FIGURE 2. Sorghum transpiration versus total leaf area for (a) water limited (WL) and well watered (WW) for 17 sorghum genotypes (points mean of 4 blocks), and (b) genotype means (points are mean of 8 points).

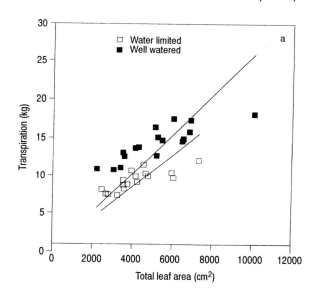

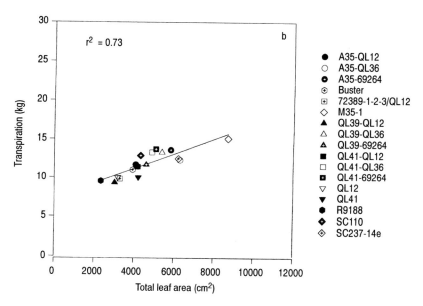

FIGURE 3. Relationship between transpiration efficiency (TE) and transpiration for water limited (WL) and well watered (WW) for 17 genotypes (points mean of 4 blocks).

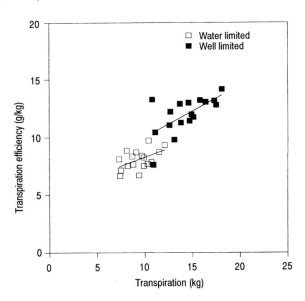

required for screening for genetic variation in TE. Also, the TE levels found in this experiment were not related to plant size, as TE was poorly related to transpiration amount (Figure 3), which was largely determined by plant size. TE did, however, correlate well with transpiration per unit leaf area (Figure 4), which is a whole plant scale indicator of conductance. Genotypes with lower levels of transpiration per unit leaf area had higher TE. TE did not correlate with biomass per unit leaf area (data not shown), which is a whole plant scale indicator of assimilation capacity.

Several variables were examined for relationships with TE (Table 4). While most variables differed significantly among genotypes, only three attributes had a significant water main effect–carbon isotope discrimination, ash in the lower leaves, and leaf C concentration in the upper leaves–and only the latter two of these had both water and genotype effects, as found for TE (Table 2). Δ did not correlate well with TE (Figure 5), confirming the similar earlier finding of Hammer, Farquhar, and Broad (1997). Of the two variables showing significant differences among the two main effects, leaf C concentration had the higher correlation with TE (Table 4, Figure 6). No relationship was evident between either specific leaf nitrogen (SLN) and TE or

FIGURE 4. Relationship between transpiration per unit leaf area and transpiration efficiency (TE) for water limited (WL) and well watered (WW) for 17 genotypes (points mean of 4 blocks).

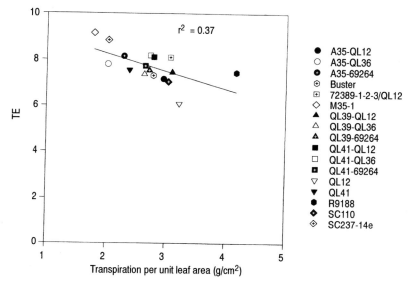

TABLE 4. The significance of water (W), genotype (G) and the interaction of W × G for a range of variables measured on leaves and the correlation with transpiration efficiency (using 34 means of W × G). Significance levels denoted as +, *, **, and *** for P < 0.10, P < 0.05, P < 0.01 and P < 0.001, respectively.

Variable	Water	Genotype	W × G	Correlation with TE
Specific Leaf Area (upper)	ns	***	**	0.05
SLA (lower)	ns	***	+	
Specific Leaf Nitrogen (upper)	ns	***	ns	−0.22
SLN (lower)	ns	***	ns	0.07
Ash (upper) (%)	ns	**	ns	−0.03
Ash (lower) (%)	***	***	ns	0.46
Carbon isotope discrim. (%) (upper leaves)	***	ns	+	0.15
Leaf Greenness[1]	ns	***	ns	−
Leaf Carbon Conc. (upper) (%)	*	*	ns	−0.56
Leaf Nitrogen Conc. (upper) (%)	ns	***	ns	−0.26

[1] Similar results for leaves 4 to 13. Measurement made with Spad meter.

FIGURE 5. Carbon isotope discrimination (delta) versus transpiration efficiency (TE) for (a) water limited (WL) and well watered (WW) for 17 sorghum genotypes (points mean of 4 blocks), and (b) genotype means (points are mean of 8 points).

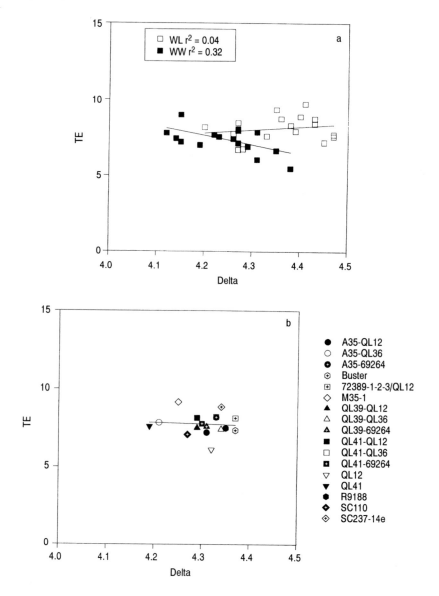

FIGURE 6. Leaf carbon concentration versus transpiration efficiency (TE) (a) water limited (WL) and well watered (WW) for 17 sorghum genotypes (points mean of 4 blocks), and (b) genotype means (points are mean of 8 points).

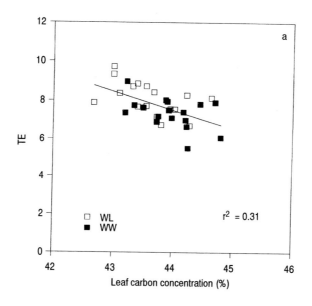

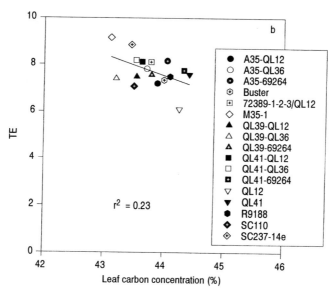

FIGURE 7. Leaf level TE (LEAFTE) versus net photosynthesis for measurements taken at 48 days after planting.

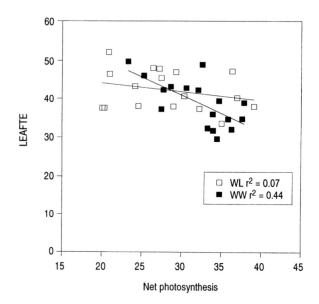

Net photosynthesis

specific leaf area (SLA) and TE in this study. Hammer Farquhar, and Broad (1997) found a negative relationship between TE and SLA, and noted a tendency for SLN to increase as SLA decreased. These results are suggestive that regulation of conductance, rather than assimilation capacity, may be responsible for the genetic variation observed in TE. The relationship with leaf C concentration warrants further attention as this attribute is easy to measure on ground leaf tissue and relatively inexpensive. Masle, Farquhar and Wong (1992) and Hammer, Farquhar, and Broad (1997) found significant correlations of TE and leaf ash. The connecting mechanism of these attributes to TE remains obscure. Further examination of leaf sections taken as part of this study may assist in interpreting this result.

Leaf level TE (LEAFTE) was calculated as A/g from photosynthetic measurements taken at 48 days after planting (22 days into the water stress period). There were relationships between both LEAFTE and net photosynthesis, A (Figure 7), and LEAFTE and stomatal conductance, g (Figure 8). Higher LEAFTE was associated with lower net photosynthesis and lower conductance. The finding for conductance is consistent with the result at the whole plant scale (Figure 4) but results for assimilation were not consistent at the two scales. The correlation between LEAFTE and TE was positive but

FIGURE 8. Leaf level TE (LEAFTE) versus stomatal conductance for measurements taken at 48 days after planting.

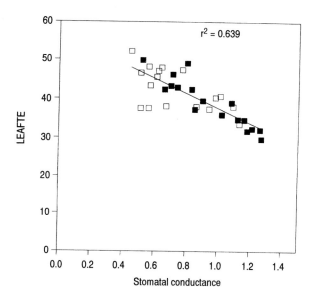

not strong (Figure 9). Many factors would cause this correlation to be poor, including the leaf level not taking into account post photosynthetic processes beyond the leaf, measurement error at both scales, and the necessarily small leaf area sample for photosynthesis measurement. The result is suggestive that issues relevant to integration to whole plant scale are not fully captured in leaf level measurement.

The underlying cause of genetic variation in TE in sorghum remains unclear. This lack of understanding hinders the search for useful selection indices. Such indices will be required before selection for this trait could be contemplated in a breeding program or before a search for molecular markers could be initiated as the difficulty of measurement of TE necessarily restricts observations to a limited number of genotypes. In addition, studies to date have concentrated on glasshouse studies. Field studies confirming these results are required. Both of these issues provide logical avenues for on-going research in this area.

CONCLUSIONS

This study has confirmed genotypic variation among 17 sorghum genotypes. The most efficient was more than 50% more efficient than the least

FIGURE 9. Transpiration efficiency (TE) versus leaf level transpiration efficiency (LEAFTE) for measurements taken at 48 days after planting (points are mean of 4 blocks).

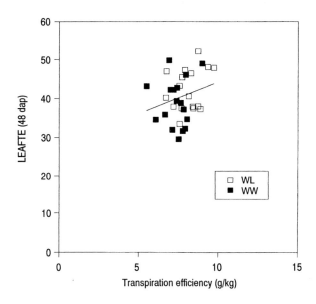

efficient and 15% more efficient than the mean. In addition, the study confirmed an increase in sorghum TE under mild water limitation. The plants in the WL treatment had a 9% increase in TE above those in the WW treatment. At the whole plant level, variation in TE was associated negatively with transpiration per unit leaf area. This association was also found between measurements of TE and conductance at the leaf level. The ash in the upper leaves and leaf C concentration (%C) was most closely correlated with variation in TE and offers promise as an avenue for development of a selection index. Before TE can be used actively in a breeding program, field studies are required to confirm these findings from glasshouse studies and more robust selection indices are needed.

REFERENCES

Briggs, L.J. and H.L Shantz. (1914). The water requirement of plants: I. Investigations in the Great Plains in 1910 and 1911. USDA *Bureau Plant Industry Bull.* 284.

Donatelli, M., G.L. Hammer, and R.L. Vanderlip. (1992). Genotype and water limitation effects on phenology, growth, and transpiration efficiency in grain sorghum. *Crop Science* 32: 781-786.

FAO, (1995). FAO Commodity Review and Outlook., 1994-5. FAO, Rome, Italy.

Farquhar, G.D. (1983). On the nature of carbon isotope discrimination in C_4 species. *Australian Journal of Plant Physiology* 10:205-226.

Farquhar, G.D., S. von Caemmerer, and J.A. Berry. (1980). A biochemical model of photosynthetic CO2 assimilation in leaves of C_3 species. *Planta* 149: 78-90.

Farquhar, G.D., M.H. O'Leary, and J.A. Berry. (1982). On the relationship between carbon isotope discrimination and the intercellular carbon dioxide concentration in leaves. *Australian Journal of Plant Physiology* 9: 121-137.

Farquhar, G.D., and R.A. Richards. (1984). Isotopic composition of plant carbon correlates with water-use efficiency of wheat genotypes. *Australian Journal of Plant Physiology* 11: 539-52.

Hammer, G.L., D. Butler, R.C. Muchow, and H. Meinke. (1996). Integrating physiological understanding and plant breeding via crop modelling and optimisation. In *Plant Adaptation and Crop Improvement*, eds. M. Cooper and G.L. Hammer, Wallingford, UK: CAB International, pp. 419-442.

Hammer, G.L., G.D. Farquhar, and I. Broad. (1997). On the extent of genetic variation for transpiration efficiency in sorghum. *Australian Journal of Agricultural Research* 48: 649-55.

Hammer, G.L., and R.C. Muchow. (1994). Assessing climatic risk to sorghum production in water-limiting subtropical environments. I Developing and testing a simulation model. *Field Crops Research* 36: 221-34.

Hammer, G.L., and R.C. Muchow. (1992). The use of simulation modelling in decision-making in sorghum production. In *Proceedings of the Second Australian Sorghum Conference*, eds. M.A. Foale, R.G. Henzell, and P.N. Vance, Melbourne, Australia: Australian Institute of Agricultural Science.

Henderson, S., S. von Caemmerer, G.D. Farquhar, L. Wade, and G.L. Hammer. (1998). Correlations between carbon isotope discrimination and transpiration efficiency in lines of the C_4 species Sorghum bicolor in the glasshouse and field. *Australian Journal of Plant Physiology* 25: 111-123.

Henzell, R.G., K.J. McCree, C.H.M. van Bavel, and K.F. Schertz. (1976). Sorghum genotype variation in stomatal sensitivity to leaf water deficit. *Crop Science* 16: 660-662.

Hubick, K.T., G.L. Hammer, G.D Farquhar, L. Wade, S. von Caemmerer, and S. Henderson. (1990). Carbon isotope discrimination varies genetically in C4 species. *Plant Physiology* 91:534-537.

Hubick, K.T., and G.D. Farquhar. (1987). Carbon isotope discrimination–selecting for water use efficiency. *Aust Cotton Grower* 8:66-68.

Hubick, K.T., and G.D. Farquhar. (1989). Carbon isotope discrimination and the ratio of carbon gained to water lost in barley cultivars. *Plant Cell and Environment* 12: 795-804.

Hubick, K.T., G.D. Farquhar, and R. Shorter. (1986). Correlation between water-use efficiency and carbon discrimination in diverse peanut *Arachis* germplasm. *Australian Journal of Plant Physiology* 13: 803-816.

Knight, J.D., N.J. Livingston, and C. van Kessel. (1994). Carbon isotope discrimination and water-use efficiency of six crops grown under wet and dryland conditions. *Plant, Cell and Environment* 17: 173-179.

Krieg, D.R. (1983). Photosynthetic activity during stress. *Agricultural Water Management* 7: 249-63.

Lal, A., and Edwards, G.E. (1996). Analysis of inhibition of photosynthesis under water stress in the C_4 species *Amaranthus cruentus* and *Zea mays:* Electron transport, CO_2 fixation and carboxylation capacity. *Australian Journal of Plant Physiology* 23: 403-12.

Lawrence, G. (1996). Marketing of grain sorghum: sorghums place in the market. In *Proceedings of the Third Sorghum Conference, Tamworth, 20 to 22 February 1996*, eds. M.A. Foale, R.G. Henzell, and J.F. Kneipp, Melbourne, Australia: Australian Institute of Agricultural Science.

Masle, J., G.D. Farquhar, and S.C. Wong. (1992). transpiration ratio and plant mineral content are related among genotypes of a range of species. *Australian Journal of Plant Physiology* 19: 709-721.

Matus, A., A.E. Slinkard, and C. van Kessel. (1996). Carbon isotope discrimination and indirect selection for transpiration efficiency in lentil (*Lens culinaris* Medikus), spring bread wheat (*Triticum aestivum* L.) durum wheat (*T. turgidum* L.) and canola (*Brassica napus* L.). *Euphytica* 87: 141-151.

Muchow, R.C., G.L. Hammer, and P.S. Carberry. (1991) Optimising crop and cultivar selection in response to climatic risk. In *Climatic Risk in Crop Production: Models and Management for the Semiarid Tropics*, eds. R.C. Muchow and J.A. Bellamy, Wallingford, UK: CAB International, pp. 235-262.

National Academy of Sciences. (1996). *Lost Crops of Africa. Volume I: Grains.* Washington, DC: National Academy Press.

Ohsugi, R., M. Samejima, N. Chonan, and T. Murata. (1988). δ^{13} C and the occurrence of suberized lamellae in some *Panicum* species. *Annals of Botany* 62:53-59.

O'Leary, M.H. (1981). Carbon isotopes fractionation in plants. *Phytochemistry* 20: 553-67.

SAS Institute Inc. (1987). *SAS/STAT TM Guide for Personal Computers.* Version 6.1. Cary, NC: SAS Institute, Inc.

Shorter, R., R.J. Lawn, and G.L. Hammer. (1991). Improving genotypic adaptation in crops–a role for breeders, phsyiologists and modellers *Experimental Agriculture* 17: 155-175.

Tanner, C.B., and T.R. Sinclair. (1983). Efficient water use in crop production: research or re-search? In *Limitations to Efficient Water Use in Crop Production*, eds. H.M. Taylor, W.R. Jordan, and T.R. Sinclair, Madison, WI: American Society of Agronomy, Crop Science Society of America, Soil Science Society of America, pp. 1-27.

Wright, G.C., K.T. Hubick, and G.D. Farquhar. (1988). Discrimination in carbon isotopes of leaves correlates with water use efficiency of field grown peanut. *Australian Journal of Plant Physiology* 15: 815-25.

Hydraulic Resistance of Sorghum (C_4) and Sunflower (C_3)

Jingxian Zhang
M. B. Kirkham

SUMMARY. It is well known that C_4 crops have a lower water requirement than C_3 crops. Reasons for the difference are not well understood. Therefore, hydraulic resistance of sorghum [*Sorghum bicolor* (L.) Moench] (C_4) and sunflower (*Helianthus annuus* L.) (C_3) was determined to see if it might be one explanation for the lower water use of crops with the C_4 photosynthetic pathway. Plants were grown under greenhouse conditions in pots with soil, which was well watered (soil matric potential of ~ 0 MPa) or allowed to dry (soil matric potential of -0.038 MPa and -0.065 MPa for sorghum and sunflower, respectively). Hydraulic resistance was calculated in two ways: (1) using the classic Ohm's law analogue, which assumes that the relation between flux (transpiration) and difference in water potentials of the soil and

Jingxian Zhang is former Graduate Research Assistant and M. B. Kirkham is Professor, Department of Agronomy, Kansas State University, Manhattan, KS 66506-5501 USA. Currently Dr. Zhang is in the Plant Molecular Genetics Laboratory, Department of Plant and Soil Sciences, Texas Tech University, Lubbock, TX 79409-2122 USA.

Address correspondence to: M. B. Kirkham, Department of Agronomy, 2004 Throckmorton Hall, Kansas State University, Manhattan, KS 66506-5501 (E-mail: mbk@ksu.edu).

This paper was presented by the junior author as an invited talk at the Wilford R. Gardner Symposium on Fundamentals of Soil Physics, held at the Soil Science Society of America Annual Meeting, St. Louis, Missouri, October 31, 1995. The full text of the talk appears in Zhang (1996). This is contribution no. 97-370-J from the Kansas Agricultural Experiment Station.

[Haworth co-indexing entry note]: "Hydraulic Resistance of Sorghum (C_4) and Sunflower (C_3)." Zhang, Jingxian, and M. B. Kirkham. Co-published simultaneously in *Journal of Crop Production* (Food Products Press, an imprint of The Haworth Press, Inc.) Vol. 2, No. 2 (#4), 1999, pp. 287-298; and: *Water Use in Crop Production* (ed: M. B. Kirkham) Food Products Press, an imprint of The Haworth Press, Inc., 1999, pp. 287-298. Single or multiple copies of this article are available for a fee from The Haworth Document Delivery Service [1-800-342-9678, 9:00 a.m. - 5:00 p.m. (EST). E-mail address: getinfo@haworthpressinc.com].

plant is linear and (2) using an equation that considers diurnal changes in leaf water content along with transpiration and difference in water potentials. Because change in leaf-water content during a day was small, hydraulic resistances calculated by the two methods resulted in similar values. Sorghum had a linear relationship between flux and difference in potentials (constant hydraulic resistance), but sunflower had a nonlinear one (variable hydraulic resistance). The hydraulic resistance of watered sunflower increased only slightly during a day and averaged about 40 MPa m^2 s mol^{-1}, which was 3.5 times less than that of watered or water-stressed sorghum ($\sim$140 MPa m^2 s mol^{-1}). The hydraulic resistance of water-stressed sunflower increased steeply during a day and by the end of the day it had a hydraulic resistance that approached that of sorghum. *[Article copies available for a fee from The Haworth Document Delivery Service: 1-800-342-9678. E-mail address: getinfo@haworthpressinc.com <Website: http://www.haworthpressinc.com>]*

KEYWORDS. Sorghum, sunflower, hydraulic resistance, Ohm's law analogue

INTRODUCTION

Water use in crop production is critically important in semi-arid regions. Therefore, crops that conserve water need to be identified and the reasons for the conservation investigated. The fact that C_3 plants have a greater water requirement than C_4 plants is well recognized (Black, 1973). The water requirement (also called transpiration ratio) is the ratio of the weight of water absorbed by a plant during its growth to the weight of dry matter or grain production (Briggs and Shantz, 1913a). The difference in water requirement was recognized even at the turn of the century, before the discovery of the different photosynthetic pathways (Briggs and Shantz, 1913a,b; Shantz and Piemeisel, 1927). Table XXI of Briggs and Shantz (1913a, p. 47) shows that the water requirement (based on dry matter produced) of plants now known to be C_4 plants is less than 370 g/g. In contrast, plants listed in this table that are now known to be C_3 plants have a water requirement that is greater than 370 g/g.

Work in semi-arid Kansas has shown that of six important row crops grown in the summer–i.e., corn (*Zea mays* L.); sorghum [*Sorghum bicolor* (L.) Moench]; pearl millet [*Pennisetum americanum* (L.) Leeke]; pinto bean (*Phaseolus vulgaris* L.); soybean [*Glycine max* (L.) Merr.]; and sunflower (*Helianthus annuus* L.)–those with the C_4 photosynthetic pathway (corn, sorghum, millet) generally have lower evapotranspiration (ET) and greater grain water use efficiency (WUE) (WUE can be considered to be the reciprocal of water requirement) than ones with the C_3 pathway (pinto bean, soybean, sunflower) (Kirkham et al., 1985; Hattendorf et al., 1988). Of the six crops, sunflower has the highest daily water use rate (6.12 mm/d) and lowest grain WUE (4.2 kg ha^{-1} mm^{-1}), and sorghum has one of the lowest daily water use rates (4.83 mm/d) and highest grain WUE (13.0 kg ha^{-1} mm^{-1}) (Hattendorf et al., 1988).

One reason for the higher WUE of C_4 plants might not only be due to their more efficient use of CO_2, but also to a difference in hydraulic resistance. Little is known about the flow of water through C_3 and C_4 plants, and we know of no work specifically comparing their hydraulic resistances. However, hydraulic conductance (reciprocal of hydraulic resistance) of two C_3 plants, one a grass and one a legume, has been compared (Gallardo et al., 1996). The grass (wheat; *Triticum aestivum* L.) had a higher hydraulic conductance than the legume (a lupine; *Lupinus angustifolius* L.). Also hydraulic resistance of two cultivars of sorghum varying in drought resistance has been compared and found to be similar (Kirkham, 1988).

Here we calculate the hydraulic resistance of sunflower (C_3) and sorghum (C_4) by two different methods: an Ohm's law analogue and a model described by Mishio and Yokoi (1991), in which changes in leaf water content are considered along with transpiration rate. We find that sorghum and sunflower differ in their hydraulic resistance.

MATERIALS AND METHODS

Seeds of sorghum [*Sorghum bicolor* (L.) Moench, cv. Funk's G522DR] and sunflower (*Helianthus annuus* L., cv. Hysun 354) were sown in a greenhouse under a natural photoperiod on 8 July 1992 in 20 plastic pots (10 for sorghum; 10 for sunflower). Each pot (17 cm diameter; 18 cm height; 5 drainage holes) was filled with a 1:1 mixture of a commercial all purpose potting mix (Fisons Horticulture, Inc., Vancouver, Canada) and fritted clay (Balcones Minerals Corp., Flatonia, Texas) (van Bavel et al., 1978). Plants were thinned to five per pot seven days after emergence. The pots were kept well watered between planting and 1 August. Between 2 August and 7 August, water was added to half of the pots while the other half was not watered (two treatments). The experimental design was a completely random one.

Soil matric potential was measured with tensiometers (Model 2100F, Soil-Moisture Equipment Corp., Santa Barbara, California), which were installed in eight pots on 28 July, four pots with sorghum and four pots with sunflower, to provide two measurements per treatment. The tips of the tensiometers were placed near the bottom of the pots. At this time, roots were visible through the drainage holes, but the roots did not grow out of the pots. Because the soil was nonsaline and because all tensiometers were placed at the same depth (reference level), the soil matric potential was assumed to be the soil water potential.

Measurements of photosynthesis and transpiration of the most recently matured leaf were made with a portable leaf photosynthesis system (Model LI-6200, Li-Cor, Inc., Lincoln, Nebraska). Water requirement was calculated by dividing leaf transpiration rate (μmol m^{-2} s^{-1}) by leaf photosynthetic rate (μmol m^{-2} s^{-1}). Water potential of a leaf was measured with a pressure chamber (Model 3000, Water Status Console, SoilMoisture Equipment Corp., Santa Barbara, California). Leaf area was measured with a meter (Model 3100, Li-Cor, Inc., Lincoln, Nebraska). Leaf water content (W, g/cm^2) was calculated according to the following formula (Mishio and Yokoi, 1991):

$$W = (\text{fresh weight} - \text{dry weight})/\text{leaf area} \qquad (1)$$

To get fresh weight, eight leaves were grouped together in one sample and one measurement was taken. After fresh weight and leaf area of the eight leaves were determined, dry weight of the leaves was obtained by drying them at 80°C for 24 h.

Hydraulic resistance was calculated using a steady-state or Ohm's law analog, as follows (Mishio and Yokoi, 1991):

$$E = (\psi_s - \psi_l)/R, \qquad (2)$$

where E is the transpiration rate (mol m^{-2} s^{-1}); Ψ_s and Ψ_l are the soil and leaf water potentials (MPa), respectively; and R is the hydraulic resistance (MPa m^2 s mol^{-1}).

Hydraulic resistance also was calculated according to the method of Mishio and Yokoi (1991). They state that leaf water content changes both with transpiration rate and water supply rate from the soil to the leaf. They give the rate of change in leaf water content as:

$$dW/dt = [(\Psi_s - \Psi_l)/R] - E, \qquad (3)$$

where W is leaf water content [(fresh weight–dry weight)/leaf area; units = mol/m^2] and t is time (s). The integrated form of Eq. (3) is:

$$W_{t2} - W_{t1} = \int [(\Psi_s - \Psi_l)/R]dt - \int E dt. \qquad (4)$$

They say that, when the sampling interval from time 1 ($t1$) to time 2 ($t2$) is within a few hours, the resistance, R, should remain constant. Therefore, they rewrite Equation (4) as follows:

$$W_{t2} - W_{t1} = (1/R) \, {}_{t1}\!\int^{t2} (\Psi_s - \Psi_l)dt - {}_{t1}\!\int^{t2} E \, dt. \qquad (5)$$

Thus, by Equation (5), Mishio and Yokoi (1991) state that R is the reciprocal of the slope obtained by plotting on a y-axis $(W_{t2} - W_{t1}) + {}_{t1}\!\int^{t2}E \, dt$, which corresponds to the water supply from the soil to the leaf, and on an x-axis, $t1\!\int^{t2}(\Psi_s - \Psi_l)dt$, the cumulative water potential difference in the soil-leaf pathway.

Hydraulic resistance was calculated separately according to Equations (2) and (5). Individual values of E and ΨY_l, used for calculating hydraulic resistance, was the mean of four and eight measurements, respectively. Tensiometer readings used to calculate Ψ_s did not change during the period of measurement (0830 to 1730 h). In watered pots, tensiometer readings read 0 MPa. In nonwatered pots, the average tensiometer readings (of two tensiometers) and standard deviation were -0.038 ± 0.013 MPa for sorghum and -0.065 ± 0.003 for sunflower. Times of measurement were 0830, 1130, 1430, and 1730 h on 7 August, and the values for times $t1$ and $t2$ (Equation 5) were sequentially taken at 0830 h ($t1$) and 1130 h ($t2$); 1130 h ($t1$) and 1430 h ($t2$); and 1430 h ($t1$) and 1730 h ($t2$).

In the figures with data plotted according to Equations (2) and (5), curves were drawn using the software GRAPHER (Golden Software, Inc., Golden, Colorado). The best polynomial fit was used. If the curves had been drawn by eye, a different fit might have been used. The software permits the same curves to be reproduced by anyone with the program and the data. Therefore, the coefficients for the curves are not given, as they have no unique significance. Raw data for all the measurements in the experiment are given by Zhang (1996), where the interested reader can obtain the data to plot the curves by eye or by using GRAPHER or another software.

In the figures for stomatal resistance, transpiration, photosynthesis, and water requirement, the value for the time period under consideration is plotted at the midpoint of the time period (1000, 1300, and 1600 h). For these latter measurements, each value plotted in the figures is the mean of four measurements.

RESULTS

Using the Ohm's law analog (Equation 2), the relation between transpiration and difference in water potentials of the soil and leaf was linear for sorghum and nonlinear for sunflower (Figure 1, top). Similar results were obtained by using the Mishio and Yokoi (1991) model (Equation 5) (Figure 1, bottom). That is, the relation between water lost (difference in water content over two time periods plus transpiration) and the difference in soil and leaf water potentials was linear for sorghum and nonlinear for sunflower. The

FIGURE 1. Data to determine hydraulic resistance of watered and water-stressed sorghum and sunflower using an Ohm's law analogue (top) and a model in which changes in leaf water content (W) are considered along with transpiration (E) to determine a total water supply from soil to leaf (bottom). The slope of a line relating transpiration rate (E) to the value of soil water potential minus leaf water potential ($\Psi_s - \Psi_l$) gives the reciprocal of hydraulic resistance (top). Similarly, the slope of a line relating ($W_{t2} - W_{t1}$) + $\int E\,dt$ versus $\int(\Psi_s - \Psi_l)dt$ gives the reciprocal of hydraulic resistance (bottom).

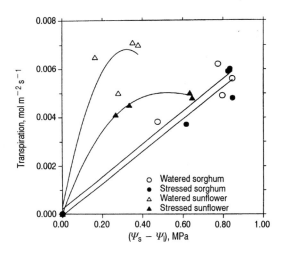

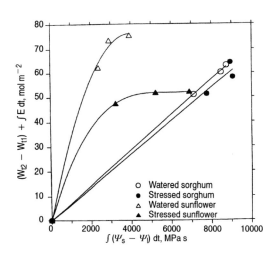

results for this experiment agree with those observed previously in separate experiments. Kirkham (1988) obtained a linear relationship for sorghum using an Ohm's law analogue, and Stoker and Weatherley (1971) and Black (1979) found a flux-dependent resistance in sunflower, in which resistance declined with an increase in transpiration rate.

Transpiration (Figure 2, top left) and stomatal resistance (Figure 2, top right) of sorghum were not affected by drought, whereas those of sunflower were changed. Watered sunflower had a higher transpiration rate and a lower stomatal resistance than water-stressed sunflower. Photosynthesis of sorghum was not affected by drought and was higher than that of sunflower (Figure 3, bottom left). Water requirement of sunflower was higher than that of sorghum

FIGURE 2. Transpiration, stomatal resistance, photosynthesis, and water requirement of leaves of watered and water-stressed sorghum and sunflower. Vertical bars indicate ± standard deviation.

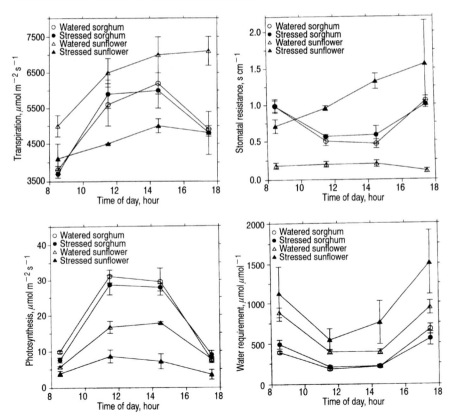

FIGURE 3. Hydraulic resistance of watered and water-stressed sorghum and sunflower during a day using an Ohm's law analogue (top) and a model in which changes in leaf water content are considered along with transpiration to determine total water supply from soil to leaf (bottom).

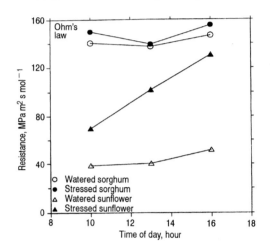

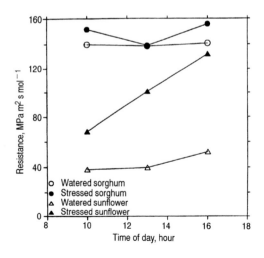

(Figure 3, bottom right). The data for transpiration, stomatal resistance, and photosynthesis agree with results from other experiments with C_3 and C_4 plants (He et al., 1992; Nie et al., 1992; Rachidi et al., 1993; Zhang and Kirkham, 1995).

Hydraulic resistances calculated using the Ohm's law analogue (Equation 2) (Figure 3, top) were similar to those calculated using the model of Mishio and Yokoi (1991) (Equation 5) (Figure 3, bottom). This is because *W*, the leaf water content on a leaf area basis, changed little (increase or decrease) during each two hour period or during the whole day (1 mol/m^2 or less). Calculations with the Ohm's law analogue gave mean hydraulic resistances of 141 and 148 MPa m^2 s mol^{-1} for watered and water-stressed sorghum, respectively. The hydraulic resistance of watered sunflower, calculated using the Ohm's law analogue, increased only slightly during the day, from 39 MPa m^2 s mol^{-1} in the morning to 52 MPa m^2 s mol^{-1} in the late afternoon. However, hydraulic resistance of water-stressed sunflower increased steeply, from 70 MPa m^2 s mol^{-1} in the morning to 130 MPa m^2 s mol^{-1} by late afternoon. Hydraulic resistance of watered sunflower ($\sim$40 MPa m^2 s mol^{-1}) was about 3.5 times less than that of watered sorghum ($\sim$140 MPa m^2 s mol^{-1}). The late afternoon value of hydraulic resistance for stressed sunflower approached the values for sorghum (Figure 3, top).

DISCUSSION

We found that during a day hydraulic resistance of sorghum was constant, but hydraulic resistance of sunflower increased. Others have found that resistances increased during a day for wheat (Jones, 1978), barley (*Hordeum vulgare* L.) (Passioura and Munns, 1984), and woody plants (Mishio and Yokoi, 1991), which indicates that woody species do not always have a constant resistance, as suggested by Jones (1983). Rüdinger et al. (1994) said that trees [spruce; *Picea abies* (L.) Karst] could have a constant resistance, because water and solutes can be taken up at a root tip which has not yet differentiated to form a Casparian strip. In this region, uptake can be all apoplastic, as no root membranes need to be traversed.

We have to consider why the hydraulic resistance is constant for sorghum but variable for sunflower. Passioura (1985) said that the linearity of the hydraulic resistance in dicotyledons has long been in doubt, but hydraulic resistance in monocotyledons often appears constant (independent of flow rate). However, the fact that monocotyledons have been shown to have both variable (e.g., wheat; Jones, 1978) and constant resistances (e.g., sorghum, Kirkham, 1988; this study) suggests that variability in hydraulic resistance may be not just be due to differences between monocotyledons and dicotyledons, although the different anatomy of the two classes of angiosperms must have an effect on hydraulic resistance. Generally, monocotyledons have a central pith in the root surrounded by vascular tissue (xylem and phloem), while dicotyledons have a central core of xylem tissue surrounded by phloem (Esau, 1965, p. 711). Wheat is an exception among the monocotyledons, because its root has a large central metaxylem element (Esau, 1965, p. 714; Passioura, 1972).

Millar et al. (1971) and Gardner (1973) suggested that an explanation for the nonlinear nature of the relation between flow and difference in water potential of the soil and plant may lie in the problem of simultaneous flow of water and ions across semipermeable plant root membranes. Solutes are taken up by the roots, which do not leak outward because of the semipermeable nature of membranes. In particular, the water and solutes must cross a membrane at the Casparian strip in the endodermis. Passioura (1984; 1988) further discussed the importance of membranes to explain nonlinearities between flow and difference in potential.

Dalton et al. (1975) provided a physically based mathematical analysis for the simultaneous uptake of water and solutes by plant roots. Their theory predicts a nonlinear or linear relationship between the flow of water and the pressure difference across root membranes, depending upon the osmotic potential outside the membrane (i.e., the osmotic potential of the nutrient solution or the soil side of a root), the reflection coefficient of the membrane, the solute permeability of the membrane, the active component for solute uptake, and the volume flux of solution across the membrane. The results of Dalton et al. (1975) suggest that the difference in hydraulic resistances of sorghum and sunflower may lie in differences in the properties of their membranes. Osmotic adjustment is observed commonly in leaf cells (Brisson et al., 1993) and is probably controlled by membrane characteristics of leaves of different species. Similarly, root membranes, under the control of root hormones (Davies and Zhang, 1991), undoubtedly regulate osmotic pressure differences across a root, and these differences, in turn, determine hydraulic resistance. Root membranes of C_3 and C_4 plants must differ fundamentally.

CONCLUSION

In sum, we found that sorghum showed a constant hydraulic resistance and sunflower showed a variable hydraulic resistance. Because sorghum also has drought-resistant characteristics, plants that show a constant hydraulic resistance may be more desirable for growth under semi-arid conditions than plants that show a variable hydraulic resistance. Under these conditions, we also want a high hydraulic resistance as shown by sorghum. Water then will be conserved and not lost via transpiration.

REFERENCES

Black, C.C., Jr. (1973). Photosynthetic carbon fixation in relation to net CO_2 uptake. *Annual Review of Plant Physiology* 24: 253-286.

Black, C.R. (1979). A quantitative study of the resistances to transpirational water movement in sunflower (*Helianthus annuus* L.). *Journal of Experimental Botany* 30: 947-953.

Briggs, L.J. and H.L. Shantz. (1913a). The water requirement of plants. I. Investigations in the Great Plains in 1910 and 1911. *Bulletin No.* 284. Washington, DC: United States Department of Agriculture, Bureau of Plant Industries, 49 pp.

Briggs, L.J. and H.J. Shantz. (1913b). The water requirement of plants. II. A review of the literature. *Bulletin No.* 285. Washington, D.C.: United States Department of Agriculture, Bureau of Plant Industries, 96 pp.

Brisson, N., A. Olioso, and P. Clastre. (1993). Daily transpiration of field soybeans as related to hydraulic conductance, root distribution, soil potential and midday leaf potential. *Plant and Soil* 154: 227-237.

Dalton, F.N., P.A.C. Raats, and W.R. Gardner. (1975). Simultaneous uptake of water and solutes by plant roots. *Agronomy Journal* 67: 334-339.

Davies, W.J. and Jianhua Zhang. (1991). Root signals and the regulation of growth and development of plants in drying soil. *Annual Review of Plant Physiology and Plant Molecular Biology* 42: 55-76.

Esau, K. (1965). *Plant anatomy.* Second ed. New York: Wiley, 767 pp.

Gallardo, M., J. Eastham, P.J. Gregory, and N.C. Turner. (1996). A comparison of plant hydraulic conductances in wheat and lupins. *Journal of Experimental Botany* 47: 233-239.

Gardner, W.R. (1973). Internal water status and plant response in relation to the external water régime. In *Plant Response to Climatic Factors,* ed. R.O. Slatyer, Paris, France: United Nations Educational, Scientific and Cultural Organization, pp. 221-225.

Hattendorf, M.J., M.S. Redelfs, B. Amos, L.R. Stone, and R.E. Gwin Jr. (1988). Comparative water use characteristics of six row crops. *Agronomy Journal* 80: 80-85.

He, H., M.B. Kirkham, D.J. Lawlor, and E.T. Kanemasu. (1992). Photosynthesis and water relations of big bluestem (C_4) and Kentucky bluegrass (C_3) under high concentration carbon dioxide. *Transactions of the Kansas Academy of Science* 95: 139-152.

Jones, H.G. (1978). Modelling diurnal trends of leaf water potential in transpiring wheat. *Journal of Applied Ecology* 15: 613-626.

Jones, H.G. (1983). Estimation of an effective soil water potential at the root surface of transpiring plants. *Plant, Cell and Environment* 6: 671-674.

Kirkham, M.B. (1988). Hydraulic resistance of two sorghums varying in drought resistance. *Plant and Soil* 105: 19-24.

Kirkham, M.B., M.S. Redelfs, L.R. Stone, and E.T. Kanemasu. (1985). Comparison of water status and evapotranspiration of six row crops. *Field Crops Research* 10: 257-268.

Millar, A.A., W.R. Gardner, and S.M. Goltz. (1971). Internal water status and water transport in seed onion plants. *Agronomy Journal* 63: 779-784.

Mishio, M. and Y. Yokoi. (1991). A model for estimation of water flow resistance in soil-leaf pathway under dynamic conditions. *Journal of Experimental Botany* 42: 541-546.

Nie, D., H. He, M.B. Kirkham, and E.T. Kanemasu. (1992). Photosynthesis of a C_3 grass and a C_4 grass under elevated CO_2. *Photosynthetica* 26: 189-198.

Passioura, J.B. (1972). The effect of root geometry on the yield of wheat growing on stored water. *Australian Journal of Agricultural Research* 23: 745-752.

Passioura, J.B. (1984). Hydraulic resistance of plants. I. Constant or variable? *Australian Journal of Plant Physiology* 11: 333-339.

Passioura, J.B. (1985). Roots and water economy of wheat. In *Wheat growth and modelling*, W. Day and R.K. Atkin, eds. NATO ASI (Advanced Sci. Inst. Ser.). New York: Plenum Press, pp. 185-198.

Passioura, J.B. (1988). Water transport in and to roots. *Annual Review of Plant Physiology and Plant Molecular Biology* 39: 245-265.

Passioura, J.B. and R. Munns. (1984). Hydraulic resistance of plants. II. Effects of rooting medium, and time of day, in barley and lupin. *Australian Journal of Plant Physiology* 11: 341-350.

Rachidi, F., M.B. Kirkham, L.R. Stone, and E.T. Kanemasu. (1993). Soil water depletion by sunflower and sorghum under rainfed conditions. *Agricultural Water Management* 24: 49-62.

Rüdinger, M., S.W. Hallgren, E. Steudle, and E.-D. Schulze. (1994). Hydraulic and osmotic properties of spruce roots. *Journal of Experimental Botany* 45: 1413-1425.

Shantz, H.L. and L.N. Piemeisel. (1927). The water requirement of plants at Akron, Colo. *Journal of Agricultural Research* 34: 1093-1190.

Stoker, R. and P.E. Weatherley. (1971). The influence of the root system on the relationship between the rate of transpiration and depression of leaf water potential. *New Phytologist* 70: 547-554.

van Bavel, C.H.M., R. Lascano, and D.R. Wilson. (1978). Water relations of fritted clay. *Soil Science Society of America Journal* 42: 657-659.

Zhang, Jingxian. (1996). *Antioxidant Status and Water Relations in Sorghum [Sorghum bicolor (L.) Moench] and Sunflower (Helianthus annuus L.) Plants Under Drought*. Ph.D. Dissertation. Manhattan, Kansas: Kansas State University, 336 pp. (Dissertation Abstract No. AAC 9629073)

Zhang, Jingxian and M.B. Kirkham. (1995). Water relations of water-stressed, split-root C_4 (*Sorghum bicolor*; Poaceae) and C_3 (*Helianthus annuus*; Asteraceae) plants. *American Journal of Botany* 82: 1220-1229.

Response of Sunflowers to Quantities of Irrigation Water, Irrigation Regimes and Salinities in the Water and Soil

Z. Plaut
A. Grava

SUMMARY. The production of sunflower grains for roasting was investigated in two soil types under different quantities of applied saline and non-saline irrigation water, different irrigation managements, soil salinity due to previous use of saline water or due to a raised water table. It was shown in one experiment, conducted in a loess type soil, that sunflowers extracted water at least to a soil depth of 120 cm, when the available water from the top layers was used up. The crop in this soil consumed all the available soil water from nearly the entire root zone, while in the clay soil limited water was consumed from deep layers, due to the high salinity and lack of aeration.

No decrease in yield was found in the loess soil when 75% of the full amount of water (which was 0.8 of Class A pan evaporation rate) was applied. When only 50% was applied a significant decrease in yield was obtained. In contrast, in the clay soil even 75% of the full amount of water decreased the yield remarkably. Under dry-land conditions approximately 65% of maximum yield was found in the loess soil but only 45% in the clay soil. These differences are all attributed to a shallow active root system in the clay soil. Residual soil salinity from previously use of saline water had no effect on grain production in the loess soil, while saline irrigation water applied during the irrigation season decreased production, but only when water supply was not rate limiting.

Z. Plaut and A. Grava are affiliated with the Institute of Soils and Water ARO, The Volcani Center, P.O. Box 6, Bet-Dagan 50250, Israel.

[Haworth co-indexing entry note]: "Response of Sunflowers to Quantities of Irrigation Water, Irrigation Regimes and Salinities in the Water and Soil." Plaut, Z., and A. Grava. Co-published simultaneously in *Journal of Crop Production* (Food Products Press, an imprint of The Haworth Press, Inc.) Vol. 2, No. 2 (#4), 1999, pp. 299-315; and: *Water Use in Crop Production* (ed: M. B. Kirkham) Food Products Press, an imprint of The Haworth Press, Inc., 1999, pp. 299-315. Single or multiple copies of this article are available for a fee from The Haworth Document Delivery Service [1-800-342-9678, 9:00 a.m. - 5:00 p.m. (EST). E-mail address: getinfo@haworthpressinc.com].

The combination of saline water and residual soil salinity had a marked effect on the decrease of grain yield under limited irrigation. In both soils a reduction in the amount of water applied per single irrigation and maintaining the entire irrigation period caused a significantly smaller decrease in yield than shortening the irrigation period and applying the full demand. *[Article copies available for a fee from The Haworth Document Delivery Service: 1-800-342-9678. E-mail address: getinfo@haworthpressinc. com <Website: http://www.haworthpressinc.com>]*

KEYWORDS. Sunflower, soil salinity, clay soil, grain yield, irrigation

INTRODUCTION

Sunflowers are known as a crop with a deep root system as compared with other field crops such as corn and sorghum. Sunflowers were found to extract soil water down to a depth of 1.8 m, while others did not extract the water from such a depth (Musick et al. 1976; Unger and Jones 1981; Unger 1990). Moreover, the penetration of sunflower roots to deep soil layers occurs during an extended part of its growing season, while in many other crops roots are active during part of the season only (Jones 1978). These phenomena are of importance under dry-land cultivation or under deficient supply of irrigation water. In such dry areas, as in the Israeli Negev, brackish or saline water is the water available for irrigation in many cases and information on sunflower's salinity tolerance is therefore desirable. Germination and early seedling growth stages of sunflowers were show to be relatively sensitive to salinity (Karami 1974; Saha & Gupta 1993; Katerji et al. 1994). Much less information is available on the response of sunflowers to salinity at advanced stages of plant development. It was considered by Maas (1986) to be moderately salinity sensitive, but this was mostly based on estimates rather than on valid evidence. Information on salinity tolerance of sunflowers is also of importance for some heavy clay soils, in which intensive irrigation and ineffective drainage, brought about a rise in the water table and soil salinization (Biniamini et al. 1991).

When water for irrigation is limiting, management will be based on a decision whether deficient irrigation should be given for an extended period of time or whether full irrigation should be given for part of the time. Full irrigation means that all the transpiration demand is added to the crop throughout the entire growing season, while deficient irrigation implies that less is being applied. One way to reduce the amount of irrigation water is by eliminating water supply during non-critical growth stages. Several studies showed that nearly full production was obtained if water stress was avoided during anthesis and grain filling, provided soil moisture content was also sufficient at early plant development (Jana et al. 1982; Unger and Jones 1981;

Pasda and Diepenbrock 1993; and Costenino et al. 1992). Reducing the quantity of water at each application and applying water throughout the entire season can also save water. This is mainly operational under frequent applications, for instance under drip irrigation. Shortening the irrigation period will induce water stress at various growth stages. The other way of limiting water supply would imply that the Crop-Irrigation-Coefficient, based on pan evaporation, would have to be decreased. This way of reducing the quantities of water per individual application will result in a smaller volume of the wetted soil and a restricted root system. The purpose of the present studies was to develop an irrigation management for sunflowers using less than the full water requirement and yet obtain high grain production.

Another purpose of this study was to determine the response of sunflowers to salinity both in the irrigation water and in the soil (either due to previous uses of saline water for irrigation or due to a raise of a saline water table). A third purpose was to examine whether sunflowers plants are able utilize free water stored in deep soil layers of heavy clay soil, due to insufficient drainage. Artificial drainage by drainage tubes is needed in such soils and the effect of such drainage in this heavy clay soil on grain production was also studied.

MATERIALS AND METHODS

Sunflowers for roasted seed consumption (cultivator of local breeding known as DY3) were grown in consecutive years at two locations in Israel. The first field trial was conducted at Kibbutz Nahal-Oz in the northern Negev, where the soil was a loess type of loamy texture (Topic Xerothert), and the second at Kibbutz Yif'at in the Jezraeel Valley where the soil was a heavy clay (Typic chromoxerert). The 2nd field was in an area of a saline water table with poor natural drainage. Bulk density varied during the season due to swelling and shrinkage. Soil physical and chemical properties of both soils are outlined in Table 1.

The first experiment was conducted on an experimental field, part of which was irrigated for several years with saline water and another part with fresh non-saline water. High rates of rainfall in the preceding winter minimized the differences between these two pre-treatments, and those were not very striking, as is shown in Table 2. The electrical conductivity of saturated paste (Eci) of the soil at planting was 1.6-4.2 dS/m, mainly increasing with depth and much less horizontally; sodium-adsorption-ratio (SAR) values were 3-6 (not presented). The field was cultivated during the fall after a previous summer crop, which was tomatoes for processing, and was finally prepared for planting in the spring, which took place on April 01. The experiment was designed in 5 blocks (replicates) and the individual plot consisted of 6 rows 1 m apart, and 12 m long. Final plant stand was 2.8 plants/m^2.

TABLE 1

A. Soil physical and chemical properties of the experimental fields.

Soil Type	Sand (%)	Silt (%)	Clay (%)	CaCO$_3$ (%)	CEC Cmol/ kg	Org. matter (%)	Field Capac. %-wt	PWP %-wt
Loess	50.00	31.0	19.0	25.1	17.6	1.49	21	10.5
Clay	12.5	22.5	65.0	8.0	54.5	2.32	48	28.0

B. Composition of the irrigation water of both sources at the loess soil experiment.

Source	Eci (dS/m)	Na (mol/m^3)	K (mol/m^3)	Ca+Mg (mol/m^3)	Cl (mol/m^3)	SAR
National Carrier	1.0	4.1	0.12	5.4	6.7	2.5
Local well	4.8	33.0	0.18	4.8	28.0	21.5

Drip lines, every alternate row and a distance of 1m between emitters and a discharge rate of 2.3 l/emitter, applied the irrigation water. Two qualities of water were used: good quality and saline water, having ECi values of approximately 1.0 and 4.8 dS/m, respectively. One experimental treatment, in which water stress was avoided throughout the entire growing period, served as a control and was irrigated for 9 weeks starting on April 27 (14 days after emergence). This covered the entire developmental period through grain filling (details on plant development in Table 3). Water was applied once a week and the quantity of water per application was based on Class A pan evaporation rate and a crop coefficient which was 0.8 during full coverage. The seasonal amount of water applied in this control treatment was 300 mm. The amount of applied water was reduced by 25% or by 50% in the rest of the treatments. This was achieved either by maintaining the duration of water application and decreasing the amount per application (Q treatments), or by shortening the period of water application to the most critical period and maintaining quantities per application unchanged (T treatments). In an additional treatment the effect of dry-land cultivation, with no water added, on sunflower production was examined. Similar treatments were conducted with both water qualities on field plots which were or were not pre-exposed to salinity; the experimental outline is presented in Table 3.

Second experiment: Soil ECi was approximately 4 dS/m in the top 60 cm, and increased gradually up to 17 dS/m at the depth of 120-150 cm. This increase in soil salinity could not be avoided in-spite of soil drainage by two systems of pipe lines. One of those was located at a depth of 1.5 m and

TABLE 2. Soil EC (in dS/m) in the loess soil type. NSS = non-saline soil (not exposed previously to salinity), SS = saline-soil (previously exposed to salinity), NSW = non-saline irrigation water, SW = saline irrigation water.

A. Loess soil type at the beginning of the experiment

Depth (cm)	NSS	SS
0-30	3.3	4.3
30-60	1.7	2.4
60-90	1.9	2.6
90-120	2.4	2.7

B. Loess soil type at termination of the experiment.

	--------------------NSS--------------------				--------------------SS--------------------			
	------NSW------		------SW------		------NSW------		------SW------	
Depth (cm)	120 mm	300 mm	120 mm	300 mm	120 mm	300 mm	120 mm	300 mm
0-30	3.5	2.0	7.0	12.5	4.0	4.0	8.0	13.5
30-60	2.6	1.6	4.8	4.3	3.3	2.5	5.5	6.9
30-90	1.9	1.8	3.3	3.9	3.1	2.7	4.1	5.3
90-120	1.8	2.0	2.7	3.5	3.3	3.0	3.5	4.1

C. Along the field of the clay soil experiment (distance in m from the lowest point, perpendicular to the drainers) at the beginning of the experiment.

Depth (cm)	---------------Distance from the bottom of the field (m)---------------			
	0	50	100	150
0-30	2.3	3.2	3.3	3.0
30-60	7.4	3.1	2.5	2.3
60-90	12.8	8.3	7.8	5.9
90-120	14.0	11.4	9.1	7.8
120-150	16.9	13.7	11.7	9.5

approximately 65 m apart and the 2nd in the center between those at the depth of 70 cm. The experiment was arranged so that each plot covered the entire distance between a drainer and the center between two adjacent drainers, namely 17 rows 1-m apart. The length of the plot was 12 m. The field was cultivated in the fall after harvest of the previous crop, which was cotton. The field was then prepared for planting in the following spring and planting had to be delayed until May 07, due to late rainfall. A similar drip system to the one used in the 1st experiment was installed and irrigation regimes were also similar (Table 4). Water of one quality only (ECi = 1.8 dS/m) was used for

TABLE 3. Stages of sunflower plants development.

Growth-stage	Loess-soil		Clay-soil	
	Date	Age (days)	Date	Age (days)
Planting	01.04	0	07.05	0
Full emergence	13.04	13	22.05	15
1st flower	06.05	32	09.06	33
Flowering–10%	21.05	47	25.06	49
Flowering–full	05.06	62	11.07	66
Maturity	15.07	102	24.08	108

TABLE 4. Design of experimental treatments, quantities of water applied and irrigation timing.

Treatment	Water applied (mm)	Crop-irrigation Coefficient	Irrigation period (days)	Designation of Treatment
Loess-soil				
Control	300	1.00	63	300
Medium quantity	230	0.75	63	230-Q
Medium time	230	1.00	42	230-T
Low quantity	120	0.50	63	120-Q
Minimal time	120	1.00	28	120-T
Dry	0	–	–	0
Clay soil				
Control	330	1.00	56	330
Medium quantity	230	0.70	56	230-Q
Medium time	230	1.00	42	230-T
Low quantity	150	0.45	56	150-Q
Minimal time	150	1.00	28	150-T
Dry	0	–	–	0

irrigation. The irrigation season started on June 19 and details on plant development are outlined in Table 3.

Soil moisture content was determined, in the loess soil type experiment from the soil surface to a depth of 120 cm and in the clay soil experiment down to 150 cm, in both by means of neutron probes, which were located between row 2 and 3 in the 1st experiment and near rows 2, 9 and 16 (at different distances from the drainer) in the 2nd experiment. All measurements were determined in quadruplicate. Plant development was very uniform within plots in the 1st experiment and therefore yields and yield components were determined together from the central 4 rows of each plot. In the 2nd experiment, however, yield and yield components were determined separately for rows 1 and 2, 5 and 6, 11 and 12, and 15 and 16 (length of 10 m).

It should be noted that there was no rainfall in both locations between planting through harvest. The rate of daily Class A pan evaporation was approximately 4.0 mm rising gradually up to 7.8 mm at harvest time in the 1st experiment and in the range of 5.6 to 7.5 in the 2nd.

RESULTS AND DISCUSSION

Depletion of soil water from the root zone may take place either by evapotranspiration or by drainage. Frequent measurements of soil water content at the bottom of the root zone never showed an increase in water content after irrigation. This was due to frequent (twice a week) applications, each with small quantities of water. Losses of soil water can thus be ascribed mainly to transpiration, and to some extent to evaporation from the surface prior to full ground coverage or at the end of the season.

In the loess soil experiment, moisture content above the permanent-wilting-point (PWP) in the top 120 cm of the control treatment, (300-NS), was never less then 90-110 mm during the first 60 days after planting (Figure 1a). When saline water was used (300-S) soil water content was higher by approximately 20 mm, suggesting a decrease in transpiration rate under saline water use, as indicated earlier (Meiri et al. 1992, Shalhevet, 1994). Soil available water content was proportional to the quantities applied and was always higher when saline water was used. The rate of water depletion was maximal when no irrigation was applied, as may be expected, and the point of 0 available water content was already obtained around 45 days after planting. This stage was delayed depending on the quantities of water applied. The depletion of soil water beyond the point of 0 available water can mainly be attributed to more extreme dehydration of the top soil layers, mainly toward the end of the season. While the rates of water losses were nearly linear in treatments of reduced quantities of water, there were fluctuations in these rates when the irrigation periods were shortened (Figure 1b). Most of the water used by the crop of the control treatments (300-NS and 300-S) was extracted from the top 30 cm of the soil during the initial 80 days after planting. In the drier treatment deeper layers were already extracted at this time. The deeper the layer the later was its water extracted, namely root activity became deeper only when the water from more shallow layers was nearly used up (data not shown).

FIGURE 1. The effect of quantities and salinity of irrigation water and of irrigation management (timing and duration, as outlined in Table 4) on available water content. Vertical bars indicate SE of the means.

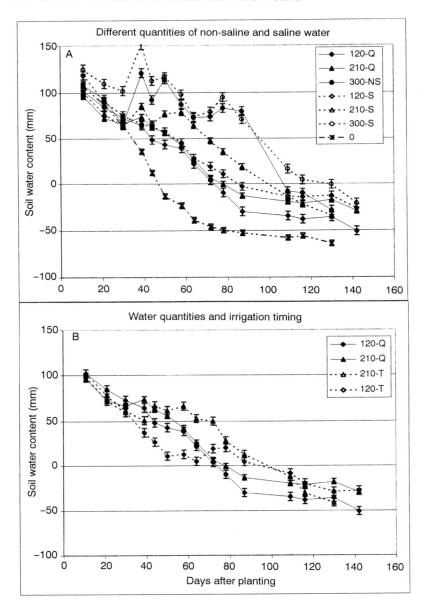

The initial higher ECi in the plots, which were in the past irrigated with saline water as compared with those which were always irrigated with non-saline water, were found again at termination of the experiment (Table 4 A and B), regardless of the quantity of water used. However, the use of saline water during the growing period raised the soil ECi up to 7.0-8.0 and 12.5-13.5 dS/m in the top layer when 120 and 300 mm of water were applied, respectively.

The irrigation period in the clay soil experiment started only 38 days after planting due to the initial high soil water content (Figure 2a). The trend in water losses during the growing season of this experiment resembled generally those outlined for the loess soil. The stage of complete utilization of available soil water was delayed in the unirrigated treatment of the clay soil up to 70 days after planting, and in the control fully irrigated treatment it was not reached even at harvest time. Soil water content was lower near the drainer as compared with locations further away (Figure 2b). This can be seen most clearly in the dry treatment, in which no irrigation water was applied, when soil water content at rows 2 and 16 are compared at all the depths.

A distinct difference between the two soil types is the extraction of water from the entire soil profile in the loess soil type in contrast to the restricted use of water from the deeper soil layers in the clay soil. The deepest soil layer tested in this experiment, which was at 120-150 cm, contributed no water to the sunflower crop (Figure 3). Soil water content of this layer was nearly the same in all treatments and throughout the season (except a decrease, which took place at the beginning of the season). The top layer of 0-30 cm contributed most of the water to the crop, and its water content decreased very significantly when no water was applied as compared with that of 150 mm application. The layers of 30-60 and 60-90 cm still contributed some water to the crop, as their water content was reduced to the PWP under partial irrigation (150-Q) after termination of the irrigation period. It thus seems that root activity below 90 cm was very poor. The effect of the raised soil water table on salinization of the clay soil is demonstrated in Table 4C. There was a gradual increase in salinity along the slope of the field toward the bottom and also from the soil surface down to the deeper layers. At the bottom of the field only the top 30 cm were free of salt, while for soil close to the peak, a depth of over 60 cm was salt free.

Grain production was much lower in the clay soil than in the loess soil. This can be attributed to the poor soil physical properties, high salt content and the late planting, under climatic conditions, which were not optimal. The yields of all plots in a given experiment relative to the maximal yield obtained in this experiment were calculated and plotted against the quantity of water applied, regardless of irrigation management, salinity or location within the field (Figure 4). Two main differences can be seen between the two soil types: (a) in the loess type soil approximately 0.65 of maximal yield was

FIGURE 2. The effect of quantities of water applied and irrigation management on available water content in the clay soil (as in Figure 1, but to depth of 150 cm); and the distribution of soil water as a function of soil depth and distance from drainer. Row 2 = next to drainer. Row 16 = maximal distance from drainer. Vertical bars indicate SE of the means.

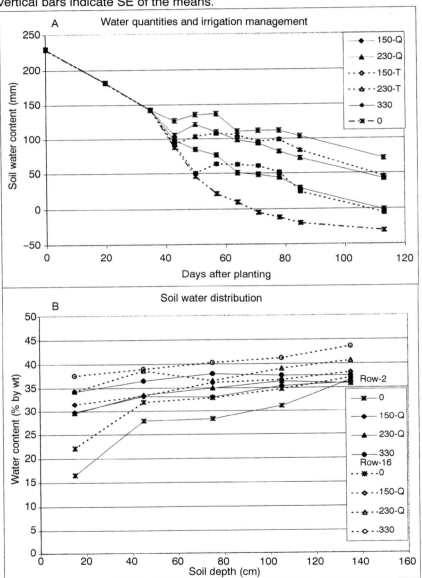

FIGURE 3. Pattern of water content in 5 soil layers (0-150 cm) throughout growing season in the clay soil. Values presented are for water content shortly before application during the irrigation period or later after termination of the irrigation period. Treatment presented were: 0, 150-Q and 330 (Details in Table 4).

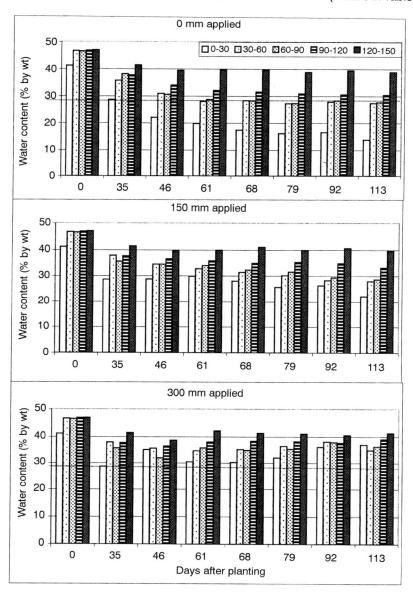

FIGURE 4. Relative grain yield obtained at both experiments as a function of the quantity of water applied. Yields were calculated per unit area and per plant. The points represent yields obtained with a given quantity of water but with different salinity levels and different irrigation management.

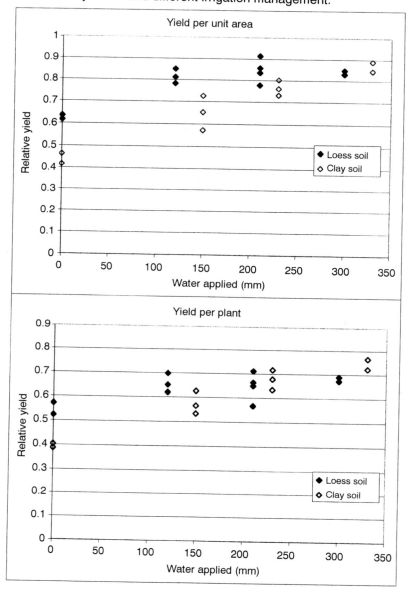

obtained without any water, while in the clay soil only 0.45. (b) In the loess soil type no increase in relative yield was found above 210 mm of water, while in the clay soil a linear increase in yield was found over the entire range of water applied. This finding, which was obtained for both: yield per unit area and yield per plant can be interpreted on the basis of restricted depth of water use by plants in the clay soils, as indicated in Figure 3. This restriction was not due to scarcity of available water, but was the result of high salt content in the deeper soil layers and the lack of aeration due to high water content and soil compaction. These conditions were most likely responsible for insufficient root development (Plaut et al. 1997) leading to limited water supply to the canopy.

Maximal sunflower grain yield of approximately 280 g/m^2 was obtained when 210 mm or more of non-saline water were applied throughout the season, regardless of pre-existing residual salinity in the soil (Figure 5). Maximal yield was significantly less when saline water was used, and was further reduced if this water was applied to previously salinized soil. Under conditions of salinized soil and saline water, a sharp decline in yield was found under limiting irrigation, possibly due to high concentration of salt in the soil. The response of grain yield to water and salinity can mainly be attributed to the number of grains per head and average grain weight. Each of those responded to the quantities of water applied and to salinity, although to a lower extent than grain yields. It was thus the additive effect, which was responsible for the decrease in yield. Number of heads per unit area is determined early in the season and was not affected by the quantity of water used if in nonsalinized soil. The number was increased when larger quantities of water were used, probably due to leaching or at least diluting the salts.

Limiting the amount of irrigation water by reducing the quantity per application was always less harmful than shortening the irrigation season to the most critical period. This was found in both experiments, although conditions were much different, and is shown for the experiment conducted in the clay soil (Figure 6). This is in contrast with some publications (Jana et al. 1982, Costenino et al. 1992, Pasda and Diepenbrock 1993). Number of heads per unit area was again hardly affected by quantities of water applied or by irrigation management, probably because sufficient water was still available at this stage in all treatments. It seems that if available soil water was sufficient up to this stage, then available water during grain filling became an important factor in the determination of yield. When water is being applied up to flowering and not later, it is probably limiting at stages of grain filling and thus reducing number and weight of grains (Figure 6).

Since soil moisture content was found to be influenced by the distance from the drainer (Figure 2), and as the top layers were mostly important for water supply to the crop, the horizontal distribution of grain yield over the field is of interest (Figure 7). A significant, although not very large rise in

FIGURE 5. Response of yield and yield components to quantities of irrigation water, salinity in the water and residual salinity in the loess soil. (NSS, SS, NSW and SW as outlined in Table 2). Vertical bars indicate SE of the means.

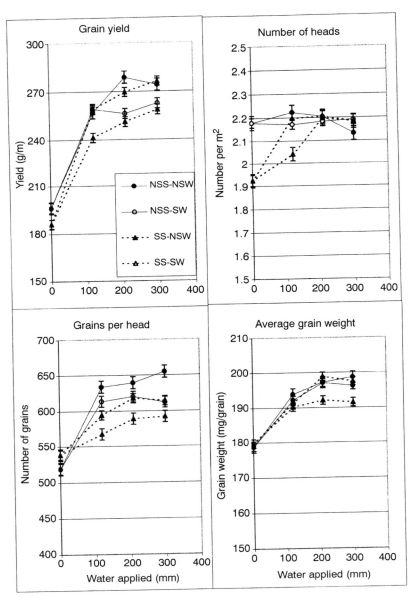

FIGURE 6. Response of yield and yield components to quantities of irrigation water, and irrigation management (timing and duration) in the clay soil. Vertical bars indicate SE of the means.

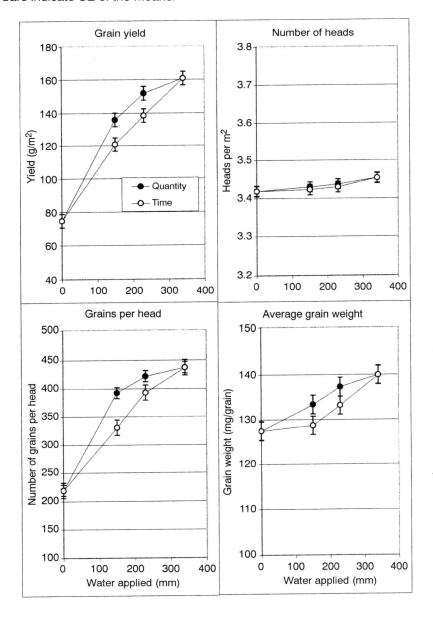

FIGURE 7. Interaction of distance from drainer and quantities of water applied and their effect on grain yield and number of grains per head in the clay soil.

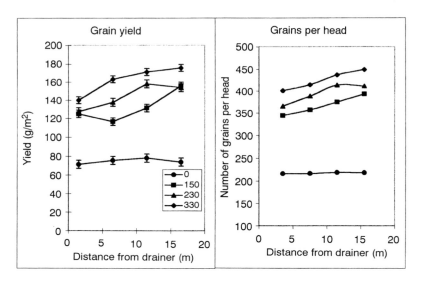

grain yields was found at increasing distance from the drainer. This was associated with a similar trend in the number of grains per head, which was shown to be the major component determining yield (Figure 6). As may be expected no such effect was found under dry-land conditions.

REFERENCES

Biniamini, A., S. Marish, A. Gafni, and M. Gutman (1991) Subsurface drainage systems as a mean for decreasing salinity. Final Report. 49 pp. (in Hebrew).

Costenino, S., O. Sartini, and P.G Litrico (1992) Yield response, canopy temperature and water status of a second crop of sunflowers (*Helianthus annuus* L.)with different irrigation regimes. *Rivieta di Agronomia* 24: 633-640.

Jana, P.K., B. Misra, and P.K. Kar (1982) Effect of irrigation at different physiological stages of growth on yield attributes, yield, consumptive use and water use efficiency of sunflower. *Indian Agriculture* 26: 39-42.

Jones, O.R. (1978) Management practices for dry-land sunflower in the U.S. southern great plains. In Proc. 8th Int. Sunflower Conf., Minneapolis, Minnesota, pp. 89-98.

Karami, E. (1974) Emergence of 9 varieties of sunflower (*Helianthus annuus* L.) in salinized soil culture. *Journal of Agricultural Science* 83:359-362.

Katerji, N., J.W. Vanhoorn, A. Hamdy, F. Karam, and M. Mastrorilli. (1994) Effect of salinity on emergence and on water stress at early seedlings growth of sunflower and maize. *Agricultural Water Management* 26: 81-91.

Maas, E.V. (1986) Salt tolerance of plants. *Applied Agricultural Research* 1: 12-26.

Meiri, A., H. Frenkel, and A. Mantel (1992) Cotton response to water and salinity under sprinkler and drip irrigation. *Agronomy Journal* 84: 44-50.

Musick, J.T., L.L. New, and D.A. Dusek (1976) Soil water depletion–yield relationships of irrigated sorghum, wheat and soybean. *Transactions of the American Society of Agricultural Engineers* (ASAE) 19:489-493.

Pasda, G. and W. Diepenbrock (1993) Physiological yield analysis of sunflower (*Helianthus annuus* L.) Part III: Agronomic factors and production techniques. *Felt Wissenschaft. Technologie* 93: 235-243.

Plaut, Z., M. Newman, E. Federman, and A. Grava (1997) Response of root growth to a combination of three environmental factors: water stress, salinity and soil compactness. In: Biology of Root Formation and Development, ed. A. Altman and Y. Waisel, New York: Plenum Press.

Saha, K. and K. Gupta (1993) Effect of LAB150978–a plant growth retardant on sunflower and mung bean seedlings under salinity stress. *Indian Journal of Plant Physiology* 36: 151-154.

Shalhevet, J. (1994) Using water of marginal quality for crop production: major issues. *Agricultural Water Management* 25: 234-269.

Unger, P.W. (1990) Sunflower. In Irrigation of Agricultural Crops, ed. B.A. Stewart and D.R. Nielsen, Agronomy Monograph No. 30. Madison, WI:American Society of Agronomy, Crop Science Society of America, and Soil Science Society of America, Madison, WI, pp. 775-794.

Unger, P.W. and O.R. Jones (1981) Effect of soil water content and growing season straw mulch on grain sorghum. *Soil Science Society of America Journal* 45: 129-134.

Turfgrass Evapotranspiration

Bingru Huang
Jack D. Fry

SUMMARY. This paper reviews factors influencing water use and methods to reduce it. Water use in turfgrass typically is quantified by evapotranspiration rate (ET), which refers to the loss of water from the soil through evaporation and from the plant through transpiration. Turfgrass ET rates vary among species and cultivars within a species. Inter- and intraspecific variations in ET rates could be explained by differences in stomatal characteristics, growth rate and habit, canopy configuration, and rooting characteristics. The ET rates also are influenced by environmental conditions and cultural practices. Environmental conditions include climatic factors such as temperature, relative humidity, solar radiation, and wind and edaphic factors such as soil temperature, water availability, and soil texture. Cultural practices include mowing, irrigation, fertility, use of antitranspirants, and plant growth regulators. *[Article copies available for a fee from The Haworth Document Delivery Service: 1-800-342-9678. E-mail address: getinfo@haworthpressinc.com <Website: http://www.haworthpressinc.com>]*

KEYWORDS. Evapotranspiration rate, turfgrass, water use

INTRODUCTION

Water is becoming increasingly limited for irrigation, especially in expanding urban areas situated in arid and semi-arid regions of the U.S. As

Bingru Huang is Assistant Professor and Jack D. Fry is Associate Professor, Department of Horticulture, Forestry, and Recreation Resources, Kansas State University, Manhattan, KS 66506.

Contribution No. 99-369-B from the Kansas Agricultural Experiment Station.

[Haworth co-indexing entry note]: "Turfgrass Evapotranspiration." Huang, Bingru, and Jack D. Fry. Co-published simultaneously in *Journal of Crop Production* (Food Products Press, an imprint of The Haworth Press, Inc.) Vol. 2, No. 2 (#4), 1999, pp. 317-333; and: *Water Use in Crop Production* (ed: M. B. Kirkham) Food Products Press, an imprint of The Haworth Press, Inc., 1999, pp. 317-333. Single or multiple copies of this article are available for a fee from The Haworth Document Delivery Service [1-800-342-9678, 9:00 a.m. - 5:00 p.m. (EST). E-mail address: getinfo@haworthpressinc.com].

317

conscientious environmental and financial managers, those employed in the turfgrass industry realize that efficient water management is a principal concern. Many research efforts over the past 25 years have been directed at obtaining a better understanding of the physiology of plant water use and determining turfgrass water requirements and factors that affect water demand.

A typical turfgrass plant has a water content ranging from 75 to 85% by weight (Beard, 1973). Only 1 to 3% of the absorbed water is utilized in plant metabolic processes; most is transported to shoots where it is transpired. Theoretically, turfgrass "water use" refers to the total amount of water needed for plant growth plus that lost through evapotranspiration (ET), defined as the sum of plant transpiration and soil evaporation. However, the terms water use and ET often used interchangeably and water use typically is quantified by measuring ET, because such a small fraction of water absorbed is metabolized by the plant. For the sake of simplicity, we will use the term ET throughout this manuscript.

Evapotranspiration usually is expressed as the depth of water lost (if measured from a flat surface) over time. Typical ET rates of turfgrasses range from 3 to 8 mm of water per day and can be as high as 12 mm per day (Beard, 1973). Transpiration may appear to be wasteful, but it is important in many physiological processes. For example, water moving through the plant in transpiration streams transports nutrients, metabolites, and products of photosynthesis. Transpiration is also an important cooling process and prevents the plant's internal temperature from reaching lethal levels. When the transpiration rate exceeds the rate at which water is absorbed by roots, wilt and desiccation of the turf occur. Without adequate irrigation, a water deficit occurs, impeding many physiological processes and restricting turfgrass growth.

Key strategies in water conservation are the selection of turfgrass species and cultivars possessing low ET rates and development or identification of cultural practices that reduce ET. Knowledge of factors that influence plant water use internally and externally is required before significant progress can be made in turfgrass water conservation. This paper provides a review of research on factors influencing turfgrass ET.

PHYSIOLOGICAL PROCESSES INFLUENCING ET

Evapotranspiration rate and internal plant water balance are determined by the interrelated processes of water absorption and transpiration. Water uptake from the soil is a crucial function of the root system and largely determines water use efficiency and water status of the plant. The uptake capacity depends on root morphological characteristics (e.g., root length density and root distribution) and physiology (e.g., hydraulic properties) (Passioura, 1988; Nobel, 1991). Most transpiration occurs through leaves, although some may occur through other plant parts exposed to the atmosphere. The relative humidity inside the stomatal cavity is almost always 100%. Hence, the magnitude of the vapor pressure gradient between the stomatal cavity and the ambient air is usually substantial, particularly in arid and semi-arid climates. Interrupting this gradient is the boundary layer, a narrow layer of water vapor that develops above and below the leaf. The presence of this layer minimizes the vapor pressure gradient between the plant and atmosphere. The transpiration rate is dependent on resistance to water movement within the leaf (internal leaf diffusion resistance), diffusion resistance of the boundary layer, and the vapor pressure gradient between leaf and ambient air.

INTER- AND INTRASPECIFIC VARIABILITIES IN TURFGRASS ET

Water absorption, translocation, and transpiration are affected by many plant morphological and physiological factors, most of which differ among turfgrass species and cultivars. Extensive research has been conducted to discern differences in ET among turfgrass species. However, relatively little information is available at the intraspecific level. Almost all of the research done has evaluated turfgrass ET under nonlimiting soil moisture conditions using small weighing lysimeters (Feldhake, 1981; Kim, 1983; Minner, 1984). Results of some of these studies are summarized below.

Typical water use rates range from 3 to 8 mm per day for cool-season grasses under non-limiting soil moisture conditions (Beard, 1994). Among cool-season species evaluated, tall fescue (*Festuca arundinacea* Schreb.), creeping bentgrass (*Agrostis stolonifera* L.), perennial ryegrass (*Lolium perenne* L.), annual bluegrass (*Poa annua* L.), and Italian ryegrass (*Lolium multiflorum* L.) have the highest ET rates; Kentucky bluegrass (*Poa pratensis* L.) and rough bluegrass (*Poa trivialis* L.) have intermediate rates; and hard fescue (*Festuca longifolia* Thuill.), chewings fescue (*Festuca rubra* L. ssp. *commutata* Gaud.), and creeping red fescue (*Festuca rubra* L. ssp. *rubra*) have the lowest ET rates (Beard, 1994).

Kentucky bluegrass had average ET rates of 5.1 mm day^{-1} in Fort Collins, Colorado, and 5.6 mm day^{-1} in Northglenn (Danielson et al., 1979). Minner (1984) reported that tall fescue used 16% more water than Kentucky bluegrass, 14% more than perennial ryegrass, and 18% more than fine fescue between June and September in Colorado. In Israel, 'Alta' tall fescue used significantly more water than 'Pennfine' perennial ryegrass (Biran et al., 1981). The ET rate of a mixture of Kentucky bluegrass, fine fescue, meadow fescue (*Festuca elatior* L.), chewings fescue, and white clover (*Trifolium repens* L.) averaged 5.1 mm day^{-1} between July and September in Nevada (Tovey, Spencer, and Muckel, 1969).

Typical ET rates range from 2 to 5 mm per day for warm-season grasses (Beard, 1994). Warm-season grasses that have relatively low ET rates include bermudagrass (*Cynodon* spp.), zoysiagrass (*Zoysia* spp.), and buffalograss (*Buchloe dactyloides* [Nutt.] Engelm.). Centipedegrass (*Eremochloa ophiuroides* [Munro] Hack.), seashore paspalum (*Paspalum vaginatum* Swartz.), bahiagrass (*Paspalum notatum* Flugge), and St. Augustinegrass (*Stenotaphrum secundatum* [Walt.] Kuntze.) are considered to have relatively high ET rates. Of 11 warm-season turfgrasses grown in Texas, 'Emerald' zoysiagrass exhibited the lowest ET rate (4.8 mm day^{-1}); 'Texas Common' St. Augustinegrass had the highest (6.3 mm day^{-1}); and 'Tifgreen' bermudagrass, common bermudagrass, and 'Meyer' zoysiagrass had intermediate ET rates (Kim, 1983). In an Arizona test, 'Tifway' bermudagrass and 'Meyer' zoysiagrass used similar amounts of water that were significantly less than those used by St. Augustinegrass when subirrigation was employed (Kneebone and Pepper, 1982). 'Meyer' zoysiagrass consumed 8 to 14% more water than 'Midlawn' bermudagrass and 'Prairie' buffalograss in Kansas (Qian, Fry, and Upham, 1997).

Generally, cool-season species exhibit higher water use rates than warm-season species. Kim (1983) reported that 'Kentucky 31' tall fescue used 16% more water than 'Texas Common' St. Augustinegrass and 47% more than 'Emerald' zoysiagrass in Texas. 'Alta' tall fescue and 'Pennfine' perennial ryegrass had higher ET rates than nine warm-season species in Israel (Biran et al., 1981). Kentucky bluegrass and tall fescue were found to use over 20% more water than 'Tifway' bermudagrass and buffalograss in Colorado (Feldhake, Danielson, and Butler, 1983). Qian, Fry, and Upham (1997) reported that ET of tall fescue was 12-25% higher than that of 'Meyer' zoysiagrass; 29-38% higher than that of 'Prairie' buffalograss; and 35% higher than that of 'Midlawn' bermudagrass in Kansas.

Evapotranspiration also can vary considerably among cultivars within a species. Shearman (1986) evaluated 20 cultivars of Kentucky bluegrass in a growth chamber and found that ET rates ranged from 3.9 mm day^{-1} for 'Enoble' Kentucky bluegrass, to 6.3 mm day^{-1} for 'Birka', 'Sydsport', and 'Merion'. Schmidt and Everton (1985) reported significantly higher water use for 'Adelphi' Kentucky bluegrass compared with four other cultivars under two nitrogen levels and two irrigation regimes. Shearman (1989) also reported significant ET variability among 12 well-watered perennial ryegrass cultivars evaluated under field conditions in Nebraska. Significant differences in ET rates among cultivars of tall fescue were reported in a Nebraska field study (Kopec and Shearman,1987) and in a greenhouse test (Bowman and Macaulay, 1991). Results showed that turf-type tall fescue cultivars had lower ET rates than forage-type cultivars.

Less variability in ET has been reported among cultivars of warm-season grasses. In Texas studies, no significant differences in ET rates occurred among 11 zoysiagrass cultivars (Green et al., 1991) or 10 St. Augustinegrass cultivars (Atkins et al., 1991).

Variability in ET suggests that proper selection of turfgrass species and cultivars are essential to maximize water conservation. However, some caution should be exercised in comparing ET results among research studies, because environmental and cultural conditions usually vary greatly from one site to the next.

SHOOT AND ROOT CHARACTERISTICS INFLUENCING ET

Variability in ET rates appears to be related to differences in shoot and root characteristics among species and cultivars. Many plant factors influence ET, including canopy configuration and growth habit, stomatal density and regulation, leaf rolling or folding characteristics, and rooting characteristics. Under well-watered conditions, boundary layer resistance and, consequently, ET may be affected by canopy structure. Stomata in St. Augustinegrass provided limited control of transpiration, and ET was controlled largely by external features of the plant, including the number and position of leaves, stem density, and leaf orientation (Johns, Beard, and van Bavel, 1983). Kneebone and Pepper (1982) and Kim (1983) demonstrated that relatively high ET rates in tall fescue and St. Augustinegrass were associated with low canopy resistance resulting from low shoot densities, high leaf surface areas, and rapid rates of vertical leaf extension. Zoysiagrass and bermudagrass did not exhibit these characteristics and had low to moderate ET rates

Species that have a prostrate growth habit and are not exposed to the turbulent air flow that might be experienced by grasses with a more upright growth habit tend to have lower ET rates (Kim, 1983). Likewise, grasses with rapid rates of vertical leaf extension commonly exhibit relatively high ET rates. Shearman (1986) observed a positive correlation between vertical extension rate and ET of 20 Kentucky bluegrass cultivars. 'Adelphi' Kentucky bluegrass, a relatively low water user, had an average leaf extension rate of 10 mm week^{-1}. 'Sydsport', possessing a high ET rate, exhibited a leaf extension rate of 15 mm week^{-1}. Bowan and Macaulay (1991) found that ET rates of 20 tall fescue cultivars increased with increasing clipping dry weights. Kim (1983) reported high ET rates for 'Texas Common' St. Augustinegrass and seashore paspalum, which had vertical leaf extension rates of 7.4 and 7.5 mm day^{-1}, respectively. A low ET rate was observed in common centipedegrass, which had a vertical leaf extension rate of 2.5 mm day^{-1}. No correlation was observed between ET rate and leaf extension rate among 24 bermudagrass cultivars (Beard, Green, and Sifers, 1992), 11 zoysiagrass cultivars (Green et al., 1991), or 10 St. Augustinegrass cultivars (Atkins et al., 1991).

Some work has been done to explain the influence of plant density on ET. Marlatt (1961) evaluated the ET rates of orchardgrass (*Dactylis glomerata* L.) at 100, 70, 50, and 30% densities. The greatest water loss occurred from plots of only 50% density. Evapotranspiration was higher at 70% density, but lower at 30% density, compared to that at full cover. Grass growing in plots of 50 and 70% density used more water because of a phenomenon referred to as the "clothesline" effect. An open canopy allowed movement of air around plants, thereby increasing water loss from advective forces. In addition, temperature measurements indicated that air directly above bare ground was warmer than that above a full-cover canopy. The warmth of the air circulating through plots of 50 and 70% covers also contributed to the increased ET rates. Below 50% cover, the plant population was so low that ET losses most likely were due primarily to evaporative water lost from fallow soil. Other researchers also have found that a dense turfgrass sward exhibits lower ET rates. When studying water use rates of 20 Kentucky bluegrass cultivars, Shearman (1986) noted that ET increased with decreasing shoot density and verdure. Kim (1983) reported that low ET rates of 12 warm-season turfgrasses were associated with high shoot density.

Results on the influence of stomatal characteristics on ET are inconsistent, which may be related to species differences and variability of soil water content during study periods. Stomatal and leaf surface characteristics that influenced transpiration varied among 12 warm-season grasses (Kim, 1987). Bermudagrass stomata were sunken and wax protected and closed rapidly following the onset of drought stress. This species also exhibited a low ET rate during drought and nonstressed conditions. A high degree of wax accumulation on leaf surfaces was associated with low ET rates of buffalograss and zoysiagrass. Bahiagrass had high ET rates under nonstressed conditions, but very low rates when subjected to drought stress, which was attributed to rapid stomatal closure. The stomata of centipedegrass and St. Augustinegrass were not protected by wax and remained open during the onset of drought stress. Consequently, these grasses exhibited relatively high ET rates regardless of whether they were well watered or entering drought stress. 'Penncross' stomatal density increased with increasing light intensity, and a positive correlation was reported between stomatal density and water use rate (Shearman and Beard, 1973).

Several other studies have suggested that plant stomatal characteristics have little influence on ET and drought resistance. Dernoeden and Butler (1978) reported no correlation between stomatal characteristics of 12 Kentucky-bluegrass cultivars and drought resistance observed in the field in Colorado. Shearman (1986) found no correlation between stomatal density or index and ET rates of 20 Kentucky bluegrass cultivars in a controlled environment. Johns, Beard, and van Bavel (1983) concluded that genetic control of stomatal resistance in St. Augustinegrass would not result in increased water savings.

Leaf rolling and folding also are employed by turfgrasses to minimize water loss at the onset of stress. Decreasing plant water potential results in collapse of bulliform cells, thereby causing the leaf to roll or fold (Beard, 1973). This reduces the amount of leaf surface area and number of stomata exposed to the ambient environment. Other xeromorphic features important in regulating plant water loss include cuticle thickness, presence of surface hairs, and amount of intercellular space. Essentially no research has been conducted with turfgrasses to determine the importance of these factors in ET regulation.

Rooting characteristics greatly influence water absorption and, thus, have a substantial impact on ET (Beard, 1973). Tall fescue, a cool-season turfgrass that develops a deep, extensive, root system, has been shown to use more water than most other species. This could be due to its ability to absorb more water by exploring a greater soil volume. Tall fescue also is considered drought resistant, indicating no direct relationship between this characteristic and ET rate (Beard, 1985). However, deep, extensive, root systems do not necessarily result in high ET rates. Bermudagrass, for example, has a deep root system, but exhibits relatively low ET rates (Youngner, 1985).

It is generally agreed that an extensive, well-branched, deep, root system improves potential for water uptake (Hurd, 1974; Kramer, 1983). The extent of rooting can be quantified by root length density (RLD), defined as root length per unit soil volume (cm root cm^{-3} soil) (Taylor, 1980). Water uptake rates of root systems generally are considered to be proportional to RLD; however, the relationship depends largely on plant species, soil water availability, and soil depth. Mason et al. (1983) reported that sunflower (*Helianthus annuus* L.) extracted as much water from irrigated clay loam as maize (*Zea mays* L.) and sorghum (*Sorghum bicolor* L.), despite having only half the total RLD of the cereals. Other researchers have found that water uptake is correlated positively with RLD when soil is moist but not when soil is dry (Gregory et al., 1978; Sharp and Davies, 1985; Al-Khafaf, Wierenga, and Philip, 1987; Shein and Pachepsky, 1995; Huang, Duncan, and Carrow, 1997). Rooting depth is correlated better with water uptake than RLD when soil is drying (Narayan, 1991).

Root length density at deep soil depths has been correlated highly with water uptake. Plants with greater RLD in deep soil layers are better able to maintain water status and stomatal conductance when soil dries than those with lower RLD (Devries et al., 1989; Fukai and Cooper, 1995; Volaire and Thomas, 1995; Carrow, 1996; Huang, Duncan, and Carrow, 1997; Qian, Fry and Upham, 1997). For example, tall fescue cultivars with a greater RLD at a depth of 60 cm were less prone to leaf wilting during periods of infrequent rainfall than those with lower RLD (Carrow, 1996).

Deep rooting enhances water uptake from deeper soil profiles where soil water is available. For example, water extraction rates at a 40 to 80 cm depth were greater for the deep-rooted seashore paspalum and buffalograss than the shallowed-rooted 'Meyer' zoysiagrass when only soil near the surface was drying (Huang, Duncan, and Carrow, 1997; Huang, 1999). However, when little water is stored deep in the soil profile, faster root extension towards these depths may be detrimental. Roots could rapidly deplete remaining water, resulting in drought stress. In contrast, some plants with sparse and poorly permeable root systems survive partly by conserving water stored more deeply in the profile (Passioura, 1988).

Hydraulic conductivity is a coefficient relating the driving force, usually a drop in water potential or in hydrostatic pressure, to water flow, such as the volume of water crossing a root's surface area per unit time. Root hydraulic conductivity can represent two-thirds of the limitations on water movement within a plant (Fiscus, Klute, and Kaufman, 1983). It is correlated proportionally to ET (Richards and Passouria, 1981) and, thus, can have a major influence on leaf water status and, in turn, on plant growth in both moist and dry soil (Blizzard and Boyer, 1980; Passioura, 1988).

Root hydraulic conductivity varies with plant species, ranging from 0.1 to 70×10^{-8} m s^{-1} MPa^{-1} when measured by applying a hydrostatic pressure gradient to the roots and from 0.5 to 8×10^{-8} m s^{-1} MPa^{-1} when measured using osmotic pressure gradients (Huang and Nobel, 1994). Water uptake or water use efficiency might be improved by genetically altering the hydraulic properties of the roots (Meyer, 1976). For example, the flow rate at a subcrown potential of -1.5 MPa would be increased by 30% in a plant with three root axes without changing the root length, if the radius (40 m) of each metaxylem vessel was increased by 3 m (Meyer, Greacen, and Alston, 1978), assuming that axial movement of water is unimpeded by cross-walls in the xylem. To date, no information is available regarding the relationship among root hydraulic conductivity, water uptake, and ET in turfgrasses.

ENVIRONMENTAL FACTORS INFLUENCING TURFGRASS ET

Climatic Factors Affecting ET

Many environmental factors influence turfgrass ET rates, including temperature (Beard, 1973; Feldhake, Butler, and Danielson, 1985), relative humidity (Beard, 1985); solar radiation (Feldhake, 1981; Aurasteh, 1983; Feldhake and Boyer, 1986); and wind (Grace, 1974; Grace and Russell, 1977; Oke, 1979). Evapotranspiration is an energy-dependent process, and the energy related to environmental factors regulates the rate at which it occurs. Beard (1985) noted that transpiration is linked closely to the energy balance equation. This is expressed as:

$$R + H + LE + G + aA = 0$$

where: R = net radiation flux
       H = sensible heat exchange with the atmosphere
      LE = latent heat exchange with the atmosphere
      G = sensible heat exchange with vegetation and soil
      aA = energy used for plant metabolic purposes

As solar radiation is absorbed, energy must be used by the plant or released in the form of heat. Plant temperature is influenced greatly by the amount of solar radiation absorbed. As plant temperature increases, ET increases (Beard, 1973). Feldhake (1981) and Aurasteh (1983) found a linear relationship between increasing solar radiation and the ET rate of Kentucky bluegrass. Beard, Green, and Sifers (1992) and Kim and Beard (1988) also reported that ET rates were associated closely with solar radiation. Researchers in Colorado evaluated the possibility of preconditioning 'Merion' Kentucky bluegrass to shade, thereby reducing its ET rate in full sun (Feldhake, Butler, and Danielson, 1985). Shade-grown turf moved to full sun exhibited ET rates similar to those of turf that had been exposed to full sun for the duration of the study.

As ambient relative humidity decreases, ET increases (Beard, 1973). The rate of water loss is dependent upon the vapor pressure gradient between the stomatal cavity and the atmosphere. For example, if the vapor pressure within the plant is high, water loss will be greater when the atmospheric relative humidity is low than when it is high. In arid and semi-arid regions, relative humidity may approach 10% or less during summer months. This, coupled with temperatures greater than 32°C, results in water use rates that are much greater than might be observed in more humid regions.

Wind increases turfgrass ET by disturbing the leaf boundary layer, thereby allowing greater water loss to occur from the leaf to the atmosphere. Grace (1974) and Grace and Russell (1977) conducted studies to evaluate the influence of wind on anatomy and water relations of tall fescue. Grace (1974) found that the transpiration rate of 'S170' tall fescue increased as wind speed above the sward increased from 1 m s^{-1} to 3.5 m s^{-1}. At lower wind speeds, disruption of the boundary layer resulted in increased water loss. At higher wind speeds, however, stomatal and cuticular resistances decreased, accelerating water loss. Leaf collisions at higher wind speeds caused abrasions, which also accelerated water loss through the plant cuticle. Grace and Russell (1977) found that tall fescue plants exposed to wind speeds of 0.5 to 1.0 m s^{-1} developed more and smaller stomata and had much lower osmotic potentials than those not exposed to wind. Wind-exposed plants were less able to restrict water loss than plants exposed to drought. Feldhake (1981) found that 30% of the water lost by 'Merion' Kentucky bluegrass could be attributed to advective forces. Beard, Green, and Sifers (1992) also reported that ET rates were associated closely with wind speed.

Edaphic Factors

Because turfgrass growth is greatly dependent upon existing soil characteristics, ET rates also are affected. Soil properties that influence water movement include temperature, saturated and unsaturated flow characteristics, and moisture content (Carrow, 1985). Feldhake and Boyer (1986) studied the effects of soil temperature on turfgrass water use rates. Two C_3 and C_4 grasses were evaluated at soil temperatures of 13, 21, and 29°C. Generally, all species exhibited higher ET rates with increasing soil temperature. At 13°C, C_4 grasses averaged 30% lower ET rates than C_3 grasses, but, the difference was reduced to only 10% at 29°C.

Other researchers have shown that ET is related directly to available soil water. Biran et al. (1981) found that the water use rates of two cool-season and nine warm-season grasses declined with decreasing soil water. Similar results were obtained in Texas with warm-season turfgrasses (Kim, 1983); ET rates declined as less soil water was available for plant absorption. Soil water content also may influence turfgrass ET by affecting soil temperature. In general, soils with greater water content tend to be cooler. Hence, in some instances, such as early spring when soils are cold and wet, ET rates may be lower in soil with greater water content.

Soil physical condition directly affects turfgrass growth and vigor and, thus, indirectly influences ET. 'Derby' perennial ryegrass grown in a silt loam soil exhibited 21 and 41% reduced ET rates under moderate and heavy soil compaction treatments, respectively. Similar results were obtained with 'Pennfine' perennial ryegrass (Sills and Carrow, 1983) and 'Ram I' Kentucky bluegrass (Agnew and Carrow, 1985).

Little information is available concerning the effect of soil texture on turfgrass ET. 'Merion' Kentucky bluegrass exhibited a 6% higher ET when growing on a sand-peat mixture than on clay during one of two summers in Colorado (Feldhake, Danielson, and Butler, 1983). This may have been due to reduced rooting in the clay soil, which has been observed in other research (Feldhake, 1979).

Essentially no work has been done to evaluate the effect of soil chemical characteristics on turfgrass ET. Soil salt levels may influence water use rates by decreasing the soil osmotic potential and thereby making water less available to the plant. Research is needed in this area.

CULTURAL PRACTICES AFFECTING ET

Research has been conducted to evaluate the influences of mowing, irrigation, fertility, antitranspirants, and plant growth regulators (PGRs) on ET. Several researchers have reported a relationship between turf mowing height and water use. Increasing mowing height increases leaf area, resulting in greater transpiration per unit ground area. High mowing also creates a rough canopy surface, allowing for more turbulent gas exchange between the canopy and bulk air, and reduces boundary layer resistance (Waddington, Carrow, and Shearman, 1992 see p. 448-449). Madison and Hagan (1962) reported that 'Merion' Kentucky bluegrass used less water when mowed at 1.2 cm than at 5 cm in California. Plants cut at the lower height also possessed a more shallow root system. Biran et al. (1981) found increased vigor and water use when the mowing height of 11 turfgrass species was raised from 3 to 6 cm in Israel. Feldhake, Danielson, and Butler (1983) reported that 'Merion' Kentucky bluegrass used 15% more water when cut at 5 cm compared to 2 cm in Colorado. A 6% increase in ET occurred for both annual bluegrass and 'Penncross' creeping bentgrass maintained at 12 mm compared to that at 6 mm mowing height in Colorado (Fry and Butler, 1989).

Despite the fact that ET decreases with mowing height, most researchers agree that a taller cut allows development of a plant that is a more efficient water user (i.e., better turf quality after a given amount of water is transpired), primarily because of a deeper root system. In general, drought resistance is greater when turf is maintained at a taller height (Dernoeden and Butler, 1978).

Mowing frequency and mower blade sharpness also can influence ET. Shearman and Beard (1973) found that the water use of Penncross creeping bentgrass increased by 15% as mowing frequency increased from 1 to 12 times every 14 days. A Kentucky bluegrass blend mowed with a sharp blade used approximately 1.3 times more water than turf cut with a dull blade in Nebraska (Steinegger et al., 1983). Turf vigor was reduced where the dull blade was used, which likely resulted in less water demand.

If soil water content influences turfgrass ET, one would suspect a response to irrigation management. Bermudagrass irrigated at 254, 540, and 808% of class A pan evaporation exhibited ET rates corresponding to 68, 109, and 119%, respectively, of pan evaporation (Kneebone and Pepper, 1984). Doss and Taylor (1970) reported that the average ET rate of coastal bermudagrass was 8 mm day^{-1} on the second day following irrigation; however, the rate decreased to 6 mm day^{-1} at 28 days after watering. In a greenhouse study, Morgan et al. (1966) found that 'Common' bermudagrass used significantly more water when irrigated with 1.2 cm of water three times weekly compared to irrigation done according to tensiometer readings. Silcock and Wilson (1981) compared transpiration rates of *Festuca* spp. during a 2-d watering period following drought stress. Transpiration was greater in plants that had been exposed to drought stress than that of well-watered ones. In addition, water use efficiency was greatest in plants exposed to prolonged water stress (383 mg shoot dry matter 100 g^{-1} H$_2$O) and least in plants suffering intermittent water stress (301 mg shoot dry matter 100 g^{-1} H$_2$O).

Most research investigating the influence of soil fertility on turfgrass ET has focused on nitrogen. Nitrogen applied at a total of 211 and 420 kg ha^{-1} over 2 years increased the ET rate of a mixture of orchardgrass, smooth bromegrass (*Bromus inermis* Leyss.), and creeping red fescue (Krogman, 1967). However, water use efficiency also increased with nitrogen applications. In a greenhouse study, nitrogen-deficient 'Merion' Kentucky bluegrass used 14% less water than turf receiving supplemental nitrogen fertilization (Sills and Carrow, 1983). Nitrogen-deficient turf also has been shown to decline more rapidly in quality when subjected to moisture stress compared to adequately fertilized turf (Feldhake, 1981). Research data concerning the influence of phosphorus and potassium on the water requirement of turfgrass are limited. Potassium increases root production, which may increase the available soil water reservoir and possibly ET (Waddington, Carrow and Shearman, 1992 see p. 451).

Certain measures can be taken to limit transpiration and thereby conserve water. Davenport (1966) evaluated two antitranspirants on creeping red fescue for effectiveness in reducing ET without phytotoxicity. Phenylmercuric acetate (PMA) gave a 20% reduction in transpiration with no growth inhibition. Stahnke (1981) examined five antitranspirants on creeping bentgrass and found that abscisic acid and a monoglycenol ester of decenyl succinic acid (a wetting agent) reduced transpiration by 59% and 26%, respectively.

Plant growth regulators (PGRs) have been evaluated extensively for use in limiting turfgrass growth and reducing mowing requirements. Plant growth regulators that contribute to the development of a short, compact turf have been shown to reduce ET rates in turfgrass. Flurprimidol { $\propto$-(1-methylethyl)-$\propto$-[4-(trifluoro-methoxy) phenyl] 5-pyrimidine-methanol} and mefluidide (N-{2,4-dimethyl-5-[[(trifuloromethyl)-sulfonyl]amino]phenyl}) acetamide reduced ET of Kentucky bluegrass for 35 to 42 days in the field, but turf rebounded with a growth flush and exhibited higher ET than untreated plants (Doyle and Shearman, 1985). Mefluidide, ethephon ((2-chloroethyl) phosphonic acid), and trinexapac-ethyl (4-(cyclopropyl-hydroxy-methylene)-3,5-dioxo-cyclohexanecarboxylic acid ethyl ester) reduced ET of tall fescue over a 6-week period in a greenhouse study (Marcum and Jiang, 1997). Flurprimidol reduced ET of St. Augustinegrass by 11 to 29% (Johns and Beard, 1981). Additional work is needed to determine the effectiveness of PGRs in reducing turf ET.

CONCLUSION

Researchers have made progress in determining the physiology of water uptake and use and the impacts of environment and culture on ET in turfgrasses. Future work likely will be directed at breeding, or genetically engineering grasses to incorporate traits that result in more efficient water use. This might include developing selections that have potential to extract water with extensive root systems or canopy characteristics that minimize ET. As new cultural practices evolve and chemicals are manufactured with potential to increase turfgrass water conservation, these also must be evaluated.

REFERENCES

Agnew, M.L., and R.N. Carrow. (1985). Soil compaction and moisture stress preconditioning in Kentucky bluegrass. I. Soil aeration, water use, and root responses. *Agronomy Journal* 77:872-877.
Al-Khafaf-S., P. Wierenga, M. Philip. (1987). Root efficiency and distribution of barley plants in irrigated and non-irrigated field plots. *Journal of Agriculture and Water Resources Research, Soil and Water Resources* 6:1-14.
Atkins, C.E., R.L. Green, S.I. Sifers, and J.B. Beard. (1991). Evapotranspiration rates of 10 St. Augustinegrass genotypes. *HortScience* 26:1488-1491.
Aurasteh, R.M. (1983). *A model for estimating lawn grass water requirement considering deficit irrigation, shading and application efficiency.* PhD. dissertation, Utah State University, Logan, Utah.
Beard, J.B. (1973). *Turfgrass: Science and Culture.* Englewood Cliffs, New Jersey: Prentice Hall, Inc. 658 pp.
Beard, J.B. (1985). An assessment of water use by turfgrasses. In *Turfgrass Water Conservation*, eds. V.A. Gibeault and S.T. Cockerham. University of California Publication 21405, Riverside, CA pp.45-62.
Beard, J.B., R.L. Green, and S.I. Sifers. (1992). Evapotranspiration and leaf extension rates of 24 well-watered turf-type *Cynodon* genotypes. *HortScience* 27:986-998.
Beard, J.B. (1994). The water-use rate of turfgrasses. *TurfCraft Australia* 39:79-81.
Biran, I., B. Bravado, I. Bushkin-Harav, and E. Rawitz. (1981). Water consumption and growth rate of 11 turfgrasses as affected by mowing height, irrigation frequency, and soil moisture. *Agronomy Journal* 73:85-90.
Blizzard, W.E., and J.S. Boyer. (1980). Comparative resistance of the soil and the plant to water transport. *Plant Physiology* 66:809-814.
Bowman, D.C. and L. Macaulay. (1991). Comparative evapotranspiration rates of tall fescue cultivars. *HortScience* 26:122-123.

Carrow, R.N. (1985). Soil/water relationships in turfgrass. In *Turfgrass Water Conservation,* eds. V.A. Gibeault and S.T. Cockerham. University of California Publication 21405. Riverside, CA pp. 85-102.

Carrow, R.N. (1996). Drought avoidance characteristics of diverse tall fescue cultivars. *Crop Science* 36:371-377.

Danielson, R.E., W.E. Hart, C.M. Feldhake, and P.M. Haw. (1979). *Water requirements for urban lawns.* Colorado completion report to OWRT Project B-035-WYO. Fort Collins CO p. 90.

Davenport, D.C. (1966). Effects of chemical antitranspirants on transpiration and growth of grass. *Journal of Experimental Botany* 18:332-347.

Dernoeden, P.H., and J.D. Butler. (1978). Drought resistance of Kentucky bluegrass cultivars. *HortScience* 13:667-668.

Devries, J.D., J.M. Bennett, S.L. Albrecht, and K.J. Boote. (1989). Water relations, nitrogenase activity and root development of three grain legumes in response to soil water deficits. *Field Crops Research* 21:215-226.

Doss, B.D., and H.M. Taylor. (1970). Evapotranspiration and drainage from the root zone of irrigated coastal bermudagrass (*Cynodon dactylon* (L.) Pers.) on coastal plains soils. *Transactions of the American Society of Agricultural Engineering* 13:426-429.

Doyle, J.M., and R.C. Shearman. (1985). Plant growth regulator effects on evapotranspiration of a Kentucky bluegrass turf. *Agronomy Abstracts* p. 115. Amer. Soc. of Agronomy, Madison, Wisc: meetings held in Chicago, all Oct. 1-6, 1985.

Feldhake, C.M. (1979). *Measuring evapotranspiration of turfgrass.* M.S. thesis, Colorado State Univ., Fort Collins, Colorado. p. 56.

Feldhake, C.M. (1981). *Turfgrass evapotranspiration and microenvironment interaction.* PhD. dissertation, Colorado State University, Fort Collins, Colorado p. 147.

Feldhake, C.M., R.E. Danielson, and J.D. Butler. (1983). Turfgrass evapotranspiration. I. Factors influencing rate in urban environments. *Agronomy Journal* 75:824-830.

Feldhake, C.M., J.D. Butler, and R.E. Danielson. (1985). Turfgrass evapotranspiration. Responses to shade preconditioning. *Irrigation Science* 6:265-270.

Feldhake, C.M., and D.G. Boyer. (1986). Effect of soil temperature on evapotranspiration by C_3 and C_4 grasses. *Agricultural and Forest Meteorology* 37:309-318.

Fiscus, E.L., A. Klute, and M.R. Kaufmann. (1983). An interpretation of some whole plant water transport phenomena. *Plant Physiology* 71:810-817.

Fry, J.D., and J.D. Butler. (1989). Annual bluegrass and creeping bentgrass evapotranspiration rates. *HortScience* 24:268-271.

Fukai, S., and M. Cooper. (1995). Development of drought-resistant cultivars using physio-morphological traits in rice. *Field Crops Research* 40:67-86.

Grace, J. (1974). The effect of wind on grasses. I. Cuticular and stomatal transpiration. *Journal of Experimental Botany* 25:542-551.

Grace, J., and G. Russell. (1977). The effect of wind on grasses. III. Influence of continuous drought or wind on anatomy and water relations in *Festuca arundinacea* Schreb. *Journal of Experimental Botany* 28:268-278.

Green, R.L., S.I. Sifers, C.E. Atkins, and J.B. Beard. (1991). Evapotranspiration rates of eleven zoysiagrass genotypes. *HortScience* 26:264-266.

Gregory, J.P., M. McGowan, P.V. Biscoe, and B. Hunter. (1978). Water relations of winter wheat. I. Growth of the root system. *Journal of Agricultural Science* 91:91-102.

Huang, B. (1999). Water relations and root activities of *Buchloe dactyloidaes* and *Zoysia japonica* in response to localized soil drying. *Plant and Soil* 28:179-186.

Huang, B., and P.S. Nobel. (1994). Root hydraulic conductivity and its components, with emphasis on desert succulents. *Agronomy Journal* 86:767-774.

Huang, B., R.R. Duncan, and R.N. Carrow. (1997). Drought-resistance mechanisms of seven warm-season turfgrasses under surface soil drying: I. Shoot response. *Crop Science* 37:1858-1863.

Hurd, E.A. (1974). Phenotype and drought tolerance in wheat. *Agricultural Meteorology* 14:39-55.

Johns, D., and J.B. Beard. (1981). Reducing turfgrass transpiration using a growth inhibitor. *Agronomy Abstracts* p. 126. Amer. Soc. of Agronomy, Madison, Wisc., meetings held in Atlanta, Georgia, Nov. 29-Oct. 4, 1981.

Johns, D., J.B. Beard, and C.H.M. van Bavel. (1983). Resistance to evaporation from a St. Augustinegrass turf canopy. *Agronomy Journal* 75:419-422.

Kim, K.S. (1983). *Comparative evapotranspiration rates of thirteen turfgrasses grown under both non-limiting soil moisture and progressive water stress conditions*. M.S. thesis, Texas A & M University, College Station, Texas. p. 64.

Kim, K.S. (1987). *Comparative drought resistance mechanisms of eleven major warm-season turf grasses*. PhD. dissertation, Texas A&M University, College Station (Dissertation Abstract 87-20914).

Kim, K.S., and J.B. Beard. (1988). Comparative turfgrass evapotranspiration rates and associated plant morphological characteristics. *Crop Science* 28:328-331.

Kneebone, W.R., and I.L. Pepper. (1982). Consumptive water use by sub-irrigated turfgrasses under desert conditions. *Agronomy Journal* 74:419-423.

Kneebone, W.R., and I.L. Pepper. (1984). Luxury water use by bermudagrass turf. *Agronomy Journal* 76:999-1002.

Kopec, D.M., and R.C. Shearman. (1987). Evapotranspiration of tall fescue turf. *HortScience* 23:300-301.

Kramer, P.J. (1983). *Water relations of plants*. Orlando, Florida: Academic Press. 489 pp.

Krogman, K.K. (1967). Evapotranspiration by irrigated grass as related to fertilizer. *Canadian Journal of Plant Science* 47:284-287.

Madison, J.H., and R.M. Hagan. (1962). Extraction of soil moisture by 'Merion' bluegrass (*Poa pratensis* 'Merion') turf, as affected by irrigation frequency, mowing height, and other cultural operations. *Agronomy Journal* 54:157-160.

Marcum, K., and H. Jiang. (1997). Effects of plant growth regulators on tall fescue rooting and water use. *Journal of Turfgrass Management* 2:13-27.

Marlatt, W.E. (1961). *The interactions of microclimate, plant cover, and soil moisture content affecting evapotranspiration rates*. Atmospheric Science Technology Paper No. 23, Colorado State University, Fort Collins, Colorado. p. 8.

Mason, W.K., W.S. Meyer, R.C.G. Smith, and H.D. Barrs. (1983). Water balance of three irrigated crops on fine-textured soils of the Riverine plain. *Australian Journal of Agricultural Research* 34:183-91.

Meyer, W.S. (1976). *Seminal roots of wheat: Manipulation of their geometry and to improve the efficiency of water use.* PhD. dissertation, University of Adelaide, Australia.

Meyer, W.S., E.L. Greacen, and A.M. Alston. (1978). Resistance to water flow in the seminal roots of wheat. *Journal of Experimental Botany* 29:1451-1461.

Minner, D.D. (1984). *Cool season turfgrass quality as related to evapotranspiration and drought.* PhD. dissertation. Colorado State University, Fort Collins, Colorado. p. 133.

Morgan, W.C., J. Letey, S.J. Richards, and N. Valoras. (1966). Physical soil amendments, soil compaction, irrigation, and wetting agents in turfgrass management. I. Effects on compactibility, water infiltration rates, evapotranspiration, and number of irrigation. *Agronomy Journal* 58:525-535.

Narayan, D. (1991). Root growth and productivity of wheat cultivars under different soil moisture conditions. *International Journal of Ecological and Environmental Sciences* 17:19-26.

Nobel, P.S. (1991). *Physicochemical and environmental plant physiology.* San Diego, California: Academic Press. 635 pp.

Oke, T.R. (1979). Advectively-assisted evapotranspiration from irrigated urban vegetation. *Boundary Layer Meteorology* 17:167-173.

Passioura, J.B. (1988). Root signals control leaf expansion in wheat seedlings growing in drying soil. *Australian Journal of Plant Physiology* 15:687-693.

Qian, Y.L., J.D. Fry, and W.S. Upham. (1997). Rooting and drought avoidance of warm-season turfgrasses and tall fescue in Kansas. *Crop Science* 37:905-910.

Richards, R.A., and J.B. Passioura. (1981). Seminal root morphology and water use of wheat. II. Genetic variation. *Crop Science* 21:253-255.

Schmidt, R.E., and L.A. Everton. (1985). Moisture consumption of Kentucky bluegrass (*Poa pratensis* L.) cultivars. pp. 373-379. In *Proceedings 5th International Turfgrass Research Conference,* ed. F. Lemaire, Avignon, France: Inst. Natl. de la Recherche Agron. pp. 373-379.

Sharp, R.E., and W.J. Davies. (1985). Root growth and water uptake by maize plants in drying soil. *Journal of Experimental Botany* 36:1441-1456.

Shearman, R.C. (1986). Kentucky bluegrass cultivar evapotranspiration rates. *HortScience* 21:455-457.

Shearman, R.C. (1989). Perennial ryegrass cultivar evapotranspiration rates. *HortScience* 24:767-769.

Shearman, R.C., and J.B. Beard. (1973). Environmental and cultural preconditioning effects on the water use rate of *Agrostis palustris* Huds., cultivar Penncross. *Crop Science* 13:424-427.

Shein-E.V., and Ya.A. Pachepsky. (1995). Influence of root density on the critical soil water potential. *Plant and Soil* 171:351-357.

Silcock, R.G., and D. Wilson. (1981). Effect of watering regime on yield, water use, and leaf conductance of seven *Festuca* species with contrasting leaf ridging. *New Phytologist* 89:569-580.

Sills, M.J., and R.N. Carrow. (1983). Turfgrass growth, N use, and water use under soil compaction and N fertilization. *Agronomy Journal* 75:488-492.

Stahnke, G.K. (1981). *Evaluation of antitranspirants on creeping bentgrass (Agrostis palustris Huds., cv. 'Penncross') and bermudagrass [Cynodon dactylon (L.) Pers.* ×

Cynodon transvaalensis Burtt-Davy, cv. 'Tifway']. M.S. thesis, Texas A & M University, College Station, Texas. p. 70.

Steinegger, D.H., R.C. Shearman, T.P. Riordan, and E.J. Kinbacher. (1983). Mower blade sharpness effects on turf. *Agronomy Journal* 75:479-480.

Taylor, H.M. (1980). *Modifying root systems of cotton and soybean to increase water absorption.* USDA-SEA-AR, Soil & Water Management Unit, Iowa State University, Ames, Iowa.

Tovey, R., J.S. Spencer, and D.C. Muckel. (1969). Water requirements of lawngrass. *Transactions of the American Society of Agricultural Engineering* 12:356-358.

Volaire, F., and H. Thomas. (1995). Effects of drought on water relations, mineral uptake, water-soluble carbohydrate accumulation and survival of two contrasting populations of cocksfoot (*Dactylis glomerata* L.). *Annals of Botany* 5:513-524.

Waddington, D.V., R.N. Carrow, and R.C. Shearman. (eds). (1992). *Turfgrass.* Madison, Wisconsin: American Society of Agronomy, Inc. Crop Science Society of America, Inc., and Soil Science Society of America, Inc. 805 pp.

Youngner, V.B. (1985). Physiology of water use and water stress. In *Turfgrass Water Conservation,* eds. V.A. Gibeault and S.T. Cockerham, University of California Publication 21405. Riverside, CA, pp. 37-43.

Wheat Yield and Water Use as Affected by Micro-Basins and Weed Control on the Sloping Lands of Morocco

Ahmed Bouaziz
Hassan Chekli

SUMMARY. Traditionally in Morocco the wooden plow was the most widely used implement for soil tillage, especially on sloping lands. Mechanization was introduced on flat areas and was extended more and more toward sloping lands, so the tillage accentuated the problems of water runoff and soil erosion. The objective of this study was to determine the effect of micro-basins on runoff, water storage by increasing infiltration, and wheat yield. In addition, the interaction with weeding was tested to reduce water losses by evapotranspiration from weeds. Four treatments were used, as follows: BW (micro-basins, weeded), NBW (no micro-basins, weeded), BNW (micro-basins, not weeded) and NBNW (no micro-basins, not weeded). Field experiments were carried out during three years using a split plot design with four replications: two at Meknès in 1990-91 and 1991-92 and a third at Khémisset (1992-93). Slopes were 2.5% for Meknès and 3.5% for Khémisset. Rainfall received was 523.3, 483.7, and 473.0 mm, respectively, for the 1990-91, 1991-92, and 1993-94 growing seasons.

The results showed that the effect of micro-basins in interaction with weeding was statistically significant in 1990-91 and 1993-94 for soil water storage, grain yield, spikes/m^2, kernels/m^2, thousand kernel

Ahmed Bouaziz is Professor at the Agronomy and Plant Breeding Department, IAV Hassan II, BP 6202 Rabat-Institute, Rabat, Morocco, and Hassan Chekli is Professor at the Machinery Department, National School of Agriculture, Meknès, Morocco.

Address correspondence to Ahmed Bouaziz at the above address.

[Haworth co-indexing entry note]: "Wheat Yield and Water Use as Affected by Micro-Basins and Weed Control on the Sloping Lands of Morocco." Bouaziz, Ahmed, and Hassan Chekli. Co-published simultaneously in *Journal of Crop Production* (Food Products Press, an imprint of The Haworth Press, Inc.) Vol. 2, No. 2 (#4), 1999, pp. 335-351; and: *Water Use in Crop Production* (ed: M. B. Kirkham) Food Products Press, an imprint of The Haworth Press, Inc., 1999, pp. 335-351. Single or multiple copies of this article are available for a fee from The Haworth Document Delivery Service [1-800-342-9678, 9:00 a.m. - 5:00 p.m. (EST). E-mail address: getinfo@haworthpressinc.com].

335

weight, and water use efficiency. In 1991-92, yield increases were not significant, because of a drought that lasted almost two months after emergence. We conclude that these technical management practices are especially important in rainy years when rainfall intensities are high and cause water runoff. *[Article copies available for a fee from The Haworth Document Delivery Service: 1-800-342-9678. E-mail address: getinfo@ haworthpressinc.com <Website: http://www.haworthpressinc.com>]*

KEYWORDS. Wheat, weeding, micro-basins, water runoff, sloping land

INTRODUCTION

Crop production, in general, and rainfed cereal crops, in particular, are characterized in arid and semi-arid regions by high, yearly fluctuations of yield. However, several studies have showed that yield improvement and maintenance of productivity at acceptable levels cannot be achieved without using adequate soil management techniques. These practices must permit better management of precipitation to limit losses, mainly from surface runoff and weed evapotranspiration. Morin and Benyamini (1988) showed that losses from water runoff can reach 30 to 50% of precipitation in certain situations. These losses from surface runoff depend especially on rainfall intensity and the physical characteristics of the land, especially degree of sloping, infiltration rate, structure of the seed bed, and roughness of the soil surface (El Haiba, 1979; Gerard et al., 1984; Lyle and Dixon, 1977). However, competition between wheat and weeds for water is often considered to be the most important factor for determining yield in rainfed agriculture (Radosevich and Holt, 1984). Thus, Zimdhal (1980) reported that reduction in wheat yield by weeds can reach 63% in the semi-arid zones of Morocco, and losses can be even more important in arid zones (Tanji, 1987).

Soil tillage techniques and weed control are considered among the best means to conserve water and soil. Papendick and Campbell (1988) stated that the sweep is one of the most desirable tools for soil management in semi-arid lands. It acts at a depth of 5 to 6 cm without turning over soil, which permits reduction of evaporation and surface runoff. It is generally used after the first rains of a season and after emergence of adventitious plants. Unger and Stewart (1983), however, reported that plowing with a mould-board plow or disks reduced runoff by creating a rough surface and numerous, unconnected depressions that temporarily store flowing water. In contrast, El Baghati (1977) in Morocco underlined the importance of using a chisel cultivator in arid and semi-arid zones for water economy. Ouattar and Ameziane (1989) stated that the chisel is recommended for arid and semi-arid zones. Several other authors reported the importance of diking, which, according to them, permits retention of rainwater on the soil surface until it infiltrates, especially on sloping land (Morin and Benyamini, 1988; Unger, 1983; Unger and Stewart, 1983; Williams et al., 1988; Krishna and Gerik, 1988). In Morocco, this technique is still in its experimental stage in the wheat-growing region of Meknès. Several authors have showed that this technique reduces or prevents runoff, if micro-basins are installed before intense and powerful rainfalls (Clark and Hudspeth, 1976; Gerard et al., 1984, cited by McFarland et al., 1991). Thus, McFarland et al. (1991) reported that this diking system improved soil water storage from 29 to 45 mm at the 150 cm depth, when compared with the control treatment.

The objective of this study was to determine the effect of making basins, in combination with chemical weeding, on water storage, yield, yield components, and water use on sloping land. These two techniques modify the crop water balance and can improve yield and water use efficiency in rainfed areas, especially on sloping land.

MATERIALS AND METHODS

We summarize data from three experiments that we conducted in Morocco using the basin technique in interaction with chemical weeding. Three experiments with a split plot design were conducted in the field: two at the experimental field site of the National School of Agriculture (ENA) carried out in 1990-91 and 1991-92 and a third on a farm of the State Society for Land Management (SOGETA) located in the Khémisset region, close to the Beht River, and done in 1993-94. For each of these experiments, we had the following four treatments: BW (micro-basins, weeded), NBW (no micro-basins, weeded), BNW (micro-basins, not weeded) and NBNW (no micro-basins, not weeded).

The durum wheat (*Triticum durum* Desf.) variety used for these experiments was 'Oum Rbiaa' with a 98% germination rate and a weight of 40.8 g for 1000 kernels. Land slopes were 2.5% and 3.5%, respectively, at Meknès and Khémisset. Soil at the first experimental site (Meknès) had more clay, silt, organic matter, and total calcareous material (Table 1). At the second experimental location, the clay content in the soil increased with depth.

Observations and measurements of parameters related to the water balance, especially the runoff coefficient, soil water storage, actual evapotranspiration, and water use efficiency, were made. To evaluate the runoff coefficient, measured soil moisture before and after rainfall was used for rain sequences where data were available. The runoff coefficient was calculated from the amount of rainfall and variation in water storage twice a year. For

TABLE 1. Soil characteristics for the 2 locations.

Location	Growing season	Soil layer (cm)	Clay (%)	Silt (%)	Sand (%)	Organic matter (%)	Total calcareous (%)
Meknés	1991-92 and 1992-93	0-90	56.2	30.0	13.8	1.6	50.2
Khémisset	1993-94	0-30	13.0	44.0	43.0	1.4	40.1
		30-60	22.6	50.3	27.1	1.2	35.2
		60-90	37.7	49.1	13.2	0.8	33.1

soil water storage, samples were taken from different treatments and blocks and oven dried at 105°C for 24 hours and moisture was calculated. Measurements also were related to crop parameters, especially grain yield and its components as well as straw yield.

RESULTS AND DISCUSSION

Climate During the Three Growing Seasons

Rainfall and temperature for the 1990-91, 1991-92, and 1993-94 growing seasons are illustrated in Figures 1, 2, and 3. They were as follows:

- For 1990-91, the total rain was 523.3 mm. It was less than 577.2 mm, which is the calculated average over 31 years (1956-86) (Ben Lhamdani, 1991). This growing season was characterized by two rainy periods: the first during the month of December with 109.7 mm and the second ranging from early February until the end of March with 173.2 mm, which is more than 50% of the rainfall for the whole year (Figure 1). The thermal regime was characterized by a weak thermal amplitude with a daily average temperature ranging between an absolute minimum of about 2°C in February and a maximum of about 34°C in September.
- For 1991-92, the total rain of the season was 483.7 mm, which was less than 577.2 mm, the average of the region (Ben Lhamdani, 1991). This season was characterized, in general, by low rainfall with an irregular distribution, a long period of drought (from the second half of December until the end of the second decade of February), and three rainy periods: the first from September to November with 141 mm, the second in March and April with 176 mm, and the last one at the end of the season in May and June with 98.5 mm (Figure 2). The thermal regime was

characterized by a strong thermal amplitude, and the average temperature ranged between 8.2°C in November and 25.2°C in September.

- For 1993-94, the recorded total rainfall was 473 mm, higher than 424 mm, which is the calculated average for Khémisset over 30 years (1933-1963). Rainfall was characterized by a good distribution during the growing season, except at the end when a water deficit occurred. Thus, 157.6 mm was recorded at the beginning of the season (October and November) and 100.9 mm in February. The last months of the season (March, April, and May) only received 9.5% of the total rainfall (Figure 3). The thermal regime was characterized by a strong thermal amplitude, and the average temperature ranged between 12.6°C in January and 20.7°C in October.

Runoff and Soil Water Storage

Runoff

For the 1990-91 experiment, runoff coefficients of treatments without the micro-basins were higher than those with micro-basins (Table 2). For the 1991-92 experiment, because of the drought that characterized the season, quantities of water that flowed overland were very low, at least during the crop growing season, and runoff coefficients for all treatments were less than 3% of the rainfall (Table 2). Thus, rain was not violent and its intensity was probably lower than the soil water infiltration rate, even with the 2.5% slope; the effect of micro-basins on the flowing quantities of water did not have a significant effect.

For the 1993-94 experiment, water runoff was low and not more than 7%

FIGURE 1. Rainfall (bars) and maximum (dot symbol), minimum (+ symbol), and average (star symbol) temperatures for the growing season 1990-1991.

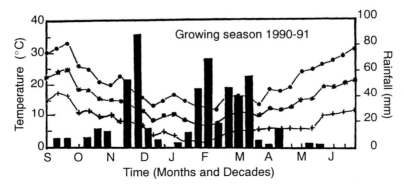

FIGURE 2. Rainfall (bars) and maximum (dot symbol), minimum (+ symbol), and average (star symbol) temperatures for the growing season 1991-1992.

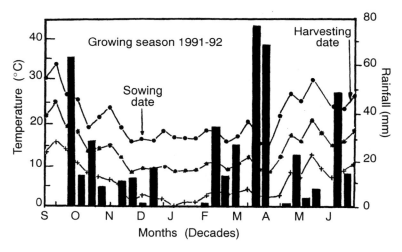

FIGURE 3. Rainfall (bars) and maximum (dot symbols), minimum (+ symbol), and average (star symbol) temperatures for the growing season 1993-1994.

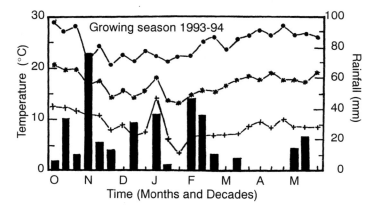

for the two rainfall sequences considered (17.8 mm and 12.0 mm). However, there was a tendency for the treatment without micro-basins to have relatively more runoff (Table 2). The low rainfall intensities did not surpass the soil water infiltration for either treatment. Field observations after rains showed signs of water retention by micro-basins.

TABLE 2. Runoff coefficient (%).

Treatment	Growing season and amounts of rainfall/sequence considered (mm)					
	1990-91		1991-92		1993-94	
	33.0	12.0	25.0	24.5	17.8	12.0
Basins	6.4	4.2	0.4	2.4	2.2	3.5
No basins	12.7	12.5	0.4	2.8	4.6	6.5

Chemical Weeding

In 1993-94, the adventitious flora was especially dominated by monocots, notably bromus and phalaris (Table 3). Dicots were dominated by *Chrysanthemum coronarium, Chenopodium mural*, and *Sinapsis arvensis*. Weeded treatments were cleaner at the heading stage than those not weeded.

Soil Water Reserve

For the 1990-91 experiment, analysis of soil water storage at seedling, heading, and dough stages (44, 93, and 140 days after planting) showed that micro-basins permitted more water to be stored (Figure 4). This difference in water reserve was due to water losses from runoff that were higher in treatments without micro-basins. In the same way, as with the absence or presence of micro-basins, the weeded situation was superior to the one not weeded because of high weed competition for water. However, in the 1991-92 experiment, there was no significant effect of the micro-basins, weeding, and their interaction on soil water reserve at any stage (seedling, heading or dough) (Table 4).

This result is due to the low runoff because of the low rainfall intensities, the low degree of slope (2.5%), and the fine structure of the surface. The effect of weeding was not significant, because of low rainfall and winter drought.

For the 1993-94 experiment, the effects of micro-basins and weeding on soil water reserve were highly significant at shoot elongation and heading stages (Table 5 and Figure 5). Thus, weeding and micro-basins procured the best soil water reserve that permitted gains of 7.9% and 9.4%, respectively. In the same way, at the heading stage, a significant effect was noticed; gains in water were, respectively, about 15.3% and 6.6% for weeding and the use of micro-basins. At the dough stage, there was no significant effect of the studied factors, but a tendency for micro-basins to improve soil water reserves was observed.

TABLE 3. Importance of adventitious species.

Species	Elongation stage				Heading stage			
	CD	SCD	CND	SCND	CD	SCD	CND	SCND
Monocotyledons	90	92	74	75	6	6	162	158
Bromus sp.	24	21	14	17	1	1	25	32
Phalaris paradoxa L.	3	3	9	1	1	0	12	11
Lolium multiflorum L.	0	0	0	1	0	0	1	1
Sub-Total	**117**	**116**	**98**	**103**	**8**	**7**	**200**	**202**
Dicotyledons								
Silybum marianum L.	1	0	2	2	0	0	2	2
Malva parvilflora L.	1	0	0	4	0	0	9	11
Sinapsis arvensis L.	0	0	13	11	0	0	17	15
Papaver rhoeas L.	1	1	7	8	1	1	9	10
Astragalus boecticus L.	0	1	7	8	0	0	9	8
Capsella busa-pastoris L.	0	0	4	3	0	0	5	5
Chrysanthemum coronarium L.	1	1	28	24	0	0	32	35
Cicorium endivia L.	1	1	3	2	0	0	4	4
Diplotaxis tenuisilique D.	1	0	4	4	0	0	5	6
Arisarum vulgare	0	1	2	3	0	0	2	2
Fumaria parvilfora L.	1	0	7	5	0	0	9	8
Glaucium corniculatum L.	1	1	3	4	0	0	4	4
Chenopodium mural L.	0	1	14	19	0	1	19	22
Polygonum aviculare L.	0	1	6	7	0	0	19	10
Convolvulus arvensis L.	1	0	8	9	0	0	16	18
Veronica agrestis L.	1	1	6	8	0	0	7	7
Reseda lutea L.	0	0	8	9	0	0	11	9
Medicago ciliaris L.	0	1	4	7	0	0	12	14
Sub-Total	**10**	**9**	**136**	**137**	**1**	**2**	**191**	**190**
Total	**127**	**125**	**234**	**240**	**9**	**9**	**391**	**392**

Yield, Evapotranspiration and Water Use Efficiency

Grain Yield

The results of the experiment conducted during the first season (1990-91) showed that grain yield improved due to the soil micro-basins and weed control (Table 6).

Thus, the micro-basin technique achieved a superior grain yield compared

FIGURE 4. Soil water storage (mm) in 1990-1991. TS: Tillering stage; SE: Stem elongation stage; H: Heading; A: Anthesis; M: Maturity.

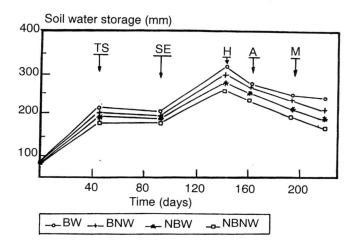

TABLE 4. Soil water reserves for 1991/92 growing season (mm).

Stage	Treatments			
	BW	BNW	NBW	NBNW
Seedling	144.3	144.3	142.7	142.7
Heading	202.7	201.9	201.9	201.9
Maturity	147.8	148.4	150.8	146.7

TABLE 5. Soil water reserves (mm) for 1993/94 growing season.

Stage	Treatments* and % gains					
	W	NW	% Gain	B	NB	% Gain
Elongation	184.5 a	170.9 b	7.9	185.7 a	169.7 b	9.4
Heading	150.4 a	130.5 b	15.3	144.9 a	135.9 b	6.6
Dough	113.5 a	111.5 a	1.8	100.5 a	97.5 a	3.8

*Reserves with a and b are significantly different.

to the one without micro-basins. Also, plant weeds caused considerable losses in all treatments. The weeded treatment with micro-basins always had better water conservation compared to the one without micro-basins and not weeded.

For the 1991-92 experiment, grain yield obtained from the different treat-

FIGURE 5. Soil water storage (mm) in 1993-1994. TS: Tillering stage; SE: Stem elongation stage; H: Heading; A: Anthesis; M: Maturity.

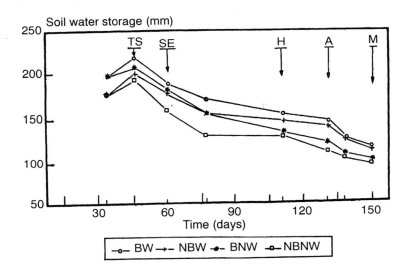

TABLE 6. Obtained yields and harvest index.

Treatments	Grain yield (kg/ha)			Straw yield (kg/ha)			Harvest Index		
	1990 -91	1991 -92	1993 -94	1990 -91	1991 -92	1993 -94	1990 -91	1991 -92	1993 -94
BW	1910	240	3030	3660	1480	6280	0.34	0.14	0.33
BNW	1390	240	1890	3040	1440	5740	0.31	0.14	0.25
NBW	1620	220	2750	3430	1210	6030	0.32	0.15	0.31
NBNW	1160	240	1100	2800	1130	4270	0.29	0.18	0.20

ments was, in general, low and was not greater than 240 kg/ha, on average. The main reason for this small yield was because most crop developmental stages occurred under unfavorable conditions of water stress, due to the long period of drought that characterized the growing season. Thus, the effect of micro-basins, prepared at sowing, was not observed; the reason was probably the absence of runoff that would have affected water storage. In addition, the effect of weeds was not statistically significant because of the bad climatic conditions.

For the 1993-94 experiment, grain yield varied between 1100 and 3030 kg/ha from one treatment to another. The statistical analysis revealed a significant difference between treatments. Thus, the weeded treatments with or without micro-basins gave higher grain yield than non-weeded treatments. The yield was lowest on the treatment that was not weeded and had no micro-basins. Weeding permitted a better utilization of water than micro-basins. The competition for water started at the tillering stage (37 days after sowing) and affected the number of spikes/m^2 and the number of grains/m^2 and, thereafter, grain yield. Gains in yield were 175, 150, and 71%, respectively, for the BW, NBW, and BNW treatments. These results are similar to those reported by Saffour (1992) in the Meknès region.

Straw Yield

In the 1990-91 experiment, differences among treatments in straw yield were very highly significant. Mean comparison showed that treatments with micro-basins achieved higher harvest indexes than treatments without micro-basins (Table 6). Similarly, as far as weeding was concerned, the analysis of the variance showed a very highly significant difference between the weeded and non-weeded treatments. The positive effects of micro-basins and weeding are explained, on one hand, by micro-basins that permitted higher water storage, and, on the other hand, by elimination of weeds that reduced or eliminated the competition between the crop and adventitious plants for water. However, during the 1991-92 growing season, because of the long period of drought to which the crop was submitted, the analysis of variance revealed no significant effect on straw yield due to the studied factors. Values obtained during this experiment were, otherwise, very low in relation to those obtained in the first experiment. In the third experiment, we noted a very highly significant effect of the studied factors. The mean ranking revealed, once again, the superiority of the treatment with micro-basins and no weeds (BW) followed by the treatment that was weeded but had no micro-basins (NBW) and then by the non-weeded treatment with micro-basins (BNW), and, finally, by the non-weeded treatment without micro-basins (NBNW).

Actual Evapotranspiration (ETR)

For the 1990-91 experiment, the analysis of variance and the mean comparison showed that the treatment without micro-basins had several low ETR's. The reason is due to the reduction in water loss from runoff and to the low soil water storage. Water consumption (ETR) was higher in the presence of micro-basins (Table 7). Yadav and Singh (1981) and Haudoufe (1988) found similar results. As far as weeds are concerned, Table 7 shows that the ETR's are high in non-weeded situations. This is due to the strong contribution of the adventitious plants to the total transpiration of the plant population cover.

TABLE 7. Actual evapotranspiration (ETR) and water use efficiencies (WUE).

Treatments	ETR (mm)			WUE (Kg of grain/mm)			WUE (Kg of straw/mm)		
	1990-91	1991-92	1993-94	1990-91	1991-92	1993-94	1990-91	1991-92	1993-94
BW	315.0	342.6	246.6	6.1	0.7	12.3	11.6	4.3	25.5
BNW	321.0	344.0	225.9	4.3	0.7	11.3	9.5	4.2	25.4
NBW	302.0	335.5	242.6	5.4	0.7	8.4	11.4	3.6	24.9
NBNW	312.0	334.6	229.7	3.7	0.7	5.1	8.9	3.4	18.6

Similarly, for the 1991-92 experiment, the analysis of variance revealed a significant effect of micro-basins on the actual ETR. The effect of controlling the adventitious plants was not evident because chemical weeding was not effective.

For the 1993-94 experiment, the analysis of variance showed a highly significant effect of weeding on actual ETR of the crop. Mean ranking showed that the ETR was very important for the treated cases. The mean ETR was 232.9 mm and was lower than that obtained in the first experiment.

Water Use Efficiency

For the 1990-91 experiment, the analysis of variance showed a very highly significant effect of micro-basins and weeding on WUE. Mean ranking revealed the superiority of the treatments with micro-basins compared to the ones without them (Table 7), and it was essentially due to the high soil water storage generated by micro-basins. Similarly, both in the presence and absence of micro-basins, WUE was always low in non-weeded situations. This difference was due to the contribution of weeds to the water use through a strong transpiration. Weeds were an important part of the plant population cover.

For the 1991-92 experiment, in contrast to the 1990-91 results, we did not notice an effect of micro-basins on WUE of either the grain or straw yields, because of the weak intensity of rains that did not generate runoff. The effect of adventitious plants also was not noticed. The values found were lower than those of the first experiment.

For the 1993-94 experiment, a highly significant effect was observed for the studied factors on WUE of both grain and straw yields. Mean ranking showed the superiority of weeded treatments. However, the WUE was low in the non-weeded situation without micro-basins. The WUE of the non-weeded treatment with micro-basins occupied an intermediate position. The results showed again that the treatment with the best storage of water permitted a high WUE.

Grain Yield Components

Plant Population

For the 1991-92 experiment, the drought had a negative effect on the plant population. Indeed, the lack of water and the competition between wheat plants, on the one hand, and the weeds on other hand, generated a very strong reduction in plant population for the different treatments (Table 8). Plant populations were 156 and 157 plants/m^2 for the cases with and without micro-basins, respectively.

In the same way, in the 1993-94 experiment, the drought that occurred during the end of the growing season reduced the plant population. The reduction was stronger for the control treatment (non-weeded one without micro-basins). The non-weeded treatments were heavily infested with *Bromus* sp., which eliminated many wheat plants through competition. These reductions of plant populations were 18.4, 33.3, 38.2, and 136.7%, respectively, for the BW, NBW, BNW, and NBNW treatments.

Spike Population

For the 1990-91 experiment, the analysis of variance showed that micro-basins and weeding had a highly significant effect on the number of ears/m^2. Ear populations were higher in the treatments with micro-basins than in those

TABLE 8. Yield components.

Yield components	Growing season	Treatments			
		BW	BNW	NBW	NBNW
Number of plants/m^2	90-91	130	130	130	130
	91-92	156	156	157	157
	93-94*	171	145	165	93
Spikes/m^2	90-91	195	140	165	105
	91-92	179	173	165	160
	93-94	215	173	209	115
Kernels/spike	90-91	38.0	37.0	35.0	38.1
	91-92	12.0	12.9	10.4	9.9
	93-94	39.6	38.8	39.5	39.2
Thousand Kernel Weight (g)	90-91	42.7	36.8	40.5	33.0
	91-92	34.8	33.4	32.6	33.4
	93-94	37.0	29.4	34.4	25.9

*Records at harvesting.

without micro-basins (Table 8). In the same way, plots treated for weeds, either with or without micro-basins, resulted in a higher spike population than those not treated. The BW treatment gave the best spike population (195 spikes/m^2).

For the 1991-92 experiment, no significant difference was noted between treatments with or without micro-basins. Similarly, weeding did not have a significant effect on spike population because of the failure of chemical weeding. The interaction between the two factors was not significant. However, in 1993-94, the studied factors had a very highly significant effect on the spike population. Mean ranking revealed that the weeded situations (BW and NBW) gave the best spike populations followed by the non-weeded situation with micro-basins, and, finally, the lowest spike population occurred in the control treatment, which was not weeded and had no micro-basins. We noted that treatments that had high plant populations also had high spike populations.

Number of Kernels Per Spike

For the 1990-91 experiment, we observed a highly significant effect of micro-basins and weeding on the number of kernels/m^2. Variability in this component is explained mainly by the number of spikes/m^2, because the number of kernels/spike is almost always nearly constant for a given climatic year. This is why the mean ranking was essentially the same as the one for spike populations. Thus, we noted a superiority of weeded treatments compared to those not weeded because of weed competition.

For the 1991-92 experiment, the analysis of variance did not reveal a significant effect of micro-basins, weeding, or their interaction on the number of kernels/m^2 or on the number of kernels/spike. The number of kernels/m^2 was determined, in general, by the number of spikes/m^2. The absence of a significant effect of weeding was always explained by the failure of chemical weeding due to the drought.

For the 1993-94 experiment, a superiority of weeded treatments was noted in relation to those not weeded for the number of kernels/m^2. The non-weeded treatment without micro-basins had the lowest number of kernels/m^2. We always observed that the number of kernels/spike was nearly constant from one treatment to the other (Table 8).

Thousand Kernel Weight

In the first experiment (1990-91), the statistical analysis revealed the existence of a very highly significant effect of micro-basins in combination with weeding (Table 8). Thus, the Newman and Keuls mean ranking showed that treatments with micro-basins had a higher 1000 kernel weight (TKW) than treatments without micro-basins. Similar results were observed for the weeded and non-weeded cases. Micro-basins, by limiting runoff and by collecting flowing water locally, permitted the preservation of more water. Even without micro-basins, the effect of the weed control was positive and very highly significant. The non-weeded treatment without micro-basins had the lowest TKW (33 g).

For the 1991-92 experiment, in contrast to the first experiment, there was no significant effect of the micro-basins or of weeding. This is mainly explained by shriveling, which occurred at the end of the crop cycle, as well as by the occurrence of water deficit that coincided with the grain filling period.

For the 1993-94 experiment, there was a highly significant effect of weeding on the 1000 kernel weight. Mean comparisons showed the superiority of weeded treatments in relation to those not weeded with a 22.6% difference. Similarly, micro-basins exercised a significant effect on the TKW and permitted a gain of 9.2%; however, we noted the absence of an interaction between the two factors.

SYNTHESIS AND CONCLUSION

The main objective of this work was to study the effect of micro-basins in combination with weeding on rainwater conservation and on the improvement of WUE of cultivated wheat on sloping lands. Main results are summarized as follows:

Concerning soil water storage, micro-basins generated a positive effect on soil water reserve. Similarly, weed control permitted maintenance of the soil water reserve at higher levels than non-weeded treatments. The effect of micro-basins and weeding on water conservation was not noted for the 1991/92 experiment due to the climatic conditions, which were characterized by a drought that lasted nearly two months at the beginning of the crop cycle.

With regard to the grain yield and its components, three components were the most important for grain yield: the spike population, the number of kernels/m^2, and the 1000 kernel weight. Thus, plots, in which micro-basins had been made, had an improved yield compared to those with no micro-basins (1650 against 1400 kg/ha and 2460 against 1930 kg/ha, respectively, for the 1990/91 and 1993/94 experiments). An effect of micro-basins on yield was not observed for the 1991/92 experiment because of the drought and limited runoff.

At the end of the experiments, we noted a positive effect of the interaction 'weeding × micro-basins' on water use efficiency (WUE). However, for the 1991/92 experiment, because of the drought, this effect was not significant due to the poor climatic conditions.

Future methodology should follow-up on the water balance, through a more detailed analysis of the soil profile, especially at the soil surface. A thorough study of the mechanisms involved (rainfall intensity, infiltration, slope effect, surface roughness, and texture effect) would permit better a better understanding of micro-basins.

Finally, more farmers should become aware of these techniques that better conserve water under rainfed conditions.

REFERENCES

Ben Lhamdani, A 1991. Etude par simulation des disponibilités en eau du blé tendre et conséquences sur la conduite technique de la culture: cas de trois régions. à climats contrastés (Tanger, Méknes et Marrakech). *Mémoire de 3ème Cycle Agronomie*. I.A.V. Hassan II, Rabat, Morocco.

Clark, R.N. and E.B. Hudspeth. 1976. *Runoff Control for Summer Crop Production in the Southern Plains*. Paper No.76-2008. American Society of Agricultural Engineers, St. Joseph, Michigan.

El Baghati, H. 1977. Compte rendu d'essais sur les technique de culture de l'orge et du blé tendre. *Doc Ronéo, DPV-MAMVA*, Rabat, Morocco.

El Haiba, M. 1979. Précipitations et ruissellement dans le bassin de Sidi Salah. Région du Tangérois. *Mémoire de 3ème Cycle Agronomie* I.A.V. Hassan II, Rabat, Morocco.

Gerard, C.J., P.D. Sexton and D.M. Conover. 1984. Effect of furrow diking, subsoiling, and slope position on crop yields. *Agronomy Journal* 76:945-950.

Handoufe, A. 1988. *Réponse à l'Eau et à l'Azote d'un Blé Tendre (Triticum aestivum L.) Sous un Climat Semi-Aride.* Thèse de Docteur Ingénieur Université Paul Sabatier. Toulouse, France

Krishna, J.H. and T.J. Gerik. 1988. Furrow-diking technology for dryland agriculture. In *Challenges in Dryland Agriculture. A Global Perspective*, eds. P.W. Unger, T.V. Sneed, W.R. Jordan and R.W. Jensen. Proceedings of the International Conference on Dryland Farming. August 15-19, 1988. Amarillo/Bushland, Texas, U.S.A. Texas Agric. Exp. Sta., College Station, Texas, pp. 258-260.

Lyle, W.M. and D.R. Dixon. 1977. Basin tillage for rainfall retention. *Transactions of the American Society of Agricultural Engineers* 20:1013-1017.

Morin, J. and Y. Benyamini. 1988. Tillage method selection based on runoff modeling. In *Challenges in Dryland Agriculture. A Global Perspective*, eds. P.W. Unger, T.V. Sneed, W.R. Jordan and R.W. Jensen. Proceedings of the International Dryland Farming. August 15-19, 1988. Amarillo/Bushland, Texas, U.S.A. Texas Agric. Exp. Sta., College Station, Texas, pp. 251-254.

McFarland, M.L., F.M. Hons and V.A. Saladino. 1991. Effects of furrow diking and tillage on corn grain yield and nitrogen accumulation. *Agronomy Journal* 83: 382-386.

Ouattar, S. and T.E. Ameziane. 1989. *Les Céréales au Maroc. De la Recherche à l'Amélioration des Techniques de Production.* Edition Toubkal. Maroc.

Papendick, R.I. and G.S. Campbell. 1988. Concepts and management strategies for water conservation in dryland farming. In *Challenges in Dryland Agriculture. A Global Perspective*, eds. P.W. Unger, T.V. Sneed, W.R. Jordan and R.W. Jensen. Proceedings of the International Conference on Dryland Farming. August 15-19, 1988. Amarillo/Bushland, Texas, U.S.A. Texas Agric. Exp. Sta., College Station, Texas, pp. 119-127.

Radosevich, S.R. and J.S Holt. 1984. *Weed Ecology Implications for Vegetation Management*. John Wiley and Sons, New York, 265 pp.

Saffour, K. 1992. Concurence entre le blé dur et les mauvaises herbes dans le Saïs. *Mémoire de 3éme Cycle Agronomie*. I.A.V. Hassan II, Rabat, Morocco.

Tanji, A. 1987. *On Farm Evaluation of Wheat Production as Affected by Three Weeding Systems and Top-Dressed Nitrogen in Chaouia (Semi-Arid Zone of Morocco)*. M.S. Thesis. Kansas State University. Manhattan, Kansas, U.S.A.

Unger, P.W. 1983. Water Conservation: Southern Great Plains. In *Dryland Agriculture*, eds. H.E. Dregne and W.O. Willis. American Society of Agronomy, Crop Science Society of America, and Soil Science Society of America, Madison, Wisconsin, U.S.A., pp. 35-55.

Unger, P.W. and B.A. Stewart. 1983. Soil management for efficient water use: An overview. In *Limitations to Efficient Water Use in Crop Production*, eds. H.M. Taylor, W.R. Jordan and T.R. Sinclair. Amererican Society of Agronomy, Crop Science Society of America, and Soil Science Society of America, Madison, Wisc. U.S.A., pp. 419-460.

Williams, J.R., G.H. Wistrand, V.W. Benson and J.H. Krishna. 1988. A model for simulating furrow dike management and use. In *Challenges in Dryland Agriculture. A Global Perspective*, eds. P.W. Unger, T.V. Sneed, W.R. Jordan and R.W. Jensen. Proceedings of the International Conference on Dryland Farming. August 15-19, 1988. Amarillo/Bushland, Texas, U.S.A. Texas Agric. Exp. Sta., College Station, Texas, pp. 255-257.

Yadav, S.K. and D.P. Singh. 1981. Effect of irrigation and antitranspirants on evapotranspiration, water use efficiency and moisture extraction patterns of barley. *Irrigation Science* 2: 177-184.

Zimdhal, R.L. 1980. *Weed Crop Competition, A Review*. International Plant Protection Center. Corvallis. Oregon State University. U.S.A.

Risk Assessment
of the Irrigation Requirements
of Field Crops in a Maritime Climate

S. R. Green
B. E. Clothier
T. M. Mills
A. Millar

SUMMARY. The need for more-efficient agricultural use of irrigation water arises out of the increased competition for water resources and the greater pressure on irrigation practices to be environmentally friendly. Here, we use a simple water-balance approach to model soil-water storage changes for the purpose of better estimating the irrigation requirements for a range of field crops growing in the Auckland region of New Zealand. In this humid, maritime climate irrigation requirements can very greatly in time and with crop type. Such information is needed by the local authorities, for planning purposes to determine appropriate water right allocations for local growers.

S. R. Green, B. E. Clothier, and T. M. Mills are affiliated with the Horticulture and Food Research Institute of New Zealand, Ltd., Palmerston North, New Zealand, and A. Millar is affiliated with the Auckland Research Council, Auckland, New Zealand. Dr. Clothier is also Adjunct Lecturer in the Institute of Natural Resources, Massey University, Palmerston North, New Zealand, and is Fellow of the Royal Socity of New Zealand, a signal honor.

This research was supported by the New Zealand Foundation for Research, Science and Technology under Contract No. CO6617. Additional financial support was obtained from the Auckland Regional Council under their Water Use Efficiency programme.

[Haworth co-indexing entry note]: "Risk Assessment of the Irrigation Requirements of Field Crops in a Maritime Climate." Green, S.R. et al. Co-published simultaneously in *Journal of Crop Production* (Food Products Press, an imprint of The Haworth Press, Inc.) Vol. 2, No. 2 (#4), 1999, pp. 353-377; and: *Water Use in Crop Production* (ed: M. B. Kirkham) Food Products Press, an imprint of The Haworth Press, Inc., 1999, pp. 353-377. Single or multiple copies of this article are available for a fee from The Haworth Document Delivery Service [1-800-342-9678, 9:00 a.m. - 5:00 p.m. (EST). E-mail address: getinfo@haworthpressinc.com].

353

The model we have developed considers the root zone to be one dimensional, comprising a uniform soil of known hydraulic properties, and having plants with roots extending vertically to a known depth. Model output consists of daily values of the soil moisture stored in the root zone. Crop water use was calculated via the Penman-Monteith model, using a generalized coefficient for each crop. A threshold moisture level, which depends on a combination of soil and crop factors, is used trigger the irrigation events. Water drainage below the root zone is calculated from easily determined soil hydraulic properties, and the amount of water stored in the profile.

We use a statistical analysis based on 25 years of weather data to provide answers to the questions of 'how much?' and 'how often?', at any level of given risk of exceedence. Irrigation requirements were considered for a wide range of crops that grow in the Auckland region. Variation in rainfall and drainage are described using a gamma probability density function (PDF), while the variation in irrigation requirement was found to be capable of description using a Gaussian PDF. This general model of irrigation requirements is easily parameterised, and can be run for any crop-soil combination using the historical weather data. It can be used to consider requirements for any level of prescribed risk. It could even be developed further to quantify specific water right allocations, and subsequently it could be turned into a Decision Support Tool for defining good irrigation practice, even in real time. *[Article copies available for a fee from The Haworth Document Delivery Service: 1-800-342-9678. E-mail address: getinfo@haworthpressinc.com <Website: http://www.haworthpressinc.com>]*

KEYWORDS. Modelling, soil water balance, crop transpiration, irrigation, rainfall, drainage

INTRODUCTION

In a temperate, maritime climate, such as the Auckland isthmus in New Zealand, the uncertain and erratic nature of rainfall over the summer months means that irrigation of field crops might sometimes not be necessary. This difficulty of irrigation in a maritime climate is because during summer periods evaporative demands are always at their peak, but rainfall is often small but always erratic.

The greater Auckland area currently supports a large number of horticultural enterprises. Most of the field crops require some irrigation at a given time during the growing season. However, because the supply of irrigation water within the Auckland region is limited, there is a need for more-efficient agricultural use of irrigation water. This need arises because of increased competition for the water resources, and a greater pressure on irrigation practices to be environmentally friendly so that irrigation water is not wasted nor leads to unnecessary leaching of root zone chemicals.

Under New Zealand's Resource Management Act (1991) all growers in the Auckland region must apply to the Auckland Regional Council (ARC) for a water right to extract water from either surface or ground water bodies. In their application, each grower is requested to provide some justification for the volumes of water they seek, both on a maximum daily basis, and the total annual requirement. The ARC has found that, in general, the monitored water usage does not justify the volumes of water requested in the resource consent applications. To improve allocation of irrigation water, the ARC established a simple set of guidelines for horticultural users. However, these guidelines, which were developed some time ago, were based on limited data, and limited scientific understanding. At the time, no consideration was given to specific crop types and phenology. Variations associated with the climate and the soils around Auckland were not considered. Each of these factors will have an important influence on the actual water requirements of specific horticultural crops.

Knowing the total planted area for each horticultural enterprise, the ARC has calculated that almost all of the available water resource has now been allocated to irrigation. Hence, any further development of land for horticultural production in the Auckland region appears to be constrained by a shortfall in the amount of water. The Water Rights currently allocated to growers may well exceed actual water requirements! If this were the case, Water Rights could sensibly be reduced so that more irrigation water could be available for a further expansion of horticulture. Furthermore, encouraging the efficient use of irrigation water should help to minimise deleterious effects on the environment and groundwater caused by wasteful drainage of water and nutrients beyond the root-zone.

As part of a water-use efficiency programme, ARC Environment sought to update the guidelines on crop water requirements. We carried out a modelling study to calculate the likely irrigation requirements of selected field crops, with particular reference to the climate and soils of the Auckland region. The model uses long-term weather records, *in situ* measurements of important soil properties, and generalised factors to represent the main soils and crops of the Auckland region. The calculations run on a daily time-step and the model output consists of a daily water balance for the root-zone of a specified field crop. The irrigation requirements are then determined by applying a specified amount of irrigation whenever the soil becomes so dry that crop production is likely to be affected. In this paper we report some results from this modelling study, with particular reference to field crops growing in the Pukekohe area. The Council now uses this information for planning purposes to determine the appropriate level of Water Right allocations for the local growers.

MATERIALS AND METHODS

Soil Water Balance

The model we developed to permit these risk assessments considers the root zone to be one dimensional. The uniform soil possesses known hydraulic properties, and the plants have roots extending vertically to a depth of z_R [mm]. The total amount of water stored in the soil profile to a depth z_R, $W = z_R\theta$ [mm], is a dynamic property of the root-zone. It changes daily to reflect the 'inputs' of rainfall and irrigation that are received, and the 'outputs' or 'losses' of water that are due to plant uptake, surface evaporation and runoff, plus any deep drainage of water beyond the root-zone. We calculate a simple water balance for the root-zone, by summing all the inputs and outputs of water, *viz.*

$$\Delta W = (I + P) - (R + D + ET) \tag{1}$$

where ΔW [mm] represents the change in the total amount of stored water. Inputs to this store are: the total amount of irrigation, I [mm], and the total amount of precipitation, P [mm]. Some of the surface-applied irrigation, and/or precipitation, may be lost as surface runoff, R [mm], and some may eventually be lost due to deep drainage, D [mm], beyond the root-zone. The other component of this water balance is the evapotranspiration, ET [mm]. This represents the amount water removed from the soil by the roots, that is then lost evaporatively as transpiration. To ET we add in that water lost as evaporation directly from the soil surface.

Meteorological Data

The model calculations require historical weather records. We have obtained these records from the NIWAR (National Institute of Water and Atmospheric Research, New Zealand) data base, using the on-line MetSeek 'Search Engine,' developed by HortResearch Institute (M. Laurensen, Orchard 2000™). The weather data consists of daily values of global solar radiation, maximum and minimum air temperature, relative humidity, wind speed at 2 m height, and rainfall. These data were measured at a climate station located near the centre of Pukekohe, which is the major crop growing area in Auckland. A few missing values from the weather data set were simply replaced by corresponding values recorded at a second climate station located some 10 km away. The daily weather record was continuous for a period of 25 years.

Crop Water Use

We calculate plant uptake of water from the long-term weather records using a generalized crop factor to account for the stage of plant development. A two-step procedure is used, and this was based on guidelines given by the Food and Agriculture Administration (FAO) of the United Nations (Doorenbos and Pruitt, 1977; Smith, 1990). Measured data for global radiation from the sun, air temperature, humidity and windspeed are first used to calculate a reference evaporation rate, ET_0 [mm d^{-1}] from the modified Penman equation

$$ET_0 = \left(\frac{s}{s + \gamma}\right)R_N + \left(\frac{\gamma f(U)}{s + \gamma}\right)D_A, \qquad (2)$$

where R_N [mm d^{-1}] is the net radiation expressed in units of an equivalent depth rate, D_A [kPa] is the difference between the saturation vapour pressure at mean air temperature and the mean actual vapour pressure of the air, s [Pa °C^{-1}] is the slope of the saturation vapour-pressure versus temperature curve, and γ [66.1 Pa] is the psychrometeric constant. The aerodynamic resistance of the surface is expressed via a wind-related function, $f(U)$, given by

$$f(U) = 2.7(1 + U/100). \qquad (3)$$

where U [km d^{-1}] is the 24-hr wind run at 2-m height. This reference value, ET_0, defines the rate of evaporation expected from an extensive surface of green grass cover of short, uniform height, actively growing, completely shading the ground, and not short of water.

The actual crop water use, ET, is related to this reference evaporation rate via a generalised crop coefficient, K_C, which expresses the physiological development of the crop,

$$ET = f_C\,(\theta).K_C.ET_0 \qquad (4)$$

For this study we chose values of K_C according to the guidelines of Doorenbos and Pruit (1977). The seasonal course of K_C for a lettuce crop planted in November, and harvested in mid-January, is shown in Figure 1. The time course of K_C for other field crops can be deduced from the data of Table 1.

FIGURE 1. Seasonal course of the crop coefficient, Kc, for a lettuce crop planted in November and harvested in mid-January. The respective growth stages of the crop are: (I) initiation, (II) rapid growth, (III) mid-season, and (IV) late season.

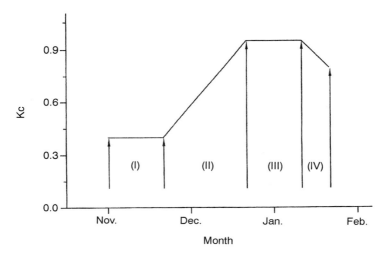

These crop coefficients are appropriate for a disease-free crop grown in a large field, under optimum soil water and fertility conditions and achieving full production potential.

Water uptake was adjusted to allow for the effects of water stress using the reduction function, $f_C(\theta)$. This takes into account the soil characteristics, rooting depths and ET demand. Crops can tolerate a certain level of water deficit in their root zone. However, once the soil dries below a given threshold then crop water use is reduced. This threshold moisture level, which depends on a combination of soil and crop factors, is used to trigger the irrigation events.

Irrigation is vital to sustain crop productivity over the summer months by maintaining 'enough' water in the root-zone. We determine the trigger point for irrigation as set out below. Figure 2 helps to define the meaning of some of these variables.

For a given soil, the water content at field capacity (θ_F at a matric potential of -0.02 M Pa) and the water content at the permanent wilting point (θ_W at a matric potential of -1.5 M Pa), are defined from the soil water retention curve. This demands that the water retention curve of the soil is known, or can be estimated. The maximum available water content, once gravity drainage has materially ceased, is equal to $\theta_A = \theta_F - \theta_W$. Not all of this water is 'readily available' to the crop, however. We assume each crop has an accept-

TABLE 1. Planting date, crop development stages, length of growing season and crop coefficient (Kc) for selected field crops (adapted from Doorenbos and Pruitt, 1977). The development stages are indicated by I/G/M/L that stand respectively for the initial, growth, mid-season and late season development stage, and (T) stands for the total growing period from planting through to harvest in days. Vegetable 1 and 2 represent continuous rotation of vegetable crops, which have either a low tolerance (1) or a high tolerance (2) to a short-term water deficit (see the *P*-factor of Table 2). Orchard 1 represents a mature, deciduous orchard (e.g., apple) with a grassed inter-row. Orchard 2 represents the same orchard with a bare inter-row.

Field crop	Planting date	Plant development (d)	Stage and crop coefficient
Vegetable 1	All year	90/90/90/95/ (365) IV (1.0)	III (1.0)
Vegetable 2	All year	90/90/90/95/ (365)	III (1.0) IV (1.0)
Pasture	All year	90/90/90/95/ (365)	III (1.0) IV (1.0)
Potato	275	30/35/50/30/ (145)	III (1.05) IV (0.70)
Onion	180	20/35/110/45/ (210)	III (0.95) IV (0.75)
Lettuce	305	20/30/15/10/ (75)	III (0.95) IV (0.90)
Cabbage	305	20/30/20/10/ (80)	III (0.95) IV (0.80)
Squash	210	25/35/35/25/ (120)	III (0.90) IV (0.70)
Orchard 1	180	150/90/90/35/ (365)	III (1.10) IV (0.80)
Orchard 2	180	150/90/90/35/ (365)	III (0.85) IV (0.60)

able depletion level, $\Delta W_C = p.\theta_A.z_R$ (mm). The value of p is the proportion of θ_A below which drought stress first occurs and begins to affect both ET and crop production. The model then applies a set irrigation amount (e.g., 15 or 25 mm) we use here, as soon as the soil water content falls below this trigger point. This irrigation aims to maximise production by maintaining just enough water in the root zone.

FIGURE 2. Water retention characteristics of the Patumahoe clay loam soil (Jim Watt, pers. comm.). Here θ_F and θ_W stand, respectively, for the water content at field capacity (-0.1 MPa) and at wilting point (-1.5 MPa) and $\theta_A = \theta_F - \theta_W$ represents the soil water content available for crop water use. Only a fraction, p of θ_A can be extracted by the crops before symptoms of water stress are exhibited and crop production is curtailed (Table 2).

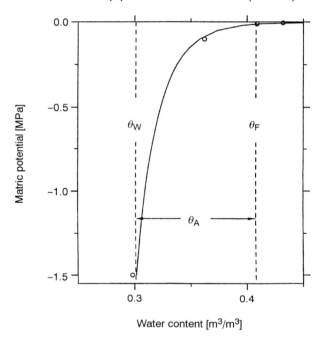

Water content [m³/m³]

Table 2 presents the values we assumed for the factor p and for the maximum rooting depth of the selected crops. Shallow-rooted field crops, such as onions and potatoes, have the lowest acceptable depletion levels. These crops are the most sensitive to water stress and are therefore most likely to require regular amounts of irrigation or free draining soils to maintain an acceptable amount of water within the root-zone. In contrast, deeper-rooted orchard trees have the highest acceptable depletion levels. These crops are less dependent on irrigation to maintain enough water in their root-zone.

Drainage Fluxes of Water

Drainage of water below the bottom of the root zone, D [mm d⁻¹], was calculated from the analytical model of Sisson et al. (1980) which is based on

TABLE 2. Assumed rooting depth (z_R), fraction of available water (P) and readily available water ($p\theta_A$) for selected field crops growing on a Patumahoe clay loam soil in the Pukekohe region.

Field crop	Root depth z_R (mm)	Fraction of available soil water (p)	Readily available soil water (mm)
Vegetable 1	0.5	0.25	14.5
Vegetable 2	0.5	0.45	24.5
Pasture	0.5	0.50	27.5
Potato	0.5	0.25	13.5
Onion	0.5	0.25	13.5
Lettuce	0.5	0.25	13.5
Cabbage	0.5	0.45	24.5
Squash	1.0	0.35	38.5
Orchard 1	1.5	0.50	82.5
Orchard 2	1.5	0.50	82.5

known values for the soil's hydraulic properties. The drainage model is derived from Richards' equation, which describes the vertical movement of water through a uniform soil:

$$\frac{\partial \theta}{\partial t} = \frac{\partial}{\partial z}\left[K\frac{\partial H}{\partial z}\right] \qquad (5)$$

where θ is the volumetric water content [$m^3\ m^{-3}$], K is the hydraulic conductivity [$m\ s^{-1}$], H is the total hydraulic head [m] which comprises the soil's matric potential, h [m], minus the total gravitational head, z [m], and t is the time [s]. For a 'unit gradient' in the total potential head, i.e., $dH/dz = -1$, and a suitable $K(\theta)$ relation, Sisson et al. (1980) showed that we can expect a linear relationship between the total depth of water in the soil profile, W [m], and the log of time [t],

$$W = C - Y\log_e(t) \qquad (6)$$

Both Y and C are soil-dependent factors. Equation (6) approximates the free drainage from a uniform soil profile only under the influence of gravity and in the absence of a shallow water table. For this study we consider the case of a power-law hydraulic conductivity function of the form

$$K(\theta) = K_s\left(\frac{\theta}{\theta_s}\right)^{1/\beta} \tag{7}$$

where K_S and θ_S represent maximum values of the hydraulic conductivity and soil water content, respectively, and β is a slope factor which dictates how fast the hydraulic conductivity declines as the soil drains. It follows from our use of a power-law conductivity function (Sisson et al. 1980), that

$$Y = \left[\frac{\beta}{\beta - 1}\right], \quad A = \left[\frac{K_s}{\beta\theta_s}\right] \text{ and } C = (1 - \beta)\theta_s z\left[\frac{A}{z}\right]^Y \tag{8}$$

For practical purposes, we are interested in the total amount of water stored in the root zone, so we set $z = z_R$, the depth of the roots. The amount of water draining below the root zone, D [m/d], over a period of one day, is then calculated as

$$D = Y\ln[1 + \exp[(W_i - C)/Y]] \tag{9}$$

where W_i [m] is the amount of water stored in the soil profile, between the soil surface and the depth of z_R, on the previous day. The 'lumped parameters' needed to estimate the drainage of water quitting the root zone are the three constants C, A and Y. Chappaz (1987) showed Equation (9) provides a good description of drainage below the root zone of both pasture and kiwifruit vines growing on the Pukekohe soil.

Soil Properties

Two soil properties are needed to estimate this drainage of water beyond the root zone. Firstly, a disk permeameter (Perroux and White, 1988) was used to obtain an *in situ* characterisation of the soil's hydraulic conductivity in the pressure head range -20 to about -150 mm, which is close to saturation. The disk permeameter data provide a measure of the soils saturated hydraulic conductivity, K_S [m s^{-1}]. This property is critical in controlling water entry into the soil (*c.f.* runoff), as well as the downward movement of water by gravity. We also need the soil's water retention characteristic since this property governs the soils ability to 'hold water' in the soil pores (Figure 2). Data for the soil's water-retention characteristics were determined using a combination of Haine's apparatus and pressure plate measurements of soil samples (Jim Watt, pers. comm.). A Brooks and Corey relationship was fitted to the water retention curve and the closed form of the hydraulic conductivity function (van Genuchten, 1980) was used to estimate the β parameter of Equation 7.

Statistical Analysis of Water Balance Calculations

The uncertain nature of rainfall over the summer months makes the scheduling of irrigation difficult. We use a statistical approach to examine variability in each of the water balance components. We used a 4-week period, for each month of the year, to define the irrigation requirements, on an annual basis, for each of the crops growing in the Pukekohe area.

From the long-term water balance calculations, we generated a probability distribution for the total rainfall, applied irrigation, crop water use and drainage that occurs for each 28-day period throughout the year. The rainfall and drainage distributions are compared against a gamma probability density function (PDF), because of the characteristic 'waiting time' between rainfall and drainage events which is implicit in the gamma model (Larsen and Marx, 1986). The corresponding distributions of irrigation and crop water use are compared against a Gaussian probability density function. Details of the gamma and normal PDFs are consigned to an appendix.

We use the gamma and Gaussian PDFs, which can be calculated from a sample mean and standard deviation, to assess the risk associated with each of the water balance components, and to define the irrigation required to meet the crop demands in, say, 4 out of 5 cases. This definition of the risk for the irrigation requirements means there is, however, a 20% chance that not enough water will be allocated to meet the crop requirements. The results provided can, however, be used to determine water requirements for any prescribed level of risk.

RESULTS

Estimates of crop water use were generated for some field crops typical of the Auckland region. The crops we considered included pasture, deciduous orchards, plus single crops of onions, potatoes, squash, lettuce and cabbage. As well, we condisered a year-round crop of green leaf vegetables. Firstly we will present the average water balance calculations of a pasture growing in Pukekohe.

The seasonal pattern of the water balance for a pasture growing on a local soil in the climate of Pukekohe, is shown in Figure 3. The mean daily ET peaks at about 5 mm/d over the summer months. These rates drop to below 1 mm/d during the winter. The average rainfall over summer is between about 2.5 to 3 mm/d so that, on average, some 2 mm/d of irrigation is required to

FIGURE 3. Mean monthly values for the water balance of an irrigated pasture growing on a Patumahoe clay loam soil in the Pukekohe region. Here ET, P, D and I stand, respectively, for the evapotranspiration, rainfall, drainage and irrigation of the pasture in mm per day.

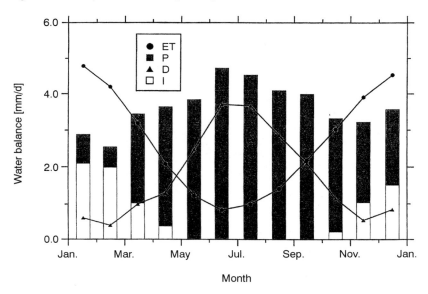

maintain the pasture production. A small amount of drainage does occur below the root zone in summer, usually following a very heavy rainfall event when the root zone is already close to field capacity. Drainage in summer is, on average, less than about 1 mm/d while runoff in summer is insignificant due to the free draining nature of the Pukekohe soil. In contrast, the average rainfall in winter peaks at just under 5 mm/d and this is much greater than the 1 mm/d rate of pasture ET. The high rainfall, coupled with the low ET, means that there is never any need to apply irrigation in the winter. Drainage rates in winter peak at about 3.5 mm/d so that, on average, only about 30% of the winter rainfall will remain within the root zone.

If we were to allocate crop water based on the 'average' irrigation require-ment shown in Figure 3 then 50% of the time the crop demands would exceed the allocation. However, due to the erratic and uncertain nature of summer rainfall, and the associated variability of crop water use caused by warm/cool or sunny/cloudy weather, we might expect a wide scatter in the irrigation requirements of particular crops on different years. In Figures 4 and 5 we demonstrate two extremes.

During a dry year (1974), the pasture required some 525 mm/yr of irriga-

FIGURE 4. Estimates of the daily water balance for a pasture at Pukekohe during a dry summer (1974). The top panel shows the average water content in the root zone, to a depth of 0.5 m. The middle panel shows inputs of water via irrigation (I) and rainfall (P). The bottom panel shows losses of water via evapotranspiration (ET) and drainage (D). Irrigation was supplied whenever $\theta <$ $\theta_F - p.\theta_A$.

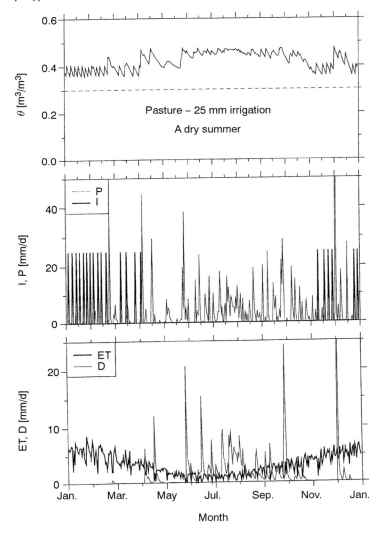

FIGURE 5. Estimates of the daily water balance for a pasture at Pukekohe during a wet summer (1992). Otherwise the same as for Figure 4.

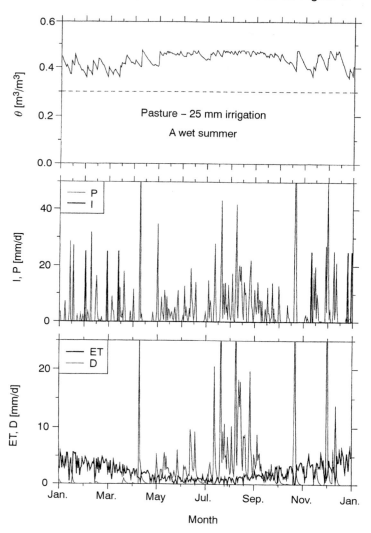

tion (Figure 4), when supplied in 25 mm units. The root zone water content remained close to the trigger point ($\theta = 0.36$ [m^3/m^3]) for much of January and February because of the infrequent and small amounts of rainfall. Summer time ET peaked at about 7.5 mm/d because of the warm, dry weather. A combination of reduced ET demands and some large rainfall events in late April conspired to return the root zone close to field capacity. This resulted in two large drainage events. However, most of the drainage occurred between the months of June to October because of increased winter rainfall that maintained the root zone soil close to field capacity. No irrigation was required between April through to October. Thereafter, another warm, dry period between October and November increased pasture water use and quickly reduced the root zone water content down to the trigger point of $\theta = 0.36$. Six irrigations of 25 mm were required in the last 2 months of the year, despite one large rainfall of some 90 mm falling at the end of November. Not all of this rainfall remained in the root zone. Rather, some 30 mm was drainage, and about 9 mm was lost as surface runoff.

During a wet year (1992), we calculated the irrigation requirements of pasture to be just 150 mm/yr, when supplied in 25 mm units (Figure 5). For this wet year, the root zone soil was already quite wet at the beginning of January. So with this initially wet soil, combined with several large rainfall events during January to March, and a much lower ET caused by the cooler, moister weather, only three irrigations were needed between January and March. In contrast, the pasture needed some 15 irrigations during the first three months of the dry summer of 1974 (Figure 4). In other words, the pasture required a total irrigation of some 375 mm for the first three months of the dry year, but only 75 mm for the wet year. This difference highlights the need to consider risk when developing irrigation requirements. On average, we expected the pasture to need a total irrigation of some 140 mm between January to March (Figure 3). A similar story, of the erratic rainfall (*c.f.* ET) having a predominant influence on the irrigation requirements of pasture, also holds true during November and December.

In contrast, the irrigation requirement of orchard trees, with their deeper root systems, is somewhat less than that of the shallower rooted pasture. Indeed, we calculate a mature, deciduous orchard would require no irrigation during the entire wet year of 1992 (Figure 6). This is because the fruit trees had enough water stored in their deep root zone to provide an effective buffer against any of the short-term water deficits that developed over the summer period.

Deciduous orchard trees on the Pukekohe soil are considered to have a storage buffer of some 82.5 mm of readily available soil water (Table 2). This buffer is enough to meet the crop water requirements for a period of about 16 days, assuming average summer ET of 5 mm/d. Summer rainfall was more

FIGURE 6. Estimates of the daily water balance for a deciduous orchard at Pukekohe during a dry summer (1974). We assume a root zone depth of 1.5 m and a bare inter-row between the trees. Otherwise the same as for Figure 4.

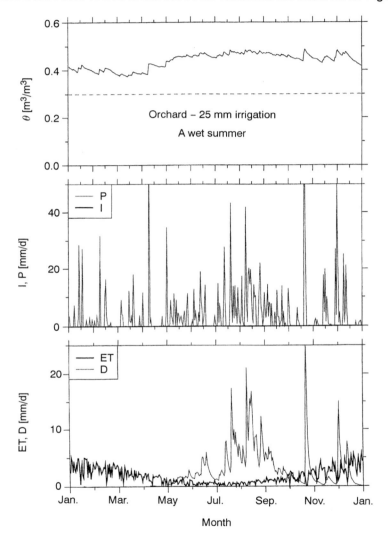

frequent than this during the wet year of 1992. In the dry year of 1974, the orchard required some 300 mm of irrigation over the entire year. This is still much less than the 525 mm of irrigation required by pasture in the same year.

It is clear from Figures 3-6 that an allocation of irrigation based on average water balance calculations will not provide enough water for every summer. Such a scheme will over-allocate half the time in the wetter years, and under-allocate during the drier years. If growers were constrained to using just the average amount of irrigation water then half the time the crops would run short of water and there would be a concomitant drop in crop production. A more conservative approach to water allocation is desirable and this we base on a probabilistic assessment of risk. Next, we use a statistical approach to examine the variability of water needs to establish the level of irrigation required to meet the crop demands in, say, 80% of years.

From the long-term water balance calculations, we can generate a probability distribution for each of the water balance components, as analysed over a 28-day period, for each month of the year. In Figure 7 we illustrate some results for pasture over summer in which we compare the rainfall and drainage distributions against a gamma probability density function (PDF). There, we also we compare the corresponding distributions of irrigation and crop ET against a normal PDF. These probability density functions, calculated from sample means and standard deviations, provide a very good description of the trend observed in each of the water balance components. We note that the PDF of drainage, as calculated from a gamma function, does not appear to be as good as the other three water balance components. This is because the drainage is often quite small during the summer. The gamma distribution does give a very good description of drainage during the winter periods when this component is a much bigger part of the total water balance.

For the purpose of water allocation it is encouraging that the irrigation PDF is well described by the Gaussian distribution. Thus, we can use the Gaussian PDF to define the irrigation required to meet the crop demands in, say 4 out of 5 cases, as shown by the circled point of Figure 7. The level of risk can of course easily be changed if required once the statistics and the PDF are established. On this 80% basis of risk we calculate a pasture at Pukekohe in the summer (January) should be allocated 78 mm/4 weeks. This allocation would meet the crop water use demands 80% of the time. This analysis is based on a 4 week period. Therefore, for the 31 days of any January we need to allocate some 86 mm for pasture irrigation. This does means, however, that there is a 20% chance that we would not allocate enough water to meet the pasture requirements across all years. But this is still much better than the 50% chance we would achieve by allocating the irrigation based on the 'average' water balance calculations of Figure 3.

The seasonal irrigation schedule for each of the selected field crops is

FIGURE 7. A statistical representation of the root-zone water balance of an irrigated pasture in Pukekohe during the summer months (Jan-Feb). The top panel shows inputs of water via rainfall (P) and irrigation (I). The bottom panel represents losses of water via evapotranspiration (ET) and drainage (D). Markers represent the cumulative probabilities as determined from a Monte-Carlo type simulation using 25 years of weather data. ET and I are described by a Gaussian distribution function (solid lines) while P and D are described by a Gamma distribution function (broken lines). The circled point defines the irrigation level that meets the crop requirements 80% of the time. This value is used for allocating irrigation over a 4 week period (see Table 3).

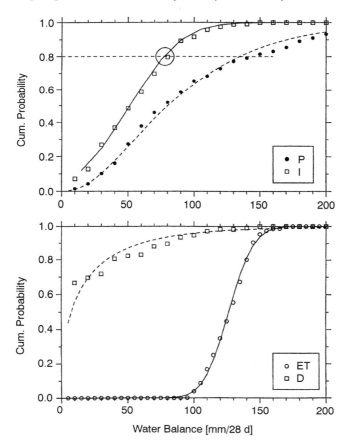

presented in Table 3. Our long-term water balance calculations assume a Gaussian PDF to describe the distribution of irrigation over a 4-week period. For a monthly schedule of irrigation, these figures need to be scaled in proportion to the number of days in each month.

The first point to note in Table 3 is that no irrigation is deemed necessary from May until October. There is always enough rainfall to more than meet the crop water demands during the winter. Continuously cropped soils and pasture will require irrigation for about 7 months of the year, from October through to early April. Mature, deciduous orchards will require irrigation for just 5 months of the year, from December through to early April. Deeper rooted trees in orchards have a far greater buffer of freely available water in their root zone than do field crops (Table 2). Hence, orchards are less reliant upon irrigation during the spring and early summer.

Potatoes require more irrigation than onions. Potatoes have a higher crop

TABLE 3. Irrigation schedule for selected field crops growing on a Patumahoe clay loam soil in the Pukekohe region. The schedule is indicated by M/S/A that stand, respectively, for the mean, the standard deviation and the allocation of irrigation in mm per 4 week period. Here A is calculated as the 80% probability that the irrigation requirement is less than or equal to A (see text and Figure 7).

Crop	Jan.	Feb.	Mar.	April	Oct.	Nov.	Dec.
Veg. 1	63/29 (87)	68/35 (98)	45/25 (67)	21/17 (36)	6/12 (18)	26/27 (49)	45/25 (66)
Veg. 2	54/32 (80)	62/39 (95)	41/28 (65)	18/18 (32)	2/6 (12)	20/25 (41)	36/27 (59)
Pasture	54/32 (78)	60/39 (89)	39/28 (59)	17/18 (28)	2/6 (10)	18/25 (31)	36/26 (54)
Potato	75/31 (101)	65/35 (95)	12/17 (27)	–	1/1 (2)	21/26 (43)	53/27 (76)
Onion	41/25 (61)	18/22 (37)	–	–	6/11 (16)	27/26 (49)	40/24 (61)
Lettuce	59/29 (84)	23/27 (46)	–	–	–	–	18/21 (36)
Cabbage	54/31 (80)	32/31 (59)	1/3 (3)	–	–	–	12/18 (27)
Squash	33/33 (61)	35/34 (64)	3/9 (11)	–	–	–	6/13 (17)
Orch. 1	39/41 (74)	59/52 (104)	38/34 (66)	10/17 (26)	–	1/4 (4)	6/18 (22)
Orch. 2	11/22 (29)	25/36 (55)	17/24 (37)	6/13 (17)	–	–	1/4 (4)

coefficient, and therefore use more water (Table 1). Of the selected crops, we find potatoes have the highest irrigation demand through the month of January. Potatoes have a shallow root system, and low tolerance to water stress. A field crop such as lettuce or cabbage, planted in early summer, has a much shorter total growing period, and will need irrigation for just 3 months of the year. Nevertheless these crops also have a shallow root system (z_R = 0.5 m) and only a mild tolerance to a water deficit. So they will still need to be allocated a significant amount of irrigation to get them through the driest summer months. We note that a single crop of squash is less reliant upon irrigation than are the other crops, because of its deeper root system.

Our so-called Orchard-1, with its grassed inter-row, consumes more water than Orchard-2 which has a bare inter-row. In terms of water use efficiency, this may not be a good thing since irrigating the inter-row will probably not increase the marketable yield of apples. However, with the emphasis on more efficient irrigation practices, it is likely that the grassed inter-row will not be irrigated. Rather, the inter-row will probably be left to dry off through the summer months. In this case the irrigation requirements of orchards may well be less than those values indicated in Table 3. Also, for the early maturing varieties of apple that are picked in late February or early March, any irrigation scheduled after harvest will not be necessary.

The irrigation schedule in Table 3 assumes a set irrigation of 15 mm. This aliquot is applied whenever the water deficit in the root zone dropped below the trigger point of $\Delta W_C = p\theta_A$. The model was also run for irrigations of 25 mm and 35 mm. However, we found that increasing the irrigation amount resulted in only slightly more drainage ($\sim$ 10-20 mm/yr), and slightly more run off ($\sim$ 5-10 mm/yr). It did not substantially alter the overall irrigation schedule. So we have not presented the irrigation schedules from these other scenarios. We also ran the model without irrigation. All crops eventually ran short of water during some time over the 25-year simulation. Although crop production was not modelled explicitly, we can assume, and we could model, a reduction in water use that will translate to a loss in crop yield. All field crops in Pukekohe will be reliant upon some irrigation if they are to consistently reach their maximum yield potential.

Our long-term calculations allow us to generate the probability distribution associated with the water balance of the irrigation applied to each of the selected crops (Figure 8). The trend in the annual irrigation amounts is well described by a Gaussian PDF. We can therefore rely on this function to define the annual irrigation allocation for each of the field crops. We estimated pasture at Pukekohe needs to be allocated 332 mm of water per year in order to meet the grass demands for water 80% of the time (Table 4). Orchards should be allocated somewhere between 256 mm and 128 mm per year, depending on whether they have a grassed inter-row, or not. A year-round

FIGURE 8. A statistical representation of the annual irrigation requirements of selected field crops at Pukekohe. Markers represent the cumulative probabilities as determined from a Monte-Carlo type simulation using 25 years of weather data. A Gamma distribution function is shown by the lines. The broken line defines the irrigation level that meets the crop requirements 80% of the time. This value is used for allocating irrigation on an annual basis (see Table 4).

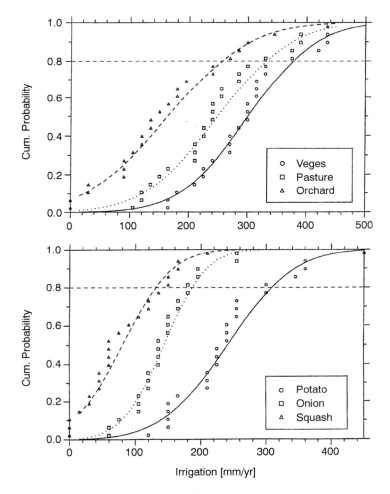

TABLE 4. Annual irrigation requirements for selected field crops growing on a Patumahoe clay loam soil in the Pukekohe region. The allocated irrigation, A, is calculated from the 80% probability that the irrigation requirement is less than or equal to A (see text and Figure 8).

Field crop	Mean irrigation (mm/yr)	Standard deviation (mm/yr)	Allocation (mm/yr)
Vegetable 1	300	91	377
Vegetable 2	248	99	332
Pasture	248	99	332
Potato	241	80	310
Onion	144	51	186
Lettuce	106	40	141
Cabbage	105	49	147
Squash	81	59	130
Orchard 1	163	110	256
Orchard 2	64	76	128

crop of leafy vegetables should be allocated between 332 and 377 mm per year, depending on the crop's tolerance to water stress. Interestingly, we found that summer rainfall was sufficient once every ten years to meet the total water use demands of the squash and orchard crops (Figure 8). Every other crop needed at least one irrigation to maintain the root zone water content above the critical soil water content that begins to affect crop production. Irrigation in a maritime climate is erratic and efficiency therefore will benefit greatly from appropriate scheduling.

CONCLUSIONS

We have used a statistical analysis that is based on 25 years of weather data. This provided answers to the question of 'how much?' and 'how often?' irrigation is required for a range of crops in the Auckland region, at any given level of prescribed risk of success. Variation in rainfall and drainage were described using a gamma probability density function (PDF), while the variation in irrigation was described using a Gaussian PDF. Our model of crop water use is easily parameterised, and can be run for any crop-soil combination using the historical weather data. The results from this study are currently being used by the Regional Council to quantify specific Water Right allocations. The model could be further developed and turned into a Decision Support Tool for defining good irrigation practice, even in real time.

REFERENCES

Chappaz, N. (1987). Development and testing of soil water regime simulation models. *Water Resources Management* 1: 293-303.

Doorenbos, J. and W.O. Pruitt. (1977). *Crop Water Requirements*. FAO Irrigation and Drainage Paper No. 24, Rome, Italy: Food and Agricultural Organisation of the United Nations, pp. 144.

Larsen, R.J. and M.L. Marx. (1986). *Mathematical Statistics and Its Application (2nd ed.)*. London, UK: Prentice-Hall International Inc., pp. 630.

Perroux, K.M. and I. White. (1988). Designs for disk permeameters. *Soil Science Society of America Journal* 52: 1205-1215.

Press, W.H., B.P. Flannery, S.A. Teukolsky, and W.T. Vetterling. (1989). *Numerical Recipes—The Art of Scientific Computing*. Cambridge, UK: Cambridge University Press, pp. 702.

Sisson, J.B., A.H. Ferguson, and M.T. van Genuchten. (1980). Simple method for predicting drainage from field plots. *Soil Science Society of America Journal* 44: 1147-1152.

Smith, M. (1990). *Report on the Expert Consultation on Revision of FAO Methodologies for Crop Water Requirements*. Rome, Italy: Food and Agricultural Organisation of the United Nations, pp 60.

van Genuchten, M.T. (1980). A closed form equation for predicting the hydraulic conductivity of unsaturated soils. *Soil Science Society of America Journal* 44: 892-989.

APPENDIX: MODEL EQUATIONS

Gamma Distribution for Rainfall and Drainage

The distribution of rainfall and drainage over a given time interval was approximated by a gamma probability density function (PDF), because of the characteristic 'waiting time' between these events which is implicit in the gamma model (Larsen and Marx, 1986). The lowest 20% quartile of the gamma distribution was used to establish the dependable level of rainfall and drainage. This dependable level represents the least amount of rainfall and drainage that we can expect 4 out of 5 times. The cornerstone of the gamma PDF is the gamma function which, for any real number $r > 0$, is defined as

$$\Gamma(r) = \int_0^\infty x^{r-1} e^{-x} dx. \tag{A1}$$

If X is a random variable such that

$$f_X(x) = \frac{\lambda^r}{\Gamma(r)} x^{r-1} e^{-x} \quad x > 0, \tag{A2}$$

then X is said to have a gamma distribution with parameters r and λ, where both r and λ must be greater than 0. The corresponding cumulative density function (CDF) may be written as

$$F_X(x) = \frac{\lambda^r}{\Gamma(r)} \int_0^R x^{r-1} e^{-x} dx. \tag{A3}$$

This function is sometimes referred to as the incomplete gamma function (Press et al., 1989). The value of $F_X(x)$ represents the probability that an event, at least as great a magnitude as x, will occur in one time interval, while the probability of exceedence during the same time interval is given by $[1 - F_X(x)]$. We can define the dependable level as the value of x when $[1 - F_X(x)] = 0.8$. Method-of-moment estimators for the gamma function are given by

$$\lambda = \frac{\bar{x}}{\sigma_x^2}, \; r = \frac{\bar{x}^2}{\sigma_x^2} \tag{A4}$$

where $\bar{x}$ and σ_x are the sample mean and standard deviation. The gamma distribution involves a number of complex integrations. However, its computation is made easy because it is usually included as a standard function in most modern spreadsheet programs. In the analysis reported here, the statistical program MINITAB (Minitab Inc., USA) has been used to calculate not only the gamma CDF, but also the inverse gamma CDF necessary to derive estimates of dependable rainfall and drainage.

Normal Probability Distribution for ET and Irrigation

The distribution of ET and irrigation over a given time interval was approximated by a Gaussian PDF. A random variable X is said to have a Gaussian distribution if its PDF is given by

$$f_X(x) = \frac{1}{\sqrt{2\pi}\sigma} e^{-(1/2)\left[(x-\mu)/\sigma\right]^2} \qquad -\infty < x < \infty \qquad [A5]$$

where μ and σ stand, respectively, for the sample mean and standard deviation. The corresponding cumulative density function may be written as

$$F_X(x) = \frac{1}{\sqrt{2\pi}\sigma} \int_{-\infty}^{x} e^{-(1/2)\left[(t-\mu)/\sigma\right]^2} dt \qquad [A6]$$

The value of $F_X(x)$ represents the probability that an event, at least as great a magnitude as x, will occur within the given time period. We define the irrigation requirements, both on a monthly and an annual basis, from the value of x when $F_X(x)] = 0.8$. The computation of $F_X(x)$ was made using the statistical program MINITAB (Minitab Inc., USA).

Index